# Genetically Modified Food Sources

# Genetically Modified Food Sources

## Safety Assessment and Control

Edited by

Prof. V. A. Tutelyan, Ph.D, D.Sc.

Institute of Nutrition, Russian Academy of Medical Sciences,
Moscow, Russia

AMSTERDAM • BOSTON • HEIDELBERG • LONDON
NEW YORK • OXFORD • PARIS • SAN DIEGO
SAN FRANCISCO • SINGAPORE • SYDNEY • TOKYO

Academic Press is an imprint of Elsevier

Academic Press is an imprint of Elsevier
32 Jamestown Road, London NW1 7BY, UK
225 Wyman Street, Waltham, MA 02451, USA
525 B Street, Suite 1800, San Diego, CA 92101-4495, USA

Copyright © 2013 Elsevier Inc. All rights reserved

*Previously published in the Russian Language by the Russian Academy of Medical Sciences, 2007.*

No part of this publication may be reproduced, stored in a retrieval system or transmitted in any form or by any means electronic, mechanical, photocopying, recording or otherwise without the prior written permission of the publisher.

Permissions may be sought directly from Elsevier's Science & Technology Rights Department in Oxford, UK: phone (+44) (0) 1865 843830; fax (+44) (0) 1865 853333; email: permissions@elsevier.com. Alternatively, visit the Science and Technology Books website at www.elsevierdirect.com/rights for further information.

**Notice**
No responsibility is assumed by the publisher for any injury and/or damage to persons or property as a matter of products liability, negligence or otherwise, or from any use or operation of any methods, products, instructions or ideas contained in the material herein. Because of rapid advances in the medical sciences, in particular, independent verification of diagnoses and drug dosages should be made.

**British Library Cataloguing-in-Publication Data**
A catalogue record for this book is available from the British Library

**Library of Congress Cataloging-in-Publication Data**
A catalog record for this book is available from the Library of Congress

ISBN: 978-0-12-405878-1

For information on all Academic Press publications
visit our website at elsevierdirect.com

Typeset by MPS Limited, Chennai, India
www.adi-mps.com

Printed and bound in United States of America

13 14 15 16   10 9 8 7 6 5 4 3 2 1

# Contents

AUTHORS ............................................................................................... ix
EDITORS ................................................................................................ xi
REVIEWERS ......................................................................................... xiii
ABBREVIATIONS .................................................................................. xv
INTRODUCTION ................................................................................. xvii

CHAPTER 1   Fundamental Concepts of Development of Genetically
            Engineered Plants ................................................................ 1
            References ......................................................................... 12

CHAPTER 2   World Production of Genetically Engineered Crops ..... 17
            References ......................................................................... 23

CHAPTER 3   Legislation and Regulation of Production and
            Sales of Food Derived From Genetically Modified
            Plants in the Russian Federation ..................................... 25
            References ......................................................................... 29

CHAPTER 4   Principles of Human Health Safety Assessment of
            Genetically Modified Plants Used in the Russian
            Federation ......................................................................... 31
            4.1  Human Health Safety Assessment ................................. 33
                 Assessment of Potential for Toxicity ............................. 33
                 Assessment of Potential Genotoxicity ........................... 36
                 Assessment of Potential for Allergenicity ...................... 36
                 Additional Studies ......................................................... 37
            4.2  Evaluation of Data Provided by Applicant ...................... 37
            4.3  Novel Approaches to Safety Assessment ....................... 37
                 Nutritional Assessment ................................................. 38
                 Assessment of Sensitive Biomarkers ............................. 39
                 Other Assessments ....................................................... 40
            Conclusion ......................................................................... 40
            References ......................................................................... 41

CHAPTER 5   Human and Animal Health Safety Assessment of Genetically Modified Plants ............................................. 43

Introduction ................................................................................. 43

5.1 Soybean .............................................................................. 44

    5.1.1 Glyphosate-Tolerant 40-3-2 Soybean Line ................... 44

        Molecular Characterization of Soybean Line 40-3-2 ....... 44

        Global Registration Status of Soybean Line 40-3-2 ....... 45

        Safety Assessment of Soybean Line 40-3-2 Conducted in the Russian Federation .......................... 45

    5.1.2 Glufosinate-Tolerant Soybean Line A2704-12 ............... 69

        Molecular Characteristics of Soybean Line A2704-12 ...... 69

        Global Registration Status of Soybean Line A2704-12 ..... 70

        Safety Assessment of Soybean Line A2704-12 Conducted in the Russian Federation .......................... 70

    5.1.3 Glufosinate-Tolerant Soybean Line A5547-127 ............ 90

        Molecular Characterization of Soybean Line A5547-127 ................................................................. 90

        Global Registration Status of Soybean Line A5547-127 ................................................................. 91

        Safety Assessment of Soybean Line A5547-127 Conducted in the Russian Federation .......................... 91

    5.1.4 Glyphosate-Tolerant Soybean Line MON 89788 ......... 108

        Molecular Characteristics of GM Soybean Line MON 89788 ..................................................... 108

        Safety Assessment of GM Soybean Line MON 89788 Conducted in the Russian Federation ........................ 110

5.2 Maize ................................................................................. 125

    5.2.1 Glyphosate-Tolerant Maize Line GA 21 ..................... 125

        Molecular Characterization of Maize Line GA 21 ...... 125

        Global Registration Status of Maize Line GA 21 ........ 126

        Safety Assessment of Transgenic Maize Line GA 21 Conducted in the Russian Federation ..... 126

    5.2.2 Transgenic Maize Line MON 810 Resistant to European Corn Borer ................................................. 142

        Molecular Characteristics of Transgenic Maize Line MON 810 ............................................................. 142

        Global Registration Status of Maize Line MON 810 ..... 143

        Safety Assessment of Transgenic Maize Line MON 810 Conducted in the Russian Federation ........ 143

    5.2.3 Transgenic Glyphosate-Tolerant Maize Line NK 603 ................................................................. 158

        Molecular Characterization of Transgenic Maize Line NK 603 ................................................................. 158

|       | Global Registration Status of Maize Line NK 603 ...... 159 |
|---|---|
|       | Safety Assessment of Transgenic Maize Line NK 603 Conducted in the Russian Federation.................. 159 |
| 5.2.4 | Diabrotica-Resistant Transgenic Maize Line MON 863 ............................................................. 175 |
|       | Molecular Characterization of Transgenic Maize Line MON 863 ............................................................. 175 |
|       | Global Registration Status of Maize Line MON 863..... 176 |
|       | Safety Assessment of Transgenic Maize Line MON 863 Conducted in the Russian Federation........ 176 |
| 5.2.5 | Transgenic Maize Line Bt11 Resistant to European Corn Borer and Tolerant to Glufosinate Ammonium................................................................. 190 |
|       | Molecular Characteristics of Transgenic Maize Line Bt11.................................................................... 190 |
|       | Global Registration Status of Maize Line Bt11........... 191 |
|       | Safety Assessment of Transgenic Maize Line Bt11 Conducted in the Russian Federation........................ 192 |
| 5.2.6 | Transgenic Maize Line T25 Tolerant to Glufosinate Ammonium................................................................. 205 |
|       | Molecular Characteristics of Transgenic Maize Line T25..................................................................... 205 |
|       | Global Registration Status of Maize Line T25 ............ 206 |
|       | Safety Assessment of Transgenic Maize Line T25 Conducted in the Russian Federation........................ 206 |
| 5.2.7 | Diabrotica-Resistant and Glyphosate-Tolerant Transgenic Maize Line MON 88017 ............................................................ 221 |
|       | Molecular Characteristics of Transgenic Maize Line MON 88017 ....................................................... 221 |
|       | Global Registration Status of Maize Line MON 88017................................................................. 222 |
|       | Safety Assessment of Transgenic Maize Line MON 88017 Conducted in the Russian Federation.... 222 |
| 5.2.8 | Diabrotica-Resistant Transgenic Maize Line MIR604 ..................................................................... 238 |
|       | Molecular Characteristics of Transgenic Maize Line MIR604 ............................................................. 238 |
|       | Global Registration Status of Maize Line MIR604 ..... 239 |
|       | Safety Assessment of Transgenic Maize Line MIR604 Conducted in the Russian Federation........... 240 |
| 5.2.9 | Maize Line 3272 Producing Alpha-Amylase Enzyme......255 |
|       | Molecular Characteristics of Maize Line 3272 ........... 255 |

Global Registration Status of Maize Line 3272 .......... 256
Safety Assessment of Maize Line 3272 Conducted
in the Russian Federation ........................................... 257
5.3 Rice ................................................................................. 272
5.3.1 Rice Line LLRICE62 Tolerant to Glufosinate
Ammonium.............................................................. 272
Molecular Characteristics of Rice Line LLRICE62 ..... 272
Global Registration Status of Rice Line LLRICE62 .... 273
Safety Assessment of Transgenic Rice Line
LLRICE62 Conducted in the Russian Federation....... 273
5.4 Potato............................................................................... 289
5.4.1 Potato Variety Superior Newleaf Resistant to
Damage Caused by Colorado Potato Beetle............... 289
Molecular Characterization of Superior NewLeaf
Potato Resistant to Damage Caused by Colorado
Potato Beetle ............................................................ 289
Global Registration Status of Transgenic Potato
Superior NewLeaf ...................................................... 290
Safety Assessment of Transgenic Potato Resistant
Against Colorado Potato Beetle Conducted in the
Russian Federation.................................................... 290

**CHAPTER 6** Control System of Food Products Derived from
Genetically Modified Organisms of Plant Origin ........ 307
6.1 Control of Food Products Containing
Genetically Modified Ingredients................................... 307
6.2 Methods to Control Food Products Containing
Genetically Modified Ingredients................................... 310
Identification of Recombinant DNA by Polymerase
Chain Reaction (PCR) Method ........................................ 310
Identification of Recombinant DNA with Biological
Microchips................................................................... 313
Quantitative Detection of Recombinant DNA by
Real-Time PCR Analysis ................................................. 315
References ................................................................................ 317

**CHAPTER 7** Monitoring of Food Products Derived from Genetically
Modified Organisms of Plant Origin in the Russian
Federation ................................................................... 319
References ................................................................................ 327

**CHAPTER 8** Information Service for the Use of Novel
Biotechnologies in the Food Industry ......................... 329
References ................................................................................ 331

**INDEX** ................................................................................................... 333

# Authors

| | |
|---|---|
| Introduction | V. A. Tutelyan[1], Prof., Ph.D., D.Sc. |
| Chapter 1 | M. P. Kirpichnikov[2], Prof., Ph.D., D.Sc. |
| | N. V. Tyshko[1] Ph.D. |
| Chapter 2 | N. V. Tyshko[1], Ph.D. |
| Chapter 3 | G. G. Onishchenko[3], Prof., Ph.D., D.Sc. |
| | E. Yu. Sorokina[1], Ph.D |
| Chapter 4 | I. N. Aksyuk[1], Ph.D. |
| | O. N. Chernysheva[1], Ph.D. |
| | L. V. Kravchenko[1], Ph.D. |
| | V. K. Mazo[1], Prof., Ph.D., D.Sc. |
| | G. G. Onishchenko[3], Prof., Ph.D., D.Sc. |
| | I. A. Rogov[4], Prof., Ph.D., D.Sc |
| | B. F. Semenov[5], Prof., Ph.D., D.Sc. |
| | K. G. Skryabin[6], Prof., Ph.D., D.Sc. |
| | V. A. Tutelyan[1], Prof., Ph.D., D.Sc. |
| | N. V. Tyshko[1] Ph.D. |
| Chapter 5 | I. N. Aksyuk[1], Ph.D. |
| | O. N. Chernysheva[1], Ph.D. |
| | N. A. Kirpatovskaya[1], Ph.D., D.Sc. |
| | L. V. Kravchenko[1], Ph.D. |
| | V. K. Mazo[1], Prof., Ph.D., D.Sc. |
| | G. G. Onishchenko[3], Prof., Ph.D., D.Sc. |
| | I. A. Rogov[4], Prof., Ph.D., D.Sc. |
| | K. E. Selyaskin[1] |
| | B. F. Semenov[5], Prof., Ph.D., D.Sc. |

|  |  |
|---|---|
|  | K. G. Skryabin[6], Prof., Ph.D., D.Sc. |
|  | E. Yu. Sorokina[1], Ph.D. |
|  | V. A. Tutelyan[1], Prof., Ph.D., D.Sc. |
|  | N. V. Tyshko[1], Ph.D. |
| Chapter 6 | O. V. Anisimova[1], Ph.D. |
|  | O. N. Chernysheva[1], Ph.D. |
|  | G. G. Onishchenko[3], Prof., Ph.D., D.Sc. |
|  | E. Yu. Sorokina[1], Ph.D. |
|  | V. A. Tutelyan[1], Prof., Ph.D., D.Sc. |
| Chapter 7 | O. V. Anisimova[1], Ph.D. |
|  | O. N. Chernysheva[1], Ph.D. |
|  | E. Yu. Sorokina[1], Ph.D. |
| Chapter 8 | N. V. Tyshko[1] Ph.D. |

[1]Institute of Nutrition, RAMS.
[2]M. V. Lomonosov Moscow State University.
[3]Chief State Sanitary Physician of the Russian Federation.
[4]Moscow State University of Applied Biotechnology, Ministry of Science and Education.
[5]I. I. Mechnikov Institute of Vaccines and Sera, RAMS.
[6]Bioengineering Center of RAS.

# Editors

**Scientific Editors:**

**E. Yu. Sorokina** (Chapters 3, 5, 6, and 7) and **N. V. Tyshko** (Chapters 1, 2, 4, 5 and 8).

**Technical editor:**

Natalia N. Bogdanova, D.V.M., Biotechnology Regulatory Solutions, LLC

**Acknowledgements:**

The authors would like to acknowledge the work on the following researches involved in the assessment of GMO safety (Chapter 5):

**Institute of Nutrition, RAMS:** L. I. Avren'eva, Yu. P. Aleshko-Ozhevsky, V. G. Baikov, E. K. Baigarin, V. A. Baturina, N. A. Beketova, L. Sh. Vorob'eva, S. V. Volkovich, O. A. Vrzhesinskya, V. G. Vysotsky, M. M. Gapparov, I. V. Gmoshinsky, N. A. Golubkina, G. V. Guseva, V. M. Zhminchenko, G. F. Zhukova, L. P. Zakharova, S. N. Zorin, V. A. Isaeva, V. M. Kodentsova, S. N. Kulakova, N. V. Lashneva, M. M. Levachev, L. G. Levin, A. B. Levitskaya, V. L. Lupinovich, N. B. Maganova, G. Yu. Mal'tsev, N. N. Makhova, F. A. Medvedev, S. V. Morozov, G. V. Nikol'skaya, V. A. Pashorina, A. I. Petukhov, A. A. Sokol'nikov, A. I. Sokolov, I. B. Sedova, Yu. A. Sysoev, I. B. Tarasova, L. A. Kharitonchik, S. A. Khotimchenko, and L. V. Shevyakova

**I. I. Mechnikov Vaccines and Sera Institute, RAMS:** M. V. Britsina, S. I. Elkina, N. S. Zakharova, and M. N. Ozeretskovskaya

**Moscow State University of Applied Biotechnology, Ministry of Science and Education:** N. V. Gurova, N. G. Krokha, I. A. Popello, and V. V. Suchkov

**Federal Research Institute of Cereals and Its Products, Russian Academy of Agricultural Sciences:** L. S. L'vova

# Reviewers

Prof. Valentin I. Pokrovsky, Ph.D., D.Sc.

Prof. Vladimir N. Yarygin, Ph.D., D.Sc.

Genetically Modified Food Sources: Safety Assessment and Control

Edited by Prof. V. A. Tutelyan, Ph.D., D.Sc.

This book reports for the first time the detailed results of the studies on human and animal food/feed safety assessment of 15 lines of genetically modified plants. The authors focused on issues of the basic legislative regulations of plant biotechnology in the Russian Federation, and approaches to the human and animal assessment of safety of food and feed, and control of the food produced from the genetically modified organisms. The book is addressed to a wide community of the specialists working in various fields of medicine and biology, to the students and postgraduates focusing on the problems of modern biotechnology and biological safety, and sanitary inspectors.

# Abbreviations

| | |
|---|---|
| AI | Anaphylactic Index |
| BV | Biological Value |
| CaMV | Cauliflower Mosaic Virus |
| DC | Diene Conjugates |
| DNA | Deoxyribonucleic acid |
| cDNA | Complementary DNA |
| T-DNA | Transferred DNA |
| ESR | Erythrocyte Sedimentation Rate |
| GM plants | Genetically Modified plants |
| GMO | Genetically Modified Organism |
| GMS | Genetically Modified Source |
| GOST | National Standards of Russian Federation |
| HEA | Hen's Egg Albumin |
| HPLC | High-Performance Liquid Chromatography |
| LPO | Lipid Peroxidation |
| MCH | Mean Cell Hemoglobin |
| MCHC | Mean Cell Hemoglobin Concentration |
| MCV | Mean Cell Volume |
| MDA | Malonic Dialdehyde |
| NPER | Net Protein Efficiency Ratio |
| PCR | Polymerase Chain Reaction |
| PER | Protein Efficiency Ratio |
| PFA | Polyunsaturated Fatty Acid |
| QMAFAnM | Quantity of Mesophilic Aerobic and Anaerobic Microorganisms |
| RAS | Russian Academy of Sciences |
| RAMS | Russian Academy of Medical Science |
| RAAS | Russian Academy of Agricultural Sciences |
| RI | Reaction Index |
| RNA | Ribonucleic acid |
| mRNA | Messenger Informational RNA |
| SE | Sheep Erythrocytes |

| | |
|---|---|
| SOD | Superoxide Dismutase |
| TLC | Thin-Layer Chromatography |
| TPD | True Protein Digestibility |
| WHO | World Health Organization |

# Introduction

The basic law of nature is maintenance of life by striving through reproduction to maintain survival of the biological species. This law applies to all living organisms from single-cell to human beings. The most important prerequisite to species preservation is the provision of sufficient amount and necessary assortment of nutritional elements (food for humans), which constitute the source of energy, building material, and biologically active regulatory substances. Deficiency or decrease in bioavailability of these nutritional substances lead to reduction or even to complete disappearance of population, while sufficient supply and availability of these substances results in the development, perfection, and expansion of the natural habitat of the living organisms.

Throughout history, mankind has tried to solve the fundamental problem of reliable food provision. While humans coped with this problem by persistent search for food and the means to preserve it, the vital problem of the settled population became not only production and preservation of the food, but also maximization of the output from natural food sources. While engaged in plant cultivation and cattle breeding, humans not only used all available means to enhance production of conventional varieties and livestock species, but they also searched for novel food sources.

Evidently, the cornerstone of many (if not all) political, socio-economic, military, and other cataclysms shaking human society during its historical development was the struggle to expand territory in order to gain access to additional food sources.

To resolve the present challenge to provide mankind with food, a wide variety of technical and technological means based on scientific achievements are being used. The most important responsibility of any state is to ensure availability and safety of food in the country, based on its own crop and cattle-breeding production in sufficient amounts to provide the necessary source of raw materials to meet the requirements of any human being in energy, food, and biologically active substances, thereby ensuring the nation's health. One of the most efficient and promising ways to increase food resources is based on application of the methods of modern biotechnology, which emerged at the interfaces between fundamental research avenues and became a powerful production force capable of contributing significantly to solution of food production challenges. Fundamental studies of the recent decades in medicine and biology, including genetics, genomics, and postgenomic technologies, opened a novel scientific field: genetic engineering. The potency of genetic engineering made feasible the replacement of the chaotic empirical search for favorable mutations by the targeted modification of genome to obtain the desired traits. First of all, it is used in plant cultivation and production of genetically modified (GM) plants with increased yield, extended shelf-life, and tolerance to various natural factors. Even now the food derived from transgenic plants, an important product of genetic engineering, significantly contributes to the global food balance.

There are several equally important aspects of the practical use of GM food sources. The first aspect relates to the technology of development and logistics of large-scale production of the new plant varieties. Until recently, it was not only an extremely sophisticated but also a very long and expensive process. However, experience acquired in the last decade, development of new methods, and improvement of the technology have recently contributed to reduction of the time and material expenditures required to bring new products to market.

At the same time, the role of the second aspect—the development and improvement of the system of human and animal health safety assessment of food derived from GM plants—has increased significantly. Currently, this aspect is the most important in decision-making about admission of GM plants to large-scale production and on their use as food and feed sources.

The third aspect concerns protection of society against intended harmful application of modern biological technologies. Any technology can be used for both welfare or detriment of man. Examples are the outstanding achievements in chemistry and microbiology that were also used to make poison gases and biological weaponry for military purposes, the use of nuclear power to provide energy as well as military applications like the atomic bomb, etc. The most important if not unique way to protect mankind from the potential unintended side effects of scientific and technological progress is to set high standards of social and industrial culture, maintain strict observance of technological requirements, and establish uncompromised control and supervisory measures.

The possibility of careless handling of projects intended to create GM food sources, and the need to assess their safety and the feasibility of obtaining genetically modified organisms (GMOs) for biological terrorism, explain the critical importance of a standardized and methodical basis for safety assessment and reliable monitoring of GMO production.

Finally, the fourth aspect unites a number of problems that seem insignificant at first glance, but become extremely important in relation to dissemination of information among professionals and the general population. Although scientific society possesses a well-developed system of scientific and technical information, the field of practical biotechnology is rather closed and does not publish broadly enough a wide variety of scientific literature on genetic engineering technology and the results of the related medical and biological studies intended to assess the safety of GMOs. Distribution of such information among the civilian population is far worse. Sad experience in Russia illustrates that insufficient attention to public relations and accessibility of information by the general population not only impedes technological progress, but also negatively impacts promising industrial applications.

The negative Russian experience of banned genetic research during the 1940s is instructive: the ignorant leadership of the country eliminated and buried genetic science at a time when it was rated highly in the world. As a result, the country was set back for decades. Now the development of some fields in this science in Russia lags behind the world level—but, one hopes, not forever. Another example is shown by the dramatic events at the end of the twentieth century. To this time, Russia had the most powerful microbiological industry in the world. Ten factories produced 1.5 million tons of fodder protein, which formed a reliable forage reserve for poultry farming and partially for cattle breeding. At this time, scientific data attesting the safety of microbiologically synthesized protein was rapidly accumulating. This problem was intensively studied in dozens of research institutes of the Soviet Union Academy of Sciences, Academy of Medical Sciences, V. I. Lenin All-Union Academy of Agriculture, Ministry of Health, Ministry of Agriculture, Central Directorate of Microbiological Industry of the USSR, and in several research institutes in East Germany. These studies yielded comprehensive data attesting to the safety of the use of microbiologically synthesized fodder protein.

However, one or two biotechnological companies producing the fodder protein identified problems related to the negative ecological effect of this production on the environment. The corresponding technological defects could be easily eliminated. Unfortunately, this was the period of election to the State Duma (Russian Parliament) characterized by especially destructive campaigns of some politicians. The problems of microbiological production became the focus of fiery speeches, which led to an absurd situation whereby a few people ignorant in microbiological science became parliamentarians and adhered to their election pledges. As a result, not only the problematic factories, but all similar production plants were closed. The country lost the entire branch of microbiological industry. Who can count the negative consequences of such forcible measures that have nothing to do with economic science and common sense? The losses of forage reserves led to persistent and progressive reduction of cattle breeding and virtually complete loss of poultry farming. Russia became dependent on food imports. Instead of the development of the domestic food industry, the country must buy foreign food. This is an example of how political ambitions can deliver a blow to food production in Russia—an event with consequences that will be experienced for a long time.

In 1994, the USA registered the first GM tomato for use as food (variety FLAVR SAVR), with enhanced resistance to rotting and increased shelf-life. Foreign countries quickly appreciated the evident advantages of agricultural GM crops and widely applied genetic engineering in plant cultivation. As a result, production of GM food sources steadily increased. At present, there is a real possibility to supply the Russian food market with GM products as well.

Consequently, the professionals of relevant ministries and departments developed the necessary legislative, normative, and methodical principles for regulation of the requirements and procedure to assess safety of GM food products and to control their presence on the food market. It is worthy of note that all the work to create the regulatory and methodological basis was prospective, as the world production of GM food was negligible at that time.

With the active participation of academician of RAS M. P. Kirpichnikov, academician of RAMS G. G. Onishchenko, academician of RAAS K. G. Skryabin, and other scientists, a system for the safety assessment of plant-derived GMOs and a system for post-market monitoring were created, and both directives are being updated in response to the requirements of modern science. For example, the medico-biological assessment of GMO safety includes the use of such modern methods as proteomic and metabolomic analyses. At present, the Russian national system of GMO safety assessment is the strictest in the world—it has more stringent requirements than those of the USA, Canada, Australia, Japan, or the European Union and is stricter than recommended by the WHO. Experience acquired during previous years has played an important role in the development of the GMO safety assessment system.

The system for monitoring of food containing GM crops secures maximum protection of the Russian food market from GMOs not registered in Russia. In 1998–2007, the Russian Consumption Inspectorate approved a number of standards and methodical directives (Sanitary Regulation, Standards, and Methodical Directives) that regulated the order and methods to control GM food products. These Directives introduced obligatory labeling of such products. Some control methods were approved as the National Standards. Due to the efforts of the Head State Sanitary Inspector of Russia, the entire system of Russian Consumption Inspectorate has the necessary instrumental and methodical basis as well as qualified specialists to efficiently monitor GMOs in all states of the Russian Federation. At present, the monitoring system for food containing GM crops carries out tens of thousands of analyses every year.

Thus, the principal problems of GMO safety assessment and control were solved, although the task of providing information for the Russian population is still pending. While the requirement to indicate the use of GM products in the label of a food product is absolutely substantiated and supported by the corresponding legislative and standard acts, the development of an adequate public education on GM sources of food is far from being achieved.

Retrospective analysis of these issues shows that pioneers of new technologies were partially "guilty" in allowing the rise of social aversion to the genetic innovations in the food industry. The public should be informed of the advent of novel technologies. Formation of public views on GM food should have been started as early as 1990s. This has nothing to do with PR actions, where persistent and annoying repetition keeps information at the subconscious level. Only open scientific information, popular science broadcasting and publications, educational programs, and explanatory work with the population allow formation of the correct social view on this issue. However, when the first GM products were placed on the market, the information on genetically modified food was confidential or highly specific, and could be used only by professionals. The reasons of confidentiality (classified know-how) and restriction of corresponding information within the limited circle of scientists (sophisticated technology, high-end scientific level) are understandable. Now it is evident that wide awareness of society about the nature of biotechnology is important; it could probably have prevented the present state of affairs in biotechnology in Russia.

Currently there are two camps in relation to biotechnology: the supporters and opponents of GM food. In those countries where the public is informed of the registration of novel GM food and placement on the agricultural market (USA, China, Australia), people have easily adopted the new technologies and relied on the state system of safety assessment of the new products. Examples of the opposite approach were shown until recently by the countries of the European Union, who limited the use of GM food for purely economic reasons, as well as some African countries. The same position is presently shared by Russia. In those countries where the public has not been sufficiently informed about the safety of GM food, people are cautious about GM food products or refuse to use them.

Unfortunately, such public opinion is unjustifiably supported by some researchers. The issue is the subject of negative propaganda that denies scientific data and arguments supporting GM food. The campaign against GM food seems profitable to some forces that seek to place barriers in the way of Russia's adoption of modern agricultural technologies. It is noteworthy that this discrediting campaign is focused only on GM crops but it does not "see" similar objects of genetic modification, the microorganisms, which for a long time have been successfully used in the pharmaceutical and food industry. In both fields, the products are meant for human consumption. Why can some GM products be used and others cannot? If the opponents care for the human genome, they should be equally concerned about GM crops and GM microorganisms. It is a good sign that, despite these opponents, the Russian Consumption Inspectorate and Russian Academy of Medical Sciences has developed an efficient system to control the safety and life-cycle of GM microorganisms. Thus, it is impossible to prevent progress in science in general and in its biotechnological branch in particular. Evidently, the future belongs to biotechnology. The next generation of GM crops is entering the market place. Some of these GM crops have improved nutritional characteristics and are able to produce higher yields under more challenging environmental conditions in the field. Biotechnology can raise the standards of human food and provide mankind with sufficient amount of vitally important minor food components such as vitamins and fatty acids, thereby improving intake of important nutrients.

More sophisticated technologies such as nanotechnology are presently being developed and introduced into modern life. To avoid the past mistakes, the developers of nanotechnologies and nanomaterials should make an effort to educate the public in these fields. The specialists should focus on the problem of safety control in nanotechnology and nanomaterials not in the future, but today. In Russia, there are some pronounced steps in this direction. However, there is a concern of potential delays in providing information to the population.

Biotechnology continues to grow and develop globally. Planting of biotechnology-derived agricultural crops has been increasing around the world. In 2010, 29 counties (the European included) planted about 148 million hectares with transgenic crops. It is expected that this figure will raise to 200 million hectares in 2015, which will account for 14% of cultivated land on the planet. Forty countries in all continents are predicted to adopt biotechnology-derived agricultural crops. In 2010, approximately one hundred lines of GM plants were registered and approved for a large-scale cultivation. Certainly, this is growing evidence indicating the considerable promise of biotechnology for the development of food and feed resources. However, in Russia we still delay the implementation of this technology and our agricultural production falls behind those countries that have adopted biotechnology. Russia has lagged the world leaders for 10–15 years. New measures should be taken to reduce this delay.

This book is an attempt to fill the informational vacuum on the safety assessment of GM crops in global scientific literature. In addition to specifics of legislative control of production and monitoring of food derived from

GM plants, as well as the principles and approaches to human and animal health and environmental assessment of their safety, this book reports for the first time the detailed results of experimental studies carried out on 15 varieties of various biotechnology-derived agricultural crops, which preceded the registration procedures in the Russian Federation. The last chapters describe important data on monitoring of the food derived from transgenic crops.

The path to production of this book met several challenges. Developers of the biotechnology-derived crops discussed herein raised confidentiality barriers which had to be overcome. The book is presented for the judgment of professionals, and we believe that it will be useful for a wide community of researchers, engineers, physicians, and biologists working in biotechnology, genetics, toxicology, hygiene, plant cultivation, etc.

The editors and authors express their gratitude to all the specialists who took part in discussion of this book, including the implacable opponents.

**Prof. V. A. Tutelyan**

# CHAPTER 1

# Fundamental Concepts of Development of Genetically Engineered Plants

At all stages of social development, ensuring the availability of food has been a prerequisite condition for survival of mankind, so the most advanced and efficient ways have been used to achieve this goal. Plant varieties and animal breeds used in agriculture have been produced through a centuries-old selection process targeted to enhance crop yield and animal productivity, adaptation to environment, increase of nutritional value, improvement of flavor, appearance, etc. By accumulation of knowledge, the researchers more and more efficiently manipulated the genome of the selected species, trying to obtain the desired traits in the shortest period of time. However, the capabilities of the methods were limited because the boundaries of selection were limited to the genome of a single species in each case. Acquisition of novel traits became possible only after overcoming interspecies barriers through the use of genetic engineering. In essence, genetic engineering uses traditional selection intended to improve the genotype of economically valuable crops, but it employs far more precise methods that significantly shorten the time needed to generate plants with desired traits [1,14,22,24,45].

At present, genetic engineering makes it possible to transfer the genes from one organism to another. Specifically, the methods of genetic engineering include synthesis of the genes, isolation of individual genes or hereditary structures from the cells, followed by rearrangement, copying, and multiplication of the isolated or synthesized genes or genetic structures, and integration of various genomes within a cell [1–3,24]. The process of generating genetically modified (GM) plants consists of several stages. The basic stages of this technique are: isolation of the target genes, insertion of the genes into a transfer vector, transformation of the plant cells, confirmation of transformation by molecular characterization of the inserted cassette, demonstrating the function of the target gene, and finally regeneration of the whole plant from the transformed cells (Figure 1.1) [5,7,61].

A genetic vector is a DNA molecule used in genetic engineering to transfer the genes from the donor to the recipient organism. A vector can be composed of a small extrachromosomal element (plasmid, phage, or viral DNA). As a

# CHAPTER 1: Fundamental Concepts of Development of Genetically Engineered Plants

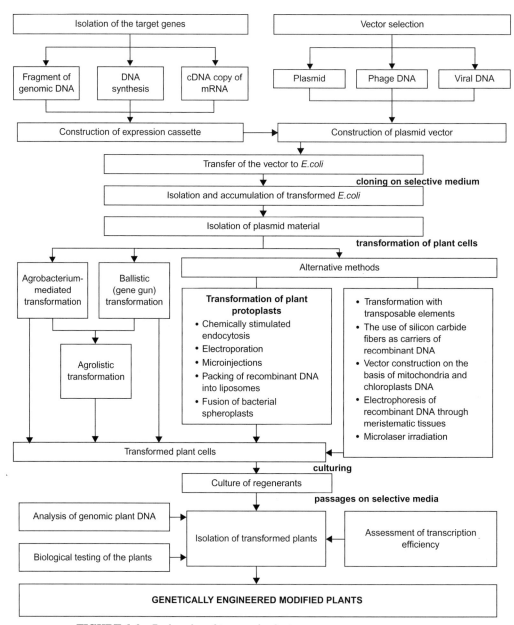

**FIGURE 1.1** Engineering of transgenic plants.

rule, plasmid vectors are used to transform the plants. The most widely used are combinations, where the role of intermediate host is given to a strain of *E. coli* and its plasmid, while *Agrobacterium tumefaciens* and its plasmid are the final hosts [39,50,53].

At the first stage of recombinant DNA production, insertions suitable for connection with the vector molecule are prepared. There are three ways to obtain such insertions:

- from genomic DNA fragmented with the use of restriction endonucleases or physical methods (such as ultrasound sonication);
- by synthesis of DNA fragments obtained by chemical or enzymatic methods or combination of thereof;
- from DNA segments (complementary DNA, cDNA) obtained through the use of enzymatic copying of RNA matrix *in vitro* [26,29,32].

In most cases, the target gene is modified, despite the universal character of the genetic code, because the codons encoding the same amino acids in prokaryotes and eukaryotes do differ significantly. Modification is necessary to exclude the gene from the sequences that potentially could be the sites of polyadenylation (premature termination of transcription) or could destabilize mRNA. In addition, the structural region of prokaryotic genes can incorporate undesirable signal sequences recognizable by splicing or degradation enzymes at mRNA level. The presence of such hidden signals results in a dramatic decrease of gene expression in the plant, so they are usually eliminated by the targeted substitution of the nucleotides [4,9,11]. There is a number of standard techniques to perform directed substitutions in the coding sequences based on the polymerase chain reaction (PCR) method. Replacement of codons does not affect the primary protein structure, although expression of a gene can be amplified up to 300-fold [42].

To express the target gene properly in plant cells, it should be placed under control of the corresponding regulatory elements that efficiently ensure transcription: promoters, transcription initiation sites, and terminators. Among the eukaryotic organisms, these elements are extremely conservative and universal. As a rule, plant cells correctly express foreign genes originated not only from the plants of different species, but also the genes of mammals, yeast, and other eukaryotes. Constitutive promoters are employed to obtain the product of the corresponding gene in significant amounts during the entire life of the plant. Examples of such promoters are the cauliflower mosaic virus 35 S promoter, the *Agrobacterium* nopaline synthase (nos1) promoter for dicotyledonous plants, the promoters of the maize alcohol dehydrogenase gene (*adh*) and the rice actin1 gene (*act*) for monocotyledonous plants. Among the presently isolated promoters, the cauliflower mosaic virus 35 S promoter has proven to be the most effective, so it is widely used for expression of target genes [11,31]. In addition to constitutive promoters, there are specific promoters active in particular tissues, cells, or at certain stages of plant ontogenesis. An example is the potato patatin gene promoter, which works in tubers only [44]. Inducible promoters that are activated by factors such as temperature, illumination, and chemical agents are being studied. Inducible promoters are rather attractive both for fundamental and applied research—specifically, in biotechnology where they can induce gene expression within a given period of time [33,40,55,61].

Therefore, the insert or expression cassette of the plasmid vector is a group of functionally related DNA fragments composed of a highly active promoter, immediately followed by the target gene and transcription terminator.

After the plasmids with inserts are obtained, it is possible to construct the plasmid vectors by digestion of the plasmid with the corresponding restriction enzymes followed by ligation with the insert. The conditions of vector ligation depend on the character of terminal parts of the vector and the insert, which can be either sticky (complementary single-strand DNA fragments at the opposite ends of the double-strand molecule) or blunt (the ends of the double-strand DNA molecule terminated with a bound pair of the complementary bases). The sticky ends of the vector and insert are connected by DNA-ligase under conditions promoting the development of hydrogen bonds between the complementary nucleotide ends. To fuse the blunt end, DNA-ligase and the fragments should be available in enhanced concentrations, because affinity of this enzyme to the blunt end is low. In addition to ligation of vector with insert, any of these fragments can bind its counterpart, producing vector-vector or insert-insert complexes and thereby decreasing the output of recombinant molecules [15].

The resulting plasmid vector is transferred into the host cell for cloning and amplification. The basic tool of molecular cloning is the two-component system of the compatible combination of vector and host, because efficient replication is possible only under optimal conditions for the plasmid vector, which employs not only the metabolites, enzymes, and other proteins of the host cell, but also its protein synthesis apparatus [24,53,63].

Penetration of isolated DNA molecules into the live cells of E. coli (transfection) proceeds most efficiently when permeability of the membrane increases—for example, due to its local breakage. The breaks in the cell membrane can be produced by exposure of the cells to certain chemicals or by the direct effect of electric current (electroporation). Thereafter the transfer of plasmid DNA is performed during few minutes (transformation efficiency determined by the number of transformed cells per 1 µg introduced DNA is $10^7$–$10^8$ or $10^6$–$10^9$, correspondingly) [3].

A prerequisite condition of successful cloning is the possibility to separate all the transfected *E. coli* cells. Usually the plasmid vectors incorporate marker genes introducing to the host cells a phenotype which can be easily used in the selection process because it indicates the presence of this vector. For example, cells sensitive to a certain antibiotic or toxin can be used in combinations with the vectors, which incorporate the genes conferring resistance to these agents. By culturing the microorganisms under conditions revealing the dependence of these microorganisms on the vector genes, it becomes possible to identify, select, and multiply the cell harboring the desired genetic material [64].

Production of recombinant DNA in amounts necessary to modify the plant genome is the final prerequisite prior to the key stage of development of a GM plant. There is a number of methods to transform the plant genome, such as *Agrobacterium*-mediated transformation, ballistic (gene gun) transfer, injection of genes into plant protoplasts, and some alternative approaches (Figure 1.1) [3,7,58,61]. Efficient and reliable transfer of vector DNA into the plant protoplasts can be performed by electroporation, microinjection, DNA packing in liposomes, fusion of bacterial spheroplasts, and chemically induced endocytosis [7]. Protoplasts (plant cells deprived of cell wall after exposure to cellulases) are the most suitable recipients of recombinant DNA, because they are bacteria-like systems where each cell assumes a competent state and possesses pronounced proliferative ability. In comparison with multicellular transformations, the use of protoplasts significantly decreases losses during selection of modified cells [58]. The alternative ways of genetic transformation are mostly experimental. One of them is the design of vectors based on plant cell organelles; another produces the genetic transformation with the use of transposed elements. In addition, there are the techniques to modify the plant genome that use silicon fibers 0.6 μm in diameter and 10–80 μm in length as carriers of the recombinant DNA and transport it across the meristematic tissues by electrophoresis, or to alter it by microlaser irradiation making microscopic pores in cell walls and membranes as a preliminary step to incubation of the perforated cells in solutions with vector DNA [7,58,61].

Ideally, the transformation system should be simple, inexpensive, and efficient. However, despite a rather wide selection of methodological approaches, no method meets all of these requirements. Presently, the large-scale production of GMOs is based mostly on the agrobacterial and ballistic technique used to modify the plant genome [20,38,49].

*Agrobacterium* (*A. tumefaciens* and *A. rhizogenes*) belong to the *Rhyzobium* genus of bacteria with the characteristic ability to induce tumors (crown-gall disease) in many dicotyledonous plants. A fragment of agrobacterial tumor-inducing (Ti)-plasmid, called transfer DNA (T-DNA), is transferred into the genome of plant cells. The delivery of T-DNA is a unique natural process of

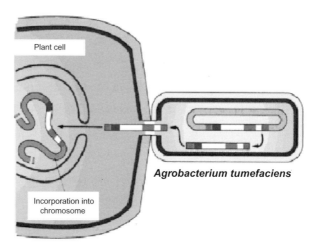

**FIGURE 1.2** *Agrobacterium*-mediated transformation of a plant cell.

genetic information exchange between prokaryotic and eukaryotic organisms (Figure 1.2) [17].

The transfer of T-DNA from the agrobacterial to the plant cell is performed with participation of the virulence locus (vir) of the Ti-plasmid and chromosomal loci chv and att, which control attachment. The chromosomal loci are expressed constitutively, while the vir-locus with its 10 operons (virA–J) is induced by plant metabolites such as acetosyringone or hydroxyacetosyringone [3,17,61]. The single-strand molecule of T-DNA, constructed by using acetosyringone-activated products of vir-genes, penetrates into the nucleus of the plant cell and incorporates itself into the plant chromosome via recombination. T-DNA is integrated through use of a 25-base-pair segment on its right flank, which probably is the recombinogenic integration site [19]. Only the sequence located to the left of this site is transferred. A similar segment, which probably marks the end of the integrated sequence, is sometimes available on the left flank of T-DNA. It is noteworthy that the vir-region of the Ti-plasmid, which determines the delivery, is not transferred into the host cell [24,48,69]. The plant cell with incorporated T-DNA produces organic substances that serve as the sources of carbon and nitrogen for agrobacteria.

Thus, the use of the Ti-plasmid as a vector of recombinant DNA is based on incorporation of the target fragment into the T-DNA region. All vectors constructed on the basis of Ti-plasmids are similarly organized and include the following elements:

- replication initiation site allowing the plasmid to replicate in *E. coli*;
- replication initiation site in *A. tumefaciens*;

- the right flanking sequence (recombinogenic integration site);
- the left flanking sequence;
- multiple cloning site, the polylinker sequence needed to incorporate the gene into the region between the boundaries of T-DNA;
- a selection marker gene controlled by eukaryotic regulatory components [3].

Because of the absence of unique restriction sites in the Ti-plasmid, it is impossible to directly insert foreign DNA into it. However, there is a method to insert the genes into T-DNA. At first, T-DNA is cloned in *E. coli* (in most cases, it is *E. Coli* plasmid pBR322). The target DNA segments are inserted into suitable restriction sites of T-DNA of the isolated plasmid, and thereafter the obtained recombinants are repeatedly cloned in *E. coli*. Then the recombinant plasmid is introduced by conjugation into *A. tumefaciens* cells which have the wild type Ti-plasmid. As a result of homologous recombination, T-DNA with inserted fragment enters the region of intact Ti-plasmid. Selection of *A. tumefaciens* cells which carry the recombinant plasmids is easily performed if the marker gene is inserted into the modified DNA.

The cloning vectors have no vir-genes; therefore, they cannot transfer and integrate T-DNA into the plant cells. However, other methods exist to solve this problem. One of them is based on the use of a binary vector system, which is a dual-plasmid *Agrobacterium* system, where the virulence genes are located in one plasmid, while the insertion fragment of T-DNA is situated on the second plasmid. Thus, the Ti-plasmid with vir-genes plays the role of a helper in integration of T-DNA into the genome of the plant cell. Efficient super-binary vectors, which contain extra copies of virulence genes, were constructed using genetic engineering. In other cases, researchers employ an integrative vector system constructed by insertion of the vector containing modified T-DNA into the "helper plasmid". Such a construct integrates T-DNA into the plant cells. It should be stressed that, in both cases, the helper plasmid contains the complete set of vir-genes, but it is deprived of T-DNA (partially or entirely), so this plasmid in non-oncogenic.

The number of the transformed cells can be increased by using a strain of *A. tumefaciens* that has enhanced virulence towards certain plant species. Nevertheless, efficiency of the agrobacterial transformation is not sufficient: only 1 of 10,000 plant cells contains the recombinant DNA [13,15,61,68,69].

The ballistic method of transformation of the plant genome, also known as microprojectile bombardment, gene gun, particle acceleration, or biolistic technique (biology + ballistics), is based on the bombardment of intact plant cells with microscopic gold or tungsten particles that are used to transfer recombinant DNA (Figure 1.3) [25]. These particles can be made of any inert metal with a high molecular weight (gold, tungsten, palladium, rhodium,

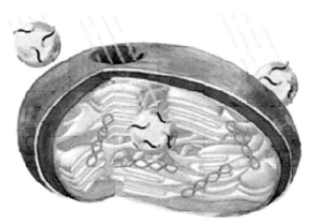

**FIGURE 1.3** Biolistic transformation of a plant cell.

platinum, indium, etc.) that do not form organometallic complexes with DNA. Importantly, their kinetic energy should be high enough to ensure efficient penetration across the cell wall [35,41,59,61].

The particles 1.5–3.0 µm in size conjugated with the recombinant DNA are accelerated to a velocity of 300–600 m/s by electric discharge or decompression in the direction of the target cells. The biolistic method is widely used to transform monocotyledonous plants; however, its efficiency is also low [59,60].

Recently a combined transformation method, called agrolistic, has been developed and successfully employed. In this technique, bombardment is used to introduce foreign DNA, which harbors the T-DNA vector with the target, marker, and agrobacterial vir-genes [7,58].

Because of the totipotency of many cells (i.e., the potency of reproduction, differentiation, and formation of whole fructiferous plants), production of GM plants from the transformed cells is not problematic. The transformation process is conducted under aseptic conditions to exclude contamination by bacteria that can lead to false-positive results during PCR testing [10,13,64].

Culturing of regenerated cells includes several series of passages in selective media. Duration of regeneration of the transformed plants amounts to several months. It is important that, during this period, the plants are exposed to selective agents present in the media at high concentrations. As a rule, two basic types of marker genes are employed: selective and reporter. Selective marker genes impart plant resistance to antibiotics and herbicides, while reporter genes determine synthesis of peptides which are neutral to the cells and can be easily detected in the tissues. These genes protect the plants and allow them to grow under the action of antibiotics and herbicides. To this end, the following agents are most frequently used:

- gene *nptII* encoding neomycin phosphotransferase which inactivates aminoglycoside antibiotics (kanamycin etc.);
- gene *hptI* encoding hygromycin phosphotransferase which inactivates antibiotic hygromycin;
- gene *bar* encoding phosphinothricin acetyltransferase which inactivates herbicide glufosinate (bialafos);
- gene *gox* encoding glyphosate oxidase which inactivates the herbicide glyphosate.

During the development of a GM plant tolerant to herbicides the introduced gene conferring herbicide tolerance is used as the target and selective gene. In contrast to selective genes, reporter genes virtually do not affect the metabolism of modified plants. In most cases, the role of reporter genes is given to the genes of β-glucuronidase (GUS), green fluorescent protein (GFP), luciferase (LUC), chloramphenicol acetyltransferase (CAT), etc. [37,46]. Frequently, it is very useful to replace the selective genes with the reporter ones during selection of GM plants, because the use of the reporter genes virtually excludes any negative unintended effects on the environment and human health.

The presence of marker genes in GM crops (especially those genes that confer resistance to antibiotics) is one of the major arguments against the use of GM plants. Therefore, novel methods have been developed to eliminate the marker genes after regeneration of GM plants. Selective elimination of the marker genes is possible with the help of special transformation systems and methods such as co-transformation system, site-specific recombination, intragenomic redistribution (transposition) of the genes with transposable elements, the use of specific promoters of the marker genes, and replacement of the original gene with the corresponding modified gene [67].

However, it should be remembered that DNA is subjected to hypermethylation in plants exposed to stress conditions during culturing in a medium with antibiotics kanamycin, hygromycin, and cefotaxime [56]. Methylation of the regulator-promoter regions in the genes of higher eukaryotes (plants included) terminates expression of these genes. As a result, the probability of "silencing" of the introduced genes is greatly enhanced. Thus, selection of GM plants in a medium with lower concentrations of kanamycin moderates the stress load and creates conditions for their normal growth and development [5].

The next stage is analysis of the genomic DNA of the plants to determine the presence of the introduced gene and the number of its copies integrated into the genome. Since the specific modification of the plant genome results from recombinant interaction of the target locus of the plant DNA with the cloned homologous (or partially homologous) fragment of foreign DNA, the "recombinant" behavior of DNA determines both quality and quantity of the transferred DNA molecules. Analysis of genomic DNA can be made with various tools, including detection with restriction analysis, Southern blot analysis, PCR of recombinant DNA, Northern blot analysis of RNA, assessment of concentration of the protein produced by the recombinant gene of interest, and the use of site-specific mutation genes in a certain combination that in the case of recombination ensures restoration of normal gene function [7].

The final stage of laboratory testing of GM plants includes the assessment of the quality and stability of the introduced trait. The necessity of such tests is mostly based on ample evidence of genetic instability which shows that the level of transgenes is largely unpredictable and greatly varies even in plants transformed by identical DNA-constructs of the clones obtained in the same experiment. It is established that expression level depends on many factors and, in particular, on the number of copies of the transgenes and their insertion sites. In addition, an insertion in genome can largely rearrange DNA structure due to duplication, inversion, etc. When a transgene incorporates into the region of active (transcribed) chromatin, its expression is usually high. To gain stable expression and reproduction of the transgenes in GM plant lines, the following should be taken into consideration.

- Gene silencing is frequently observed after integration of the complex insertions and the copies with rearrangements, duplications, and deletions, therefore, the greatest probability of stable expression results from complete single insertion of the target gene.
- The degree and length of homology of the introduced gene with the recipient genome should be minimal.
- It is preferable to integrate the transgene copies into a non-methylated region of the plant genome.

Thus, monitoring of expression level of the transgene in subsequent plant generations is absolutely necessary [3,7,15].

Based on the methods described above, over fifty varieties of GM plants have been developed and tested under field conditions globally. However, only a little over one hundred GM plant lines have been registered for mass production. Most of them are lines resistant to plant insect pests and herbicides. Resistance of GM plants to non-selective wide-spectrum herbicides results from insertion of genes encoding a corresponding protein which is insensitive to a particular class of herbicides (glyphosate, chlorsulfuron, and imidazole herbicides) or by introduction of genes determining accelerated metabolism of the herbicides in the plants (ammonium glufosinate, dalapon) [12].

At present, glyphosate is the most widely used herbicide, which explains the need for glyphosate-tolerant GM plants [16]. Glyphosate (N-[phosphomethyl] glycine isopropylamine salt) belongs to non-selective herbicides that inhibit synthesis of several essential amino acids by affecting metabolism of shikimic acid. The key stage is synthesis of 5-enolpyruvilshikimate-3-phosphate from phosphoenolpyruvate and shikimate-3-phosphate catalyzed by 5-enolpyruvilshikimate-3-phosphate synthase. This enzyme is the target for glyphosate. The described metabolic pathway of shikimic acid is common in plants, algae, bacteria, mushrooms, and unicellular protozoa. Other living organisms, including insects, fish, birds, mammals, and humans, have no such metabolic pathway [34,47,65]. To obtain glyphosate-resistant plants, the *epsps* gene encoding synthesis of 5-enolpyruvilshikimate-3-phosphate synthase is modified in such a way that the corresponding expressed enzyme becomes tolerant to the herbicide. The gene *epsps* is obtained from the DNA of *Agrobacterium* sp. strain SP4 or by modification of the native gene of the plant to be transformed [30].

The active ingredient of herbicides synthesized on the basis of ammonium glufosinate is phosphinothricin, an inhibitor of glutamine synthase in plant cells. In plants, glutamine synthase converts ammonia into glutamine. Blockage of glutamine synthase by glufosinate results in a rapid depletion of glutamine in the plant, accumulation of ammonia in the photosynthesizing tissues, and plant intoxication [18,27]. The glufosinate-tolerance gene is denoted as *pat* or *bar* depending on its source, which can be either *Streptomyces viridochromogenes* or *Streptomyces hydroscopicus*, which produce a potent tripeptide antibiotic: bialafos. The active molecule of bialafos is L-phosphinothricin [3]. The gene *pat/bar* encodes the synthesis of phosphinothricin acetyltransferase, which acetylates the free $NH_2$-group of phosphinothricin. Plants modified with this gene acquire the ability to produce phosphinothricin acetyltransferase, resulting in tolerance to glufosinate [11,27,28,54].

The insecticidal action of δ-endotoxins produced during sporulation of gram-positive soil bacteria *Bacillus thuringiensis* is based on their specific binding to the receptors of intestinal epithelium of the insects, leading to disruption of osmotic balance, swelling and lysis of the cells [43]. δ-Endotoxins are produced in bacterial cells in the form of prototoxins converted in the insect intestines into the active protein, which is toxic to insects of *Lepidoptera*, *Diptera*, and *Coleoptera* classes, while it is harmless to mammals [4,21,23]. All described genes of δ-endotoxins are divided into multiple classes currently ranging from *cry1* to *cry70* on the basis of homology in the amino acid sequences and insecticidal activity of the encoded proteins. To obtain stable expression of *cry* genes in plants the amino acid sequence of the wild-type genes is codon optimized for expression in plants [51,52]. For example, the partially modified gene *cry3A* encoding δ-endotoxin in *Bacillus thuringiensis* subsp. *tenebrionis* is used to impart resistance against Colorado potato beetle to potato plants, while partially modified gene *cry1A* from *Bacillus thuringiensis* subsp. *kurstaki* is employed in maize protected from the damage of maize insect pests [57,66].

In instances where a native plant gene needs to be blocked, an antisense RNA method is used. In this case, a copy of the target gene with antisense DNA strands is inserted into the host genome. As a result, in the transgenic plant the target gene and its complementary copy produce two complementary mRNA molecules, which mutually bind and therefore block synthesis of the target protein. This method has been used to produce GM tomato plants with improved quality of fruits. The tomato genome was supplemented with an "inversed" copy of the *pg* gene, which controls synthesis of polygalacturonase involved in pectin depolymerization. In non-transgenic plants, polygalacturonase content is elevated during the ripening period, which results in fruit softening and dramatic shortening of the storage time. The use of antisense mRNA made it possible to produce tomato plants with the new trait resulting in longer shelf life of the fruits [6,8].

Between 1994 and 2004, GM food sources in the world market were represented by GM plants of the first generation, which are resistant to damage by larvae of insect pests, viruses, fungal infections, tolerant to herbicides, and possessing improved qualities. It can be predicted that GM plants of the second generation (2005–2015) will be resistant to insect pests damage, pathogens, and tolerant to herbicides, unfavorable climatic factors, high levels of salt in soil, etc. In addition, they will have a longer shelf life, enhanced nutritional quality, improved flavor properties, potential decrease in allergenicity, and be capable of producing pharmaceutical compounds. The third generation of GM plants (after 2015) will additionally have controllable periods of plant development. They also will be characterized by altered size, shape, and the number of fruits in the plant. The efficiency of photosynthesis will be enhanced, and these future plants will be capable of producing food with an enhanced ability to withstand environmental stress [36,62].

## REFERENCES

[1] Vorob'ev AI. Biotechnology and genetic engineering: high-priority courses of scientific and technical progress [in Russian]. Vest RAMS 2004(10):8–11.

[2] Glazko VI, Glazko GV. Russian-English-Ukrainian explanatory dictionary on applied genetics, DNA-Technologies, and bioinformatics [in Russian]. Kiev: 2000.

[3] Glik BR, Pasternak DD. Molecular biotechnology: principles and application [in Russian]. Moscow: 2002.

[4] Gulina IV, Shul'ga OA, Mironov MV, et al. Expression of partially modified gene of δ-endotoxin from *Bacillus thuringiensis* var. *tenebrionis* in transgenic potato plants [in Russian]. Mol Biol (Moscow) 1994;28(5):1166–75.

[5] Zakharchenko NS, Kalyaeva MA, Bur'yanov Ya. I. Methods of genetic transformation of white beet varieties [in Russian]. Fiziol Rast 2000;47(1):79–85.

[6] Kulaeva ON. How plant life is regulated [in Russian]. Soros Obrazovat Zh 1995(1):20–7.

[7] Kurochkina SD, Kartel' NA. Genetic transformation of the plants, recombination, and control of gene expression in the transgenic plants [in Russian]. Mol Gen Microbiol Virusol 1998(4):3–12.

[8] Lutova LA. Genetic engineering of plants: achievements and hopes [in Russian]. Soros Obrazovat Zh 2000;6(10):10–17.

[9] Maniatic T, Fritsch EF, Sambrook J. Molecular cloning: a laboratory manual. Cold Spring Harbor, N.Y; 1982.

[10] Padegimas L, Shul'ga OA, Skryabin KG. Creation of transgenic plants *Nicotiana tabacum* and *Solanum tuberosum* resistant to herbicide phosphinothricin [in Russian]. Mol Biol (Moscow) 1994;28(2):437–43.

[11] Padegimas L, Shul'ga OA, Skryabin KG. Testing of transgenic plants with polymerase chain reaction [in Russian]. Ibid 1993;27(4):947–51.

[12] Pokrovskaya SF. Genetic engineering technologies in production of grain-crops and other cultivars abroad: a review [in Russian]. Moscow: 2001.

[13] Revenkova EV, Kraev AS, Skryabin KG. Transformation of cotton plant *Gossypium Hirsutum L.* with supervirulent strain *Agrobacterium Tumefaciens* A281 [in Russian]. Mol Biol (Moscow) 1990;24(4):1017–23.

[14] Semenova ML. Why do we need transgenic animals? [in Russian]. Soros Obrazovat Zh 2001;7(4):13–20.

[15] Singer M, Berg P. Genes and Genomes. [Russian Translation] Moscow: 1998. p. 373–391.

[16] Skryabin KG., Spiridonov YY. Modern trends in combating agricultural pests with novel classes of herbicides and herbicide-resistant transgenic plants [in Russian]. Moscow: vol. 2; 2001.

[17] Chumakov MI. T-DNA transfer from agrobacteria to plant cells across cell walls and membranes [in Russian]. Mol Gen Microbiol Virusol 2001(1):13–28.

[18] Chumikov IA. Libertylink technology of aventis cropscience–a novel tool for successful combating agricultural pests. Modern trends in combating agricultural pests with novel classes of herbicides and herbicide-resistant transgenic plants. In: Skryabin KG, Spiridonov YY, editors. [in Russian]. Moscow: vol. 2; 2001. p. 77–79.

[19] Shestakov SV. Genomics of pathogenic bacteria [in Russian]. Vest RAMS 2001(10):18–25.

[20] Bagyan IL, Revenkova EV, Pozmogova GE, et al. 5'-Regulatory region of *Agrobacterium tumefaciens* T-DNA gene 6b directs organ-specific, wound-inducible and auxin-inducible expression in transgenic tobacco. Plant Mol Biol 1995;29(6):1299–304.

[21] Brunke KJ, Meeusen RL. Insect control with genetically engineered crops. Trends Biotechnol 1991;9:197–200.

[22] Chassy BM. Food safety evaluation of crops produced through biotechnology. J Am Coll Nutr 2002;21(3, Suppl):166S–73S.

[23] Cheng J, Bolyard MG, Saxena RC, Sticklen MB. Production of insect resistant potato by genetic transformation with a δ-endotoxin from *Bacillus thuringiensis* var. *kurstaki*. Plant Sci 1992;81:83–91.

[24] Chrispeels MJ, Sadava DE. Plants, genes, and crop biotechnology. 2nd ed. Sudbury, Massachusetts; 2002.

[25] Christou P. Genetic transformation of croup plants using microprojectile bombardment. Plant J 1992;2:275–81.

[26] Climie S, Santi DV. Chemical synthesis of the thymidylate synthase gene. Proc Natl Acad Sci USA 1990;87(2):633–7.

[27] Cremer J. The performance of liberty in GM libertylink crops (maize, osr and sugar beet) as a broad spectrum herbicide in Europe (experience from 10 years testing). Modern trends in combating agricultural pests with novel classes of herbicides and herbicide-resistant transgenic plants. In: Skryabin KG, Spiridonov YY, editors. [in Russian]. Moscow: vol. 2; 2001. p. 89–96.

[28] Dekker J, Duke SO. Herbicide-resistant field crops. Adv Agron 1995;54:69–116.

[29] Di Donato A, de Nigris M, Russo N, et al. A method for synthesizing genes and cDNAs by the polymerase chain reaction. Anal Biochem 1993;212(1):291–3.

[30] Fichet Y, Brants I. Glyphosate-tolerant sugar beet, an overview. Modern trends in combating agricultural pests with novel classes of herbicides and herbicide-resistant transgenic plants. In: Skryabin KG, Spiridonov YY, editors. [in Russian]. Moscow: vol. 2; 2001. p. 68–75.

[31] Finnegan H, McElroy D. Transgene inactivation: plants fight back!. Biotechnology 1994;12:883–8.

[32] Fox DK, Westfall B, Nathan M, et al. Striding new distances with 5'RACE: long 5'RACE of human APS and TSC-2 cDNA. Focus 1996;18:33–7.

[33] Gatz C, Lenk I. Promoters that respond to chemical inducers. Trends Plant Sci 1998;3:352–8.

[34] Giesy JP, Dobson S, Solomon KR. Ecotoxicological risk assessment for Roundup® herbicide. Rev Environ Contam Toxicol 2000;167:35–120.

[35] Hagio T. Optimizing the particle bombardment method for efficient genetic transformation. JARQ 1998;32:239–47.

[36] Harlander SK. The evolution of modern agriculture and its future with biotechnology. J Am Coll Nutr 2002;21(3, Suppl.):161S–5S.

[37] Harper BK, Mabon SA, Leffel SM, et al. Green fluorescent protein as a marker for expression of a second gene in transgenic plants. Nat Biotechnol 1999;17(11):1125–9.

[38] Hooykaas PJ, Schilperoort RA. Agrobacterium and plant genetic engineering. Plant Mol Biol 1992;19(1):15–38.

[39] Ishida Y, Saito H, Ohta S, et al. High efficiency transformation of maize (*Zea mays* L.) mediated by *Agrobacterium tumefaciens*. Nat Biotechnol 1996;14(6):745–50.

[40] Jepson I, Martinez A, Sweetman JP. Chemical-inducible gene expression systems for plants—a review. Pesticide Sci 1998;54:360–7.

[41] Klein TM, Wolf ED, Wu R, Sanford JC. High velocity microprojectiles for delivering nucleic acids into living cells. Nature 1987;327:70–3.

[42] Koziel MG, Carozzi NB, Desai N, et al. Transgenic maize for the control of european maize borer and other maize insect pests. Ann NY Acad Sci 1996;792:164–71.

[43] Kumar PA, Sharma RP, Malik VS. The insecticidal proteins of *Bacillus thuringiensis*. Adv Appl Microbiol 1996;42:1–43.

[44] Liu XL, Prat S, Willmitzer L, Frommer WB. Cis regulatory elements directing tuber-specific and sucrose-inducible expression of a chimeric class I patatin promoter GUS Gene Fusion. Mol Gen Genet 1990;223:401–6.

[45] Madden D. Food Biotechnology. Brussels: ILSI Press; 1995.

[46] Malik VS, Saroha MK. Marker gene controversy in transgenic plans. J Plant Biochem Biotechnol 1999;8:1–13.

[47] Mensink H, Janssen P. Glyphosate. Geneva: World Health Organization; 1994.

[48] Mitsuhara I, Ugaki M, Hirochika H, et al. Efficient promoter cassettes for enhanced expression of foreign genes in dicotyledonous and monocotyledonous plants. Plant Cell Physiol 1996;37(1):49–59.

[49] Newell CA, Rozman R, Hinchee MA, et al. Agrobacterium-mediated transformation of *Solanum tuberosum* L. cv. "Russet Burbank". Plant Cell Rep 1991;10:30–4.

[50] Pauls KP. Plant biotechnology for crop improvement. Biotechnol Adv 1995;13(4):673–93.

[51] Perlak FJ, Deaton RW, Armstrong TA, et al. Insect resistant cotton plants. Biotechnology (NY) 1990;8(10):939–43.

[52] Perlak FJ, Fuchs RL, Dean DA, et al. Modification of the coding sequence enhances plant expression of insect control protein genes. Proc Natl Acad Sci USA 1991;88(8):3324–8.

[53] Potrykus I. Gene transfer to plants: assessment of published approaches and results. Annu Rev Plant Physiol 1991;42:205–25.

[54] Quinn JP. Evolving strategies for the genetic engineering of herbicide resistance in plants. Biotechnol Adv 1990;8(2):321–33.

[55] Rupp HM, Frank M, Werner T, et al. Increased steady state mRNA levels of the *STM* and *KNAT1* homeobox genes in cytokinin overproducing *Arabidopsis thaliana* indicate a role for cytokinins in the shoot apical meristem. Plant J 1999;18(5):557–63.

[56] Schmitt F, Oakeley EJ, Jost JP. Antibiotics induce genome-wide hypermethylation in cultured *Nicotiana tabacum* plants. J Biol Chem 1997;272(3):1534–40.

[57] Shan DM, Rommens CMT, Beachy RN. Resistance to diseases and insects in transgenic plants: progress and applications to agriculture. Trends Biotechnol 1995;13:362–8.

[58] Songstad DD, Somers DA, Griesbach RJ. Advances in alternative DNA delivery techniques. Plant Cell Tissue Organ Cult 1995;40:1–15.

[59] Southgate EM, Davey MR, Power JB, Marchant R. Factors affecting the genetic engineering of plants by microprojectile bombardment. Biotechnol Adv 1995;13(4):631–51.

[60] Vain P, De Buyser J, Bui Trang V, et al. Foreign gene delivery into monocotyledonous species. Biotechnol Adv 1995;13(4):653–71.

[61] van den Eede G, Aarts H, Buhk HJ, et al. The relevance of gene transfer to the safety of food and feed derived from genetically modified (GM) plants. Food Chem Toxicol 2004;42(7):1127–56.

[62] Vasil IK. The science and politics of plant biotechnology—a personal perspective. Nat Biotechnol 2003;21(8):849–51.

[63] Walden R, Schell J. Techniques in plant molecular biology—progress and problems. Eur J Biochem 1990;192(3):563–76.

[64] Walden R, Wingender R. Gene-transfer and plant regeneration techniques. Trends Biotechnol 1995;13:324–31.

[65] Williams GM, Kroes R, Munro IC. Safety evaluation and risk assessment of the herbicide roundup and its active ingredient, glyphosate, for humans. Regul Toxicol Pharmacol 2000;31(2. Pt 1):117–65.

[66] Williams S, Friedrich L, Dincher S, et al. Chemical regulation of *Bacillus thuringiensis* σ-endotoxin expression in transgenic plants. Biotechnology 1992;10:540–3.

[67] Yoder JI, Goldsbrough AP. Transformation systems for generating marker-free transgenic plants. Biotechnology 1994;12(3):263–7.

[68] Zambryski P. Basic processes underlying Agrobacterium-mediated DNA transfer to plant cells. Annu Rev Genet 1988;22:1–30.

[69] Zambryski P, Tempe J, Schell J. Transfer and function of T-DNA genes from agrobacterium Ti and Ri plasmids in plants. Cell 1989;56(2):193–201.

# CHAPTER 2

# World Production of Genetically Engineered Crops

The large-scale production of GM crops started in 1996, when they were planted on 1.7 million hectares worldwide. Between 1996 and 2011 the planted area of GMOs increased 96-fold to 160 million hectares (Figure 2.1) [4]. In 2011, GM crops were grown in 29 countries, although the leading states remained the same: USA (43%), Argentina (15%), Brazil (19%), and Canada (7%) (Tables 2.1–2.3) [1,4].

As of October 2012 in the European Union, two GM maize lines (MON 810, T25) and potato variety EH92-527-1 [www.gmo-compass.org] are approved for cultivation and use as food and feed. The MON 810 line, which is protected from damage by the European corn borer, is the main GM maize line cultivated in the EU. Romania, prior to joining EU in 2007, had extensive experience with GM crops: during the period from 1996 to 2006, GM soybean was successfully grown in a planted area that increased by more than 9-fold, reaching 143.9 thousand hectares (a production increase of 33%) [5]. However, after joining the EU, planting of GM soybean in Romania was banned.

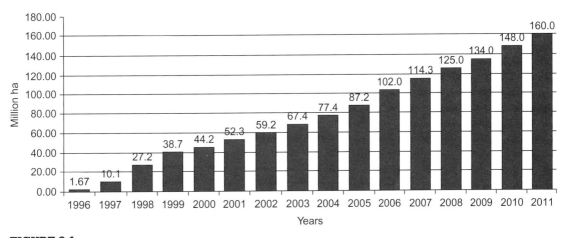

**FIGURE 2.1**
Growth of areas planted with GM crops [4].

Genetically Modified Food Sources. DOI: http://dx.doi.org/10.1016/B978-0-12-405878-1.00002-1
© 2013 Elsevier Inc. All rights reserved.

**Table 2.1** Growth of the areas planted with GM crops during 1996–2011

| Country | 1996 | 1997 | 1998 | 1999 | 2000 | 2001 | 2002 | 2003 | 2004 | 2005 | 2006 | 2007 | 2008 | 2009 | 2010 | 2011 |
|---|---|---|---|---|---|---|---|---|---|---|---|---|---|---|---|---|
| USA | 1.449 | 7.46 | 19.26 | 26.25 | 28.25 | 33.02 | 37.53 | 40.72 | 44.79 | 47.40 | 54.6 | 57.7 | 62.5 | 64.0 | 66.8 | 69.0 |
| Argentina | 0.037 | 1.76 | 4.82 | 6.84 | 9.61 | 11.78 | 13.59 | 14.90 | 15.88 | 16.93 | 18.0 | 19.1 | 21.0 | 21.3 | 22.9 | 23.7 |
| Brazil | – | 0.10 | 0.50 | 1.18 | 1.30 | 1.31 | 1.74 | 3.00 | 5.00 | 9.00 | 11.5 | 15.0 | 15.8 | 21.4 | 25.4 | 30.3 |
| Canada | 0.139 | 0.65 | 2.16 | 3.53 | 3.33 | 3.21 | 3.25 | 4.43 | 5.07 | 5.86 | 6.1 | 7.0 | 7.6 | 8.2 | 8.8 | 10.4 |
| China | – | 0.03 | 0.26 | 0.65 | 1.22 | 2.17 | 2.10 | 2.80 | 3.70 | 3.30 | 3.5 | 3.8 | 3.8 | 3.7 | 3.5 | 3.9 |
| India | – | – | – | – | – | – | 0.04 | 0.10 | 0.50 | 1.30 | 3.8 | 6.2 | 7.6 | 8.4 | 9.4 | 10.6 |
| Paraguay | – | – | – | – | 0.09 | 0.34 | 0.48 | 0.74 | 1.20 | 1.80 | 2.0 | 2.6 | 2.7 | 2.2 | 2.6 | 2.8 |
| Australia | 0.040 | 0.06 | 0.10 | 0.13 | 0.19 | 0.20 | 0.16 | 0.17 | 0.25 | 0.28 | 0.2 | 0.1 | 0.2 | 0.2 | 0.7 | 0.7 |
| South Africa | – | – | 0.00008 | 0.0008 | 0.09 | 0.15 | 0.21 | 0.30 | 0.53 | 0.60 | 1.4 | 1.8 | 1.8 | 2.1 | 2.2 | 2.3 |
| Pakistan | – | – | – | – | – | – | – | – | – | – | – | – | – | – | 2.4 | 2.6 |
| Uruguay | – | – | – | – | <0.1 | <0.1 | <0.1 | <0.1 | <0.1 | 0.30 | 0.4 | 0.5 | 0.7 | 0.8 | 1.1 | 1.3 |
| Other countries | 0.0009 | 0.02 | 0.06 | 0.07 | 0.09 | 0.01 | 0.02 | 0.11 | 0.43 | 0.39 | 0.5 | 0.5 | 1.3 | 1.7 | 2.2 | 2.4 |
| TOTAL | 1.67 | 10.08 | 27.16 | 38.71 | 44.26 | 52.28 | 59.21 | 67.36 | 77.45 | 87.16 | 102.0 | 114.3 | 125.0 | 134.0 | 148.0 | 160.0 |

## Table 2.2 GMO-Planted Area in 2011

|  | Country | Area (Mha) | GM Crops |
|---|---|---|---|
| 1 | USA | 69.0 | Soybean, maize, cotton, oilseed rape, papaya, vegetable marrow, alfalfa, sugar beet |
| 2 | Brazil | 30.3 | Soybean, maize, cotton |
| 3 | Argentina | 23.7 | Soybean, maize, cotton |
| 4 | India | 10.6 | Cotton |
| 5 | Canada | 10.4 | Oilseed rape, maize, soybean, sugar beet |
| 6 | China | 3.9 | Cotton, tomato, poplar, papaya, sweet bell red pepper |
| 7 | Paraguay | 2.8 | Soybean |
| 8 | Pakistan | 2.6 | Cotton |
| 9 | South Africa | 2.3 | Maize, soybean, cotton |
| 10 | Uruguay | 1.1 | Soybean, maize |
| 11 | Bolivia | 0.9 | Soybean |
| 12 | Australia | 0.7 | Cotton, oilseed rape |
| 13 | Philippines | 0.6 | Maize |
| 14 | Myanmar | 0.3 | Cotton |
| 15 | Burkina Faso | 0.3 | Cotton |
| 16 | Mexico | 0.2 | Cotton, soybean |
| 17 | Spain | 0.1 | Maize |
| 18 | Colombia | <0.1 | Cotton |
| 19 | Chile | <0.1 | Maize, soybean, oilseed rape |
| 20 | Honduras | <0.1 | Maize |
| 21 | Portugal | <0.1 | Maize |
| 22 | Czech Republic | <0.1 | Maize, potato |
| 23 | Poland | <0.1 | Maize |
| 24 | Egypt | <0.1 | Maize |
| 25 | Slovakia | <0.1 | Maize |
| 26 | Romania | <0.1 | Maize |
| 27 | Sweden | <0.1 | Potato |
| 28 | Costa Rica | <0.1 | Soybean, maize |
| 29 | Germany | <0.1 | Potato |

International Service for the Acquisition of Agribiotech Applications (ISAAA) reported that soybean was the major GM crop in 2011, occupying 47% (75.4 Mha) total GMO-planted area [4]. The corresponding area in 2010 was 50% (73.3 Mha); in 2009, 52% (69.2 Mha); in 2008, 53% (65.8 Mha). Maize was second in this rating: 23% (37.3 Mha) in 2011; 31% (46.8 Mha) in 2010; 31% (41.7 Mha) in 2009; 30% (37.3 Mha) in 2008. Cotton was third with 11% (17.9 Mha) in 2011, 14% (21.0 Mha) in 2010, 12% (16.1 Mha in 2009), and 12% (15.5 Mha) in 2008. Rapeseed occupied fourth position with planting area remaining at 5% (8.2 Mha) in 2011, 5% (7.0 Mha) in 2010, 5% (6.4 Mha) in 2009, 5% (5.9 Mha) in 2008.

**Table 2.3** GMO-Planted Area in the European Union in 2006–2011

| Country | 2006 | 2007 | 2008 | 2009 | 2010 | 2011 |
|---|---|---|---|---|---|---|
| Spain | ~60 kha | ~70 kha | ~100 kha | ~90 kha | ~90 kha | ~100 kha |
| Germany[a] | + | + | + | − | + | + |
| France | + | + | − | − | − | − |
| Portugal | + | + | + | + | + | + |
| Poland | − | + | + | + | + | + |
| Romania | − | + | + | + | + | + |
| Slovakia | + | + | + | + | + | + |
| Czech Republic | + | + | + | + | + | + |
| Sweden | − | − | − | − | + | + |
| Total excluding Spain | ~8.5 kha | ~8.7 kha | ~7.7 kha | ~4.8 kha | ~1.6 kha | ~17.2 kha |

[a]In 2009, Germany refused to cultivate GMOs (although GMOs continued to be used in food and feed production), but in 2010 the cultivation of GMOs was resumed.

Total area of soybean in the world is ~90 million hectares, of maize is ~158 million hectares, of cotton is ~33 million hectares, of rapeseed is ~31 million hectares. Thus in 2011, GM soybean crops accounted for 84% of the total area occupied by soy; cultivation of GM maize, cotton, and rapeseed accounted for 32%, 75%, and 26% of the respective totals (Figure 2.2).

In agriculture, the most popular of the transgenic varieties were those tolerant to herbicides. In 2011, plants with this trait occupied 59% or 93.8 million of total GMO hectares, 60% (89.3 Mha) in 2010, 62% (83.6 Mha) in 2009, and 63% (79 Mha) in 2008.

Adoption of transgenic plants with combined herbicide tolerance and insect-resistance traits (containing two or more transformed events, so called "stacked-gene varieties") took second place with 26% (42.2 Mha) in 2011, 22% (32.3 Mha) in 2010, 21% (28.7 Mha) in 2009, and 22% (26.9 Mha) in 2008. In 2011, Bt-crops (crops genetically modified to express genes from *Bacillus thuringiensis* providing protection from damage by insect pests) occupied third place with 15% (23.9 Mha), compared with 18% (26.3 Mha) in 2010, 16% (21.7 Mha) in 2009, and 15% (19.1 Mha) in 2008.

The major developers of GM plants are research centers focused on production of chemicals for the agricultural sector. The parallel researches in biotechnology and chemistry produce a kind of tandem of herbicide and plant varieties tolerant to this herbicide. In addition to efficient combat of

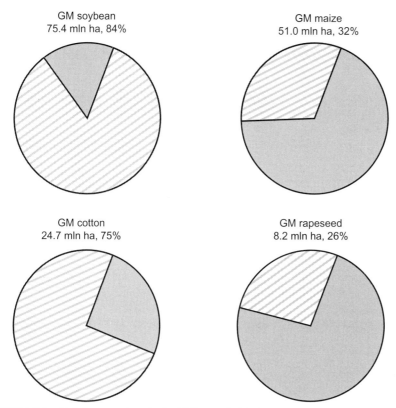

**FIGURE 2.2** Global area under principal GM crops in 2012.
Percentages show proportion of total (GM + non-GM) area for each crop.

agricultural pests, the positive effect of these complementary pairs includes a pronounced decrease in the use of pesticides in agriculture [1,2,3].

As of October 2012, over one hundred GM crop lines were registered for use as food and feed globally, including 30 lines of combined trait (staked) products (22 maize hybrids, 7 cotton, and 1 rapeseed line) [www.cera-gmc.org]. In the USA, 84 lines were approved according to the database of the US Food and Drug Administration (Table 2.4), and 45 lines were registered in the EU (Table 2.5).

In the Russian Federation as of October 2012 only 18 lines are approved for human food (4 soybean, 10 maize, 2 potato varieties, 1 rice, and 1 sugar beet line), and 13 lines are approved for use as animal feed (4 soybean and 10 maize lines) (Table 2.6).

One of the worldwide trends early in the 21st century is expansion of transgenic crops on the market. According to analysts, in 2015 the total planted area of GM crops will rise in the world to 200 million hectares or 13–14% total tillage on the planet [5].

**Table 2.4** GM Crops Approved for Food and Feed Use in the USA as of October 2012

| Crop | Number of Lines (Not Including Stacked Products) |
|---|---|
| Alfalfa | 1 |
| Cantaloupe | 1 |
| Cotton | 14 |
| Creeping bentgrass | 1 |
| Flax | 1 |
| Maize | 27 |
| Oilseed rape (canola) | 11 |
| Papaya | 2 |
| Plum | 1 |
| Potato | 4 |
| Radicchio | 1 |
| Rice | 1 |
| Soybean | 8 |
| Squash | 2 |
| Sugar beet | 3 |
| Tomato | 5 |
| Wheat | 1 |
| Total | 84 |

*Source:* http://www.accessdata.fda.gov/scripts/fcn/fcnNavigation.cfm?rpt=bioListing

**Table 2.5** GM Crops Approved for Food and Feed Use in the EU (as of October 2012)

| Crop | Number of Lines (Including Stacked) | Number of Stacked Hybrids |
|---|---|---|
| Soybean | 3 | – |
| Maize | 22 | 10 |
| Oilseed rape | 3 | 1 |
| Cotton | 6 | 2 |
| Carnation | 3 | – |
| Sugar beet | 1 | – |
| Potato | 1 | – |
| Total | 39 | 13 |

*Source:* http://www.gmo-compass.org/eng/gmo/db/ (search included lines with "valid authorization" and "Notified as existing product").

**Table 2.6** GMOs Registered in the Russian Federation as of October 2012

| GM Crop | Registered Use | |
|---|---|---|
| | Food | Feed |
| Soybean line 40-3-2 tolerant to glyphosate | 1999 | 2003 |
| Soybean line A2704-12 tolerant to ammonium glufosinate | 2002 | 2007 |
| Soybean line A5547-127 tolerant to ammonium glufosinate | 2002 | 2007 |
| Soybean line MON 89788 tolerant to glyphosate | 2010 | 2010 |
| Maize line MON 810 resistant to European maize borer *Ostrinia nubilalis* | 2000 | 2003 |
| Maize line GA21 tolerant to glyphosate | 2000 | 2003 |
| Maize line NK603 tolerant to glyphosate | 2002 | 2003 |
| Maize line T25 tolerant to ammonium glufosinate | 2001 | 2006 |
| Maize line MON 863 resistant to rootworm *Diabrotica* spp. | 2003 | 2003 |
| Maize line Bt-11 tolerant to ammonium glufosinate and resistant to European maize borer *Ostrinia nubilalis* | 2003 | 2006 |
| Maize line MON88017 tolerant to glyphosate and resistant to maize rootworm *Diabrotica* spp. | 2007 | 2008 |
| Maize line MIR604 resistant to rootworm *Diabrotica* spp. | 2007 | 2008 |
| Maize line 3272 with α-amylase | 2010 | 2010 |
| Maize line MIR162 resistant to lepidopteran pests | 2011 | 2012 |
| Rice line LL62 tolerant to ammonium glufosinate | 2003 | – |
| Sugar beet line H7-1 tolerant to glyphosate | 2006 | – |
| Sugar beet line GTSB77 tolerant to glyphosate | 2001–2006[a] | – |
| Potato variety "Elisabeth 2904/1 kgs" resistant to Colorado potato beetle | 2005 | – |
| Potato variety "Lugovskoy 1210 amk" resistant to Colorado potato beetle | 2006 | – |
| Potato variety RUSSET BURBANK NEWLEAF (RBBT02-06) resistant to Colorado potato beetle | 2000–2008[a] | – |
| Potato variety SUPERIOR NEWLEAF (SPBT02-5) resistant to Colorado potato beetle | 2000 to 2008[a] | – |

[a]Re-registration is not intended due to termination of the project.

# REFERENCES

[1] Brookes G, Barfoot P. GM crops: the first ten years — global socio-economic and environment impacts. ISAAA Brief N 36. Ithaka, N.Y; 2006.

[2] Cockburn A. Assuring safety of genetically modified (GM) foods: the importance of a holistic, integrative approach. J Biotechnol 2002;98(1):79–106.

[3] Harlander SK. The evolution of modern agriculture and its future with biotechnology. J Am Coll Nutr 2002;21(3, (Suppl.)):161S–5S.

[4] James C. Global status of commercialized biotech/GM crops: 2011. ISAAA Brief N 43. Ithaka, N.Y; p. 30.

[5] Otiman IP, Badea EM, Buzdugan L. Roundup ready soybean, a Romanian story. Bull UASVM Anim Sci biotechnologies 2008;65(1–2):352–7.

# CHAPTER 3

# Legislation and Regulation of Production and Sales of Food Derived From Genetically Modified Plants in the Russian Federation

Use of commercialized genetically modified crops as new sources of food and feed is widespread [8,9]. While developing and using GM crops, two aspects should be considered: safety for the current and future generations of mankind, and ecological safety. To ensure safety of food derived from GM crops, three major regulatory elements are required:

- a reliable system for human health safety assessment of GM food;
- efficient monitoring of the production and sale of GM food;
- availability of public information on use of new genetic engineering technologies for food production [3–6].

In order to address these issues in Russia, a relevant legal, regulatory and methodological basis has been developed. The Ministry of Health and Social Development of the Russian Federation, Rospotrebnadzor (the Federal Service for Customer Rights Protection and Public Well-Being), the Russian Academy of Medical Sciences, the Russian Academy of Agricultural Sciences, and the Russian Academy of Sciences have developed the system for GMO safety assessment and control for Russia. This system incorporated international scientific achievements and practices in biotechnology, as well as substantial Russian experience, particularly that of the 1960s–70s, when Russia held the leading position in industrial biotechnology.

In 1996 Russia adopted the Federal Law No.86-FZ of 05.07.96 "On State Regulation of Genetic Engineering" which addresses the issues related to the use of natural resources, environmental protection, and ecological safety arising in the course of genetic engineering. According to Article 4 of this law, one of the main tasks of state regulation is to ensure human and environmental safety in the course of use of genetic engineering and products derived from it. In accordance with Article 11 of this law, "the products (services) developed by genetic engineering shall comply with the environmental safety requirements, sanitary regulations, pharmacopoeial standards, and obligatory requirements of the national standards".

Federal Law No 52-FZ of 30.03.1999 "On Sanitary and Epidemiological Welfare of the Population" determined that "food, food additives, raw materials, as well as materials and products that come in contact with those during manufacturing, storage, transportation or sale, must meet the sanitary and epidemiological requirements and standards". Hygienic quality and safety standards for food are specified in the sanitary-and-epidemiological rules and specifications "Hygienic Requirements For Safety And Nutritional Value Of Food" [SanPiN 2.3.2. 1078-01]. Article 43 of this Federal Law contains the list of products and the procedure of state registration of food products. In accordance with Paragraph 1 of this Article, "chemical and biological substances, produced for the first time and never used before, products (substances) derived from those, potentially hazardous to humans, as well as certain products, including food imported into Russia for the first time" are subject to state registration.

The Federal Law No. 29-FZ of 02.01.2000 "On Quality and Safety of Food" does not classify GM food or GMO raw materials as a separate food group which must undergo special quality and human safety assessment. However, Article 10 of this law introduces a system of state registration for food, materials, and goods. According to Paragraph 1 of this Article, novel food, materials, and goods produced in the Russian Federation are subject to registration, and GM food falls into this category. It is particularly specified that imported food, materials, and goods must be registered prior to their shipment to Russia.

Since 1995 the Ministry of Health has required that all products derived from or containing GMOs are to be declared and undergo sanitary and epidemiological expertise. In 1996–1997 the authorities started developing a GM food quality and safety assessment system. During this time the first laboratory studies were carried out. In 1999 the Decree No.7 of 06.04.1999 "On Procedure for Sanitary Assessment and Registration of Food Derived from Genetically Modified Sources" issued by the Chief State Sanitary Doctor of the Russian Federation described the procedure for registration of genetically modified food sources, which was amended later (the Decree No.14 of 08.11.2000 "On Procedure for Sanitary and Epidemiological Expertise of Food Derived from Genetically Modified Sources"). In compliance with the set procedure, each GM crop newly introduced to the Russian market is subject to sanitary and epidemiological expertise which includes three categories:

1. medical (human health) and genetic assessment (the Centre of Bioengineering of the Russian Academy of Sciences);
2. medical and biological (human health) assessment (the Research Institute of Nutrition, Russian Academy of Medical Sciences);
3. assessment of the technological parameters (Moscow State University of Applied Biotechnology, Ministry of Education and Science of the Russian Federation).

Regulatory documents and materials submitted by an applicant, as well as results of the studies conducted in Russia (Figure 3.1), are used to support determination of the safety of the GM crop and food derived from it. The report on product safety is submitted to the Federal Service for Customer Rights Protection and Public Well-Being which issues a permit for use of the

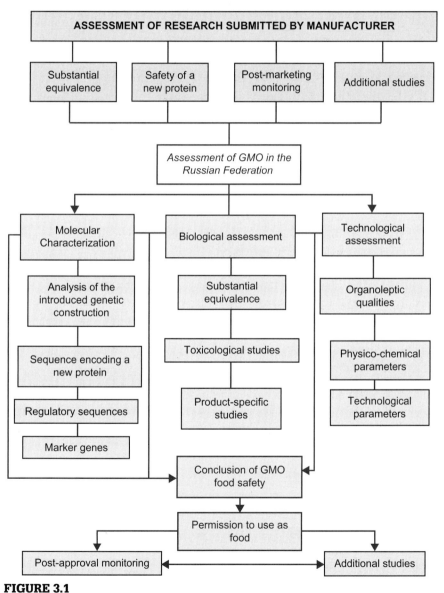

**FIGURE 3.1**
Food safety assessment of genetically modified food sources in the Russian Federation.

GMO in the food industry and for sale, otherwise the request for registration is denied. In addition, the Decree states that GM food registered in the Russian Federation must be entered into the Register of GM food.

In 2000 the Russian Ministry of Health approved methods for assessment of quality and human health safety of GMOs and food derived from them (Methodological Guidelines MUK 2.3.2. 970-00 "Medical and Biological Assessment of Food Derived from Genetically Modified Sources"). Based on the experience of registration of plant GMOs in the Russian Federation, the system of medical and biological safety assessment of GMOs was extended and improved (Methodological Guidelines MU 2.3.2.2306-07 "Medical and Biological Safety Assessment of Plant Genetically Modified Organisms"), and corresponding amendments were introduced into the Sanitary and Epidemiological Regulations and Hygienic Standards of the Russian Federation (Additions and Amendments No.6 to SanPiN 2.3.2.1078-01 "Hygienic Requirements for Safety and Nutritional Value of Food [SanPiN 2.3.2.1078-01]).

Only GM products that have been extensively studied, approved, and used without restrictions for food in other countries for several years (the USA, EU, Canada, China, etc.) can enter the Russian market.

At the same time, the GMO safety assessment system adopted in Russia requires the conduct of extensive local studies including long-term toxicological animal studies (180 days), genotoxic, genomic and proteomic analysis, assessment of potential allergenicity with model systems, etc. These assessments are considered an additional tool to ensure safety. These studies are carried out by various leading research institutions of the Federal Service for Customer Rights Protection and Public Well-Being, the Russian Academy of Medical Sciences, Russian Academy of Sciences, Russian Academy of Agricultural Sciences, and the Ministry of Education and Science of the Russian Federation.

During 16 years of the global industrial production of GM crops, more than one hundred transformation events were registered and approved for food in the USA and other countries. In Russia, only 21 events were assessed and registered. It should be noted that none of the authorities accredited to register GM crops in any country in the world reported any negative effect of GMOs, and no single GM line was ever banned or restricted to use in food for adults or children, which is specified in the documents of international organizations [10], Regulations of the European Union [12–13], statements of the US Food and Drug Administration (FDA) [7], and documents of the Codex Alimentarius Commission [11].

Another important issue is monitoring of GM food. In Russia, two methods for GMO identification have been approved by the national standards organization (GOST). For qualitative analysis, polymerase chain reaction (PCR) that allows detection by electrophoresis or biological microchips is used [1,2]. For quantitative analysis, a real-time PCR is used (Methodological Guidelines MUK 4.2.2304-07 "Methods of Identification and Quantitative Determination of Plant Genetically Modified Organisms". Rospotrebnadzor, 2008, registered with the Ministry of Justice No. 11117 of 06.02.08). Currently, all regional authorities in Russia conduct monitoring of GM food, performing up to 35,000 analyses per year.

Another issue, important from the ethical standpoint, is informing the public on new technologies used for food production. This has nothing to do with food safety, but gives the consumer a right to get reliable information on methods of food processing. This is the labeling issue. Russia has a number of laws and regulations on labeling. In 2000, Russia introduced a voluntary labeling of GM food (the Decree No.13 of 08.11.2000 "On Labeling of Food Derived from Genetically Modified Sources"). However, since 2002, GM food labeling is obligatory in Russia (SanPiN 2.3.2. 1078-01 "Hygienic Requirements for Safety and Nutritional Value of Food"). This document set a 5% threshold for all products excluding those not containing DNA or protein.

In accordance with Federal Law No.2300-1 of 07.02.1992 "On Consumer Rights Protection" (version of the Federal Law No.171-FZ of 21.12.2004), consumers must be informed of the use of genetic engineering methods for food production; in other words, "labels should contain information on GMO components in food". In 2007, in order to harmonize the Russian legislation with international practices, the government issued the Federal Law No.234-FZ of 25.10.2007 "On Amendments to the RF Law 'On Consumer Rights Protection', and amendments to Part 2 of the Civil Code of the Russian Federation", where a 0.9% threshold for GMO was set as an unintentional or technically unavoidable (adventitious) level.

To date, the Russian Federation has a functional legislative, regulatory, and methodological framework allowing the use of GMOs for food and feed purposes. Progress in development of new methods used in biotechnology requires continuous updating of the legislation and regulations for biotechnology and GMO application in Russia.

# REFERENCES

[1] National Standards of Russian Federation: Method to identify plant-derived genetically modified sources (GMS). GOST R 52173-2003. Moscow: 2004.

[2] National Standards of Russian Federation: Method to identify plant-derived genetically modified sources (GMS) with biological microchips. GOST R 52174-2003. Moscow: 2004.

[3] Onishchenko GG, Tutelyan VA, Petukhov AI, et al. Modern approaches to safety assessment of genetically modified food sources. Experience in the study of 40-3-2 line soy beans. Vopr. Pitan 1999(5-6):3–7.

[4] Tutelyan VA, Petukhov AI, Korolev AA, et al. Genetically modified food: biomedical evaluation. Doctor 2000(2):36–8.

[5] Tutelyan VA, Sorokina EY. Medical and biological evaluation of GMO foods: principles of safety assessment, teaching approaches. Food industry 2003(6):17–20.

[6] Tutelyan VA. Transgenic plants as raw food: biomedical evaluation. International scientific and practical conference "Transgenic plants–a new direction in biological plant protection", Krasnodar: 2003.

[7] U.S. Food and Drug Administration, <http:/www.fda.gov/default.htm>.

[8] Global Status of Commercialized Biotech/GM Crops: 2011. ISAAA Briefs 43-2011.

[9] GM crops: The first ten years–global socio-economic and environmental impacts: 2005. ISAAA Briefs 36-2006.

[10] Modern food biotechnology, human health and development: an evidence-based study. Food Safety Department World Health Organization: Geneva: 2005.

[11] Principles for the Risk Analysis of Foods derived from modern biotechnology. CAC/GL 44- 2003.

[12] Regulation (EC) No 1829/2003 of the European Parliament and of the Council of 22 September 2003 on genetically modified food and feed.

[13] Regulation (EC) No 1830/2003 of the European Parliament and of the Council of 22 September 2003 concerning the traceability and labelling of genetically modified organisms and the traceability of food and feed products produced from genetically modified organisms and amending Directive 2001/18/EC.

# CHAPTER 4

# Principles of Human Health Safety Assessment of Genetically Modified Plants Used in the Russian Federation

The practical implementation of novel methods to transform the genome of living organisms promoted strict regulation of the safety assessment of GMOs produced for the food market. When discussing GM food safety, one should realize that, in order to confirm the safety of GM crops, a large variety of requirements must be met during production, storage, and processing of the food products compared with the widespread view on perceived safety of traditional food sources for human health. In addition, good sanitary conditions of food products *per se* cannot prevent the development of diseases (atherosclerosis, diabetes mellitus, obesity, essential hypertension, gout, etc.) caused by imbalanced diet. The risk factors associated with food products could include deficiency of nutritional substances, the adverse effects of microorganisms and chemicals naturally present in the food (e.g., glycoalkaloids and saponins in potato), intentionally introduced into the food production chain (food additives, the residual agrochemicals), or entered the food unintentionally (environmental pollutants). Therefore, if a GM variety is tested and shown to be as safe as its corresponding non-GM traditionally bred plant, it can be reasonably considered that the GM counterpart is not dangerous for human health [9,10,12]. In this chapter the basic parameters of the GMO safety assessment system used in the Russian Federation are described.

The development of the GMO safety assessment currently used in the Russian Federation started in 1995–1996. The methodological approaches to comprehensive complex medical and biological assessment of GMOs were developed in the Russian Federation with due regard for international and national experience as well as new scientific approaches based on the achievements of contemporary fundamental science: genomic and proteomic analysis, detection of DNA damage or mutagenic activity, identification of products of free-radical modifications of DNA or other sensitive biomarkers [4,14,22–24].

GMO safety assessment is carried out for the state registration. Any novel food derived from plant GMO produced in Russia or imported into Russia for the first time is subject to the state registration [2,4]. Guidance for safety assessment is specified in MU 2.3.2.2306-07 "Medico-Biological Safety Assessment of Plant Genetically Modified Organisms" [7].

Required assessments were, for the first time, generalized in the methodological regulations "Medico-Biological Assessment of Food Products Derived from Genetically Modified Sources" (MUK 2.3.2.970-00) [7]. According to the accepted regulations, the human health assessment of a novel GMO to be placed on the domestic market includes the following (Figure 4.1):

- molecular assessment;
- human health safety assessment;
- assessment of technological parameters;
- assessment of the data presented by the applicant on the examined GMO.

Expert analysis and assessment of data describing the novel GMO includes: information enabling identification of the subject of registration (species, variety, transformation event); information on the initial parental organism and donor organisms of the introduced genetic sequences; information on the method of genetic modification, genetic structure, and gene expression; information on approval of the GMO in other countries; results of safety assessment (compositional equivalence, toxicological studies, allergenic properties, and other analyses) on the basis of which the GMO was approved in other countries.

- *Molecular assessment* includes analysis of genetic construction, genetic modification method, and the gene expression level.
- *Technological assessment* includes determination of organoleptic and functional properties, and analysis of technological characteristics of the finished products.
- *Human health safety assessment* includes several sections of required assessments: analysis of compositional equivalence and toxicological, genotoxicological, and allergological safety studies in accordance with SanPiN 2.3.2.1078-01 "Hygienic Requirements for Safety and Nutritional Value of Food" for specific food; toxicological genotoxicological and allergological studies with animals *in vivo*".
- *Methods for identification* include qualitative and quantitative assay of GMO in food (studies targeted at determination of correspondence of these methods to those used in Russia in order to provide monitoring of use and labeling of GM food).

The list and the scope of required studies is determined on the basis of analysis of information of the GMO submitted for registration; however, the above-mentioned studies are required [2,7]. If significant changes in the GMO's genome, proteome, or metabolome are shown, additional studies may be required to

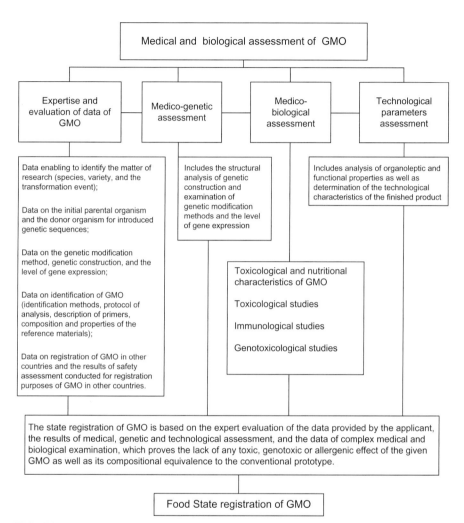

**FIGURE 4.1** The system of GMO safety assessment used in the Russian Federation.

determine: biological value and absorbency reproductive effect; gonadotoxic, embryotoxic, teratotoxic effect; potential carcinogenic effect; lifetime, etc.[7].

## 4.1 HUMAN HEALTH SAFETY ASSESSMENT

### Assessment of Potential for Toxicity

Toxicological studies (the study duration is no shorter than 6 months) are performed on laboratory animals such as Wistar rats with initial body weight of 60–80 g that are kept on a diet enriched with the examined GMO (the test group) or with its conventional counterpart (the control group) [4,5]. All the experiments on laboratory animals are carried out according to a similar protocol. The animals should be uniformly randomized between both groups: they must not have significant differences in age, body weight, appearance, etc.; the samples of GMO products and its conventional counterpart should be derived from plants grown under identical conditions; they should be similarly processed and conform to Russian Safety Requirements. Otherwise, interpretation of the obtained results would be equivocal or erroneous. The objective of such experiments require only a single input difference between the control and test groups by the presence of GMO or its conventional counterpart in the diet [7].

The control and test products are included in the diet in maximum allowed quantity to keep the diet nutritionally balanced. During these studies, dynamics of integral (appearance, body weight, etc.), hematological, biochemical, and morphological parameters are recorded.

Another feature of the Russian system for GMO safety assessment is the use of sensitive biomarkers as parameters that reflect the level of the living organism's adaptation to the environment and which are highly sensitive to foreign agents [19,25,26]. Special attention is paid to the systems responsible for the protection of organisms against exogenous and endogenous toxic compounds: the systems of xenobiotic metabolism enzymes, antioxidant protection, and apoptosis regulation (Figure 4.2).

Particular emphasis is placed on the systems protecting the organism against the effects of exogenous and endogenous toxic agents—primarily on the enzymes of phases 1 and 2 of xenobiotic degradation and on lysosomal enzymes. Taking into consideration that many physiological and metabolic functions are closely related to the free-radical oxidation processes, whose alteration is the early non-specific reaction of the organism to extreme conditions, assessments of enzymatic activity in the antioxidant protection system and the content of lipid peroxidation (LPO) products are the early informative tests employed in safety evaluation of the effects of adverse environmental factors such as food contaminants.

The samples are isolated on Day 30 and Day 180 after the start of the study. The data obtained at every stage of the study have both individual and an overall value. The samples isolated on Day 30 correspond to the stage of sexual maturity of rats, characterized by accelerated growth and cell differentiation. In this case, the integral parameters have significant diagnostic value. If the samples are isolated from the mature rats (on Day 180) kept on the diet supplied with the tested product in maximum possible dosage for one-fifth to one-quarter of the mean rat life span (2–2.5 years), the morphologic examination can be most informative. Analysis and generalization of data obtained

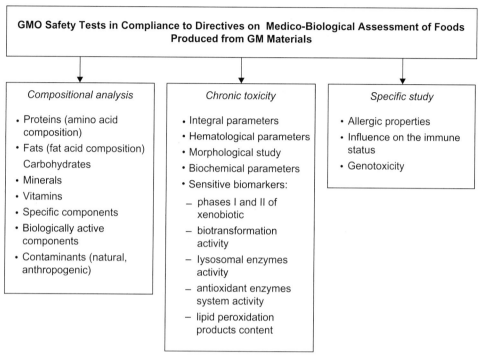
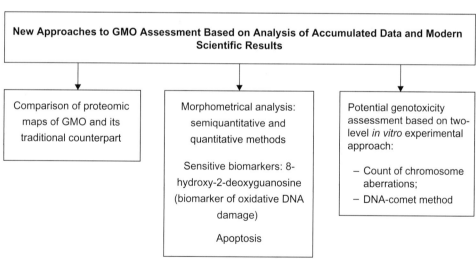

**FIGURE 4.2** Human health safety assessment of GMO.

at various time points of the experiment makes it possible to assess the condition of the animals in test and control groups at different stages of ontogenesis, which greatly improves the diagnostic value of the examination [14,19].

## Assessment of Potential Genotoxicity

In the course of human health (medico-biological) safety assessment of a plant GMO, along with general toxicological analysis, it is important to study potential specific toxic effects *in vivo* [7]. In accordance with current research practice, methods to obtain the greatest possible full and reliable information on potential genotoxic, immunotoxic, or allergic effect of the GMO, as well as to identify potential unexpected effects of the genetic modification, are used [3,4,10,22,28,30,33]. Genotoxic studies include the assessment of genetic material at various stages (DNA molecules, chromosomes); immunotoxic studies—assessment of immunomodulating and sensitizing effect in experiments with mice of oppositely reacting lines; allergenicity studies—assessment of active anaphylactic shock and the strength of humoral immune response in simulated systemic anaphylaxis in rats [7].

The genotoxicological studies are carried out *in vivo* on C57Bl/6 line mice to assess induction of chromosome mutations in the germ cells and to detect mutagenic activity. This activity is indicated by elevation of embryonic death rate in the progeny of males kept on a diet with the examined GMO product. Detection of mutagenic activity is based on the count of chromosome abnormalities in metaphase cells in the proliferating tissues. According to modern views on the development of chromosome mutations, they are always related to induction of molecular abnormalities resulting in DNA breaks caused among other factors by the action of xenobiotics.

## Assessment of Potential for Allergenicity

Allergenicity assessment studies include the following:

1. analysis of immunomodulating and sensibilizing properties of GMO in experiments on CBA and C57Bl/6 line mice in four tests:
   - potential effect on the humoral component of the immune system (assessment of the level of hemagglutinins to sheep erythrocytes;
   - potential effect on the cellular component of the immune system in delayed hypersensitivity reaction to sheep erythrocytes;
   - potential effect as sensibilizing agent in sensitivity test to histamine;
   - potential effect on natural resistance of mice to *Salmonella typhimurium*;
2. immunological studies of GMO effects on the rat model of generalized anaphylaxis, where severity of the active anaphylactic shock and the vigor of humoral immune response are examined.

**Table 4.1** Comparative Susceptibility of CBA and C57Bl/6 Mice in Allergenicity Assessment

| Agent | CBA Line | C57Bl/6 Line |
|---|---|---|
| Sheep erythrocytes | High susceptibility | Low susceptibility |
| Histamine | Non-susceptible | Susceptible |
| *Salmonella typhimurium* | Non-susceptible | Susceptible |

The comparative characteristics of CBA and C57Bl/6 line mice are described in Table 4.1.

## Additional Studies

If studies have revealed significant unintended alterations in the proteomic and metabolomic genome of the GMO, additional studies can be carried out as required. They will be focused on biological value and digestibility of GMO product, its effects on reproduction and life span, its possible gonadotoxic, teratogenic, and carcinogenic action, and so on [7].

## 4.2 EVALUATION OF DATA PROVIDED BY APPLICANT

Expert evaluation is carried out on the following information presented by the applicant:

- data enabling identification of the matter of research (species, variety, and the transformation event);
- data on the initial parental organism and the donor organisms of the introduced genetic sequences;
- data on the genetic modification method, genetic construct, and the gene expression level;
- data necessary to provide control during the GMO's lifetime (the method used to identify the transformation event, the protocol of analysis, description of the primers, composition and properties of the standard reference samples);
- data on registration of the GMO in other countries and the results of safety research that ensured registration of the GMO in other countries.

## 4.3 NOVEL APPROACHES TO SAFETY ASSESSMENT

The state registration of a GMO is based on the expert evaluation of the data provided by the applicant, the results of medical, genetic, and technological assessments, and data from complex medical and biological studies of the GMO which document the absence of any toxic, genotoxic, and allergenic effects as well as compositional equivalence of the examined GMO to its conventional counterpart.

During 1999–2012, 21 GMO products were subjected to complete safety assessment. Analysis of the accumulated data and consideration of the trends in modern research with due account of the necessity of advanced development of the safety assessment system yielded novel approaches to medical and biological examination of GMOs (Figure 4.2) [8–10,12,14,18–20,24,25,28,33,34]. These novel approaches affected virtually all the stages of GMO expertise.

## Nutritional Assessment
### Compositional Equivalence Analysis

Analysis of compositional equivalence includes comparison of the macro- and micronutrient composition, the content of specific components, the biological active substances, AND the natural and anthropogenic contaminants in the GMO and its conventional counterpart. Conclusions on similarity and differences in composition of the examined products are made with due account for the range of physiological variations characteristic of the particular parameter in the examined biological species.

### Proteomic Analysis

Compositional equivalence analysis was supplemented with the study of the GMO's proteomic profile combined with further comparison of proteomic maps of the GMO and its conventional counterpart. The methods used in proteomics enable us to compile an inventory of the proteins synthesized by the cells and to reveal their posttranslational modifications. The specificity of proteomic analysis enables discovery of not only the structural differences between the proteomes of the examined objects, but also the composition of the functionally active structures involved in various metabolic chains and the interaction of various proteins or subunits of the oligomer complexes [2,6,25,34]. Since structural and functional modifications of an individual gene affect the genomic function as a whole (therefore, they can modify the protein profile in the entire organism), there is every reason to believe that, in the nearest future, proteomic studies will be of paramount importance for tracing the unintended effects of gene modification.

Proteomic analysis compares hundreds and sometimes thousands of parameters of the examined GMO and its conventional counterpart, so there is a probability that some of these parameters will differ significantly. In many cases, comparison of literature data on proteomes yields equivocal results. For example, comparison of proteome maps of 8 GM potato lines with the map of parental conventional potato, made by S. O. Kärenlampi and S. J. Lehesranta (Institute of Applied Biotechnology) [29], showed that among 730 visually observed protein spots, only 9 (1.2%) had the statistically significant difference, while all GM-potato lines had no novel proteins. The comparison of proteome maps of 32 conventional potato varieties showed that among 1111 visually detected protein spots, 1097 (96.9%) demonstrated the statistically significant differences [18]. Therefore, the differences in protein expression result both from the genomic modification and the intraspecies differences induced by the external and internal factors.

Evaluation of differences between the proteome maps of GMOs and their conventional counterparts should take into account the natural variations in the content of a protein, induced by a number of factors. While carrying out the proteomic analysis of the GMO and its control counterpart, it should be remembered that the presence of novel proteins in GMO is much more dangerous than the absence of some proteins, because the novel protein could induce an allergic reaction or impart toxic properties to the GMO.

At present, the lack of data on plant proteomes limits interpretation of the available data. Therefore, creation of respective databases and algorithms of plant proteomic analysis is among the most important avenues of fundamental science aimed to predict the functions of genes and the properties of the products encoded by them [25,30,31].

## Assessment of Sensitive Biomarkers

The set of parameters examined during toxicologic studies is persistently widened. These studies determine the parameters of pronounced diagnostic significance, which reveal the toxic properties of the examined product.

Specifically, such biomarkers of oxidative stress as the products of free-radical modification of DNA (for example, 8-hydroxy-2-deoxyguanosine or 8-oxodG), determined in urine and in DNA, are very informative [4,5,26,27]. 8-oxodG is a product of the direct effect of free radicals on DNA, and it is the most reliable marker of antioxidant status. In addition, simultaneous assay of 8-hydroxy-2-deoxyguanosine in DNA and in urine makes it possible to evaluate the status of the reparative system, whose activity is an integral parameter of adverse influences in the organism.

The list of sensitive biomarkers is supplemented by parameters characterizing intensity of apoptosis. Control of apoptosis is affected by the internal genetic program triggered by intracellular physiological or pathogenic factors such as the effects of various damaging physical and chemical agents. Therefore, these parameters are unique in detecting weak influences before they provoke necrosis [11,15,16,21–23,32].

### Other Assessments

Morphological studies are supplemented by morphometric and immunohistochemical analysis, which can (1) examine conformity of organ structure to the norm, (2) demonstrate the absence of pathologic alterations in the organs or otherwise describe them, (3) reveal morphological correlation with changes of non-specific resistance of the organism, and (4) evaluate the state of local immune protection in an organ. These methods employ semiquantitative and quantitative analyses of the data, allowing the use of statistical criteria [1,17].

The set of obligatory genotoxicological tests was enlarged in accordance with new results of scientific research. A two-tier system based on direct assessment of integrity of the double-strand DNA polymer structure is suggested, which accounts for chromosomal aberrations in the metaphase cells of proliferating tissues in the laboratory animals kept on a diet with the examined product [3,13].

Assessment of DNA damage includes analysis of DNA integrity by the DNA-comet method. This method is based on different motility of DNA and DNA fragments of the lysed cells in agar gel. The DNA fragments migrate to the anode, where they form a structure that, after staining, looks like a comet tail. The total DNA content in the "comet", its length, and other parameters relating to DNA integrity reveal the disturbances in DNA structure.

Assessment of potential genotoxicity of GMO makes it possible to reveal both the damaged DNA fragments to be repaired later and the chromosomal aberrations fixed in the genome, which greatly enhances the diagnostic value of the data.

## CONCLUSION

It can be concluded that the system of GMO safety assessment in the Russian Federation is one of the strictest in the world. In addition, some lag in the use of novel biotechnology methods in Russia makes it possible to use data of the post-registration monitoring obtained in various countries that already use GMO food products. Thus, the state registration of any GMO in the Russian Federation is based upon complex medical and biological assessment of GMO safety, the results of similar procedures carried out in other countries, and the results of population studies.

Efficient studies aimed to improve the methodological basis, the use of novel approaches of research, and information exchange underscore the rapid and advanced development of scientific knowledge needed to ensure safety of novel food products. Currently, there is a body of scientific data on GMO safety, including the results of studies conducted in the course of GMO registration in Russia, as well as data of the worldwide scientific literature on safety assessment of biotechnology products. Of course, science is not standing still. Active development of methods, the use of high technology and global exchange of information inspire confidence that modern science is developing with increased speed, in particular to ensure safety of novel food.

Since 2007, the Russian Federation also has updated the requirements for safety assessment of GMOs contained in the MU 2.3.2.2306-07 "Medico-biological safety assessment of genetically modified organisms of plant origin".

# REFERENCES

[1] Avtandilov GG. Introduction into qualitative pathologic morphology. Moscow: Meditsina (Medicine); 1982.

[2] Archakov AI. Genomis, proteomics and bioinformatics: the sciences of XXI century. Med Kafedra 2002(3):6–13.

[3] Durnev AD, Seredenin SB. Mutagens: screening and pharmacological prevention of their effects. Moscow: Meditsina (Medicine); 1998.

[4] Zhanataev AK, Durnev AD, Seredenin SB. Perspectives of 8 hydroxy-2-deoxyguanosine essay as an oxidative stress biomarker in test and clinic. Vestn RAMN 2002(2):45–9.

[5] Zinov'eva VN, Ostrovskii OV. Free radical DNA oxidation and its oxidated guanosine biomarker (8-oxodG). Vopr Med Khim 2002;48(5):419–31.

[6] Makarov OV, Govorun VM, Taranets IN, et al. Early proteomics of the ovary cancer. Myth or reality. Biomed Khim 2003;49(1):2–7.

[7] Medico-biologic safety assessment of genetically-engineered and modified organisms of plant origin: Methodological instructive regulations MY 2.3.2.2306-07 Moscow, 2008. p. 21.

[8] Onishchenko GG. Hygienic and normative aspects of registration, marking and labeling the food produced from GM sources. Vopr Pitan 2001(2):3–7.

[9] Onishchenko GG, Tutelyan VA, Petukhov AI, et al. Genetically modified food sources: medical and biological assessment. Vrach 2000(3):35–7.

[10] Onishchenko GG, Tutelyan VA, Petukhov AI, et al. Modern approaches to safety assessment of genetically modified food sources. Experience in the study of 40-3-2 line soy beans. Vopr Pitan 1999(5–6):3–7.

[11] Pal'tsev MA, Ivanov AA, Severin SE. Intracellular interaction. Moscow: Meditsina (Medicine); 2003.

[12] Pokrovskii VI. Medical problems of biosafety. Vestn RAMN 2002(10):6–9.

[13] Application of alkali gel electrophoresis of isolated cells in assessment of genotoxic features of natural and synthetic compounds: Methodical guidelines. Moscow, 2006.

[14] Rakhmanin A, Mikhailova RI, Zaitseva NV, Vaisman Ya I. Methods of prenosological diagnostics of ecologically conditioned diseases. Gig San 2001(5):58–61.

[15] Samuilov VD. Biochemistry of programmed cell death (apoptosis) in animals. Open Soc J Educ 2001;7(10):18–25.

[16] Spiridonov VK, Vorob'eva VF, Tolochko ZS, et al. Effector effect of stimulation and damages of capsaicin-sensible afferent neurons. Byull SO RAMN 2004;2(112):135–40.

[17] Stefanov SB. Visual classification in qualitative comparison of images. Arkh Anat Gist Embr 1985;88(2):78–83.

[18] Tutelyan VA. Food and safety. Vestn RAMN 2002(10):14–19.

[19] Tutelyan VA, Bondarev GI, Martinchik AN. Nutrition and biotransformation of foreign substances. Itog Nauk Tekhn Ser Toksikologiya 1987;15:212.

[20] Tutelyan VA, Kravchenko LV, Lashneva NV, et al. Medical and biological safety assessment of protein concentrate produced from genetically modified soybean: biochemical studies. Vopr Pitan 1999(5–6):9–12.

[21] Fil'chenkov AA. Caspases: regulators of apoptosis and other cellular functions. Biokhimiya 2003;68(4):453–66.

[22] Khotimchenko SA, Alexeeva IA. Approaches to assess the alimentary load with foreign substances. Gig San 2001(5):25–7.

[23] Yudina TV, Rakitskii VN, Egorova MV, Fedorova NE. Antioxidant status parameters in prenosological diagnostics. Gig San 2001(5):61–2.

[24] Abbott A. The promise of proteomics. Nature 1999;402:715–20.

[25] Cellini F, Chesson A, Colquhoun I, et al. Unintended effects and their detection in genetically modified crops. Food Chem Toxicol 2004;42(7):1089–125.

[26] Collins A, Griffiths H, Poulsen H, et al. Markers of oxidative damage and antioxidant protection. Brussels: ILSI Press; 2000.

[27] Gackowski D, Speina E, Zielinska M, et al. Products of oxidative DNA damage and repair as possible biomarkers of susceptibility of lung cancer. Cancer Res 2003;63(16):4899–902.

[28] Huggett AC, Schilter B, Roberfroid M, et al. Comparative methods of toxicity testing. Consensus document following an international life sciences Institute-ILSI Europe Workshop held in May 1995. Food Chem Toxic 1996;34(2):183–92.

[29] Kärenlampi SO, Lehesranta SJ. Proteomic profiling and unintended effects in genetically modified crops. ISB News Report <http://www.bioinform.ru>.

[30] König A, Cockburn A, Crevel RWR, et al. Assessment of the safety of foods derived from genetically modified (GM) crops. Food Chem Toxocol 2004;45:1047–88.

[31] Kuiper HA, Kok EJ, Engel KN. Exploitation of molecular profiling techniques for GM food safety assessment. Cur Opin Biotechnol 2003;14(2):238–43.

[32] Lee S-H, Chang K-T, Kwon O-Yu, Kwon T-K. Etoposide activates caspase-3 via a caspase-1 independent mechanism in cancer cells. Exp Oncol 2002;24:105–8.

[33] Millstone E, Brunner E, Mayer S. Beyond 'substantial equivalence'. Nature 1999;401:525–6.

[34] van Hal NL, Vorst O, van Houwelingen AM, et al. The application of DNA microarrays in gene expression analysis. J Biotechnol 2000;78(3):271–80.

# CHAPTER 5

# Human and Animal Health Safety Assessment of Genetically Modified Plants

## INTRODUCTION

This chapter reports the examination of 15 genetically modified lines for their safety as human food and animal feed: four soybean, nine corn, one rice, and one potato. Out of these lines, nine were developed to be tolerant to herbicides, four to be resistant to damage from main insect pests, two to be resistant to insect pests and tolerant to herbicides, and one to have improved nutritional qualities. The results of the safety assessments for all of these products were reviewed by international regulatory agencies, and conclusions on food and feed safety were reached. Data for the product characterization and human and animal safety assessment produced by the developers of the technology are cited. Studies conducted in the Russian Federation according to the regulations described in Chapter 4 are summarized.

These studies were conducted by qualified independent academic laboratories and institutions sponsored by the Russian government according to approved protocols. Analyses of potential for chronic toxicity, assessment of parameters indicative of the ability of the organism to respond to foreign agents (sensitive biomarkers), potential for geno- and immunotoxicity, as well as additional assessment for allergenicity for all products, confirmed the conclusions of previously conducted studies of the safety of the GM crops to human and animal health. In addition, the nutritional equivalence of the examined GM crops to conventional varieties was found and confirmed to be in agreement with the assessments conducted by the technology providers.

The results of previously published regulatory safety assessments confirm the safety of the introduced protein(s) and the GM plants for humans and animals. For example, results of the No Observed Effect Levels (NOEL) obtained in acute toxicity and chronic studies in mammals accumulated for all proteins used in GM crops confirmed the absence of any toxicity[1]. Put in perspective, assessment of acute dietary exposure demonstrated for Cry proteins

---

[1] Food safety of Proteins in Agricultural Biotechnology, Ed. Bruce G. Hammond. CRC Press, 2008. P. 263–265.

expressed in insect-protected corn showed that a human adult weighing 70 kg will need to consume over 900 metric tons of grain in one day to attain the same acute dosage (4000 mg/kg) of Cry1Ab protein to achieve the dose given to mice that had no toxic effect[2].

These results are not surprising, as crops produced by methods of biotechnology have been used globally as human food and animal feed for over a decade with not a single reliable indication of any undesired effects. Development of a new variety of genetically modified crop is a very lengthy and expensive process, so the safety criteria used by the developers of GM crops are very strict and take potential for allergenicity and toxicity of introduced proteins very seriously. The line selection process takes years and numerous studies, so only lines that are confirmed to have a stable trait and are safe advance to registrations and commercialization.

## Subchapter 5.1

# Soybean

### 5.1.1 GLYPHOSATE-TOLERANT 40-3-2 SOYBEAN LINE

#### Molecular Characterization of Soybean Line 40-3-2
*Recipient Organism*
Family legumes (*Leguminosae*), the genus *Glicine*, species *max* has a long history of cultivation and safe use as human food (about 4000 years) [18].

*Donor Organism*
The donor of the *cp4 epsps* gene responsible for tolerance to glyphosate, *Agrobacterium tumefaciens* strain CP 4, is a gram-negative soil bacterium. There are no data on the adverse effects of this microorganism on humans or animals. The gene *cp4 epsps* encodes 5-enolpyruvilshikimate-3-phosphate synthase, CP4 EPSPS, a key enzyme in the synthesis of aromatic amino acids in plants and microorganisms. In *Agrobacterium* sp. strain CP4, this enzyme

---

[2] Food safety of Proteins in Agricultural Biotechnology, Ed. Bruce G. Hammond. CRC Press, 2008. P. 272–273.

is naturally resistant to inhibition by glyphosate, so glyphosate-containing herbicides cannot inhibit this enzyme [19,34,35,22,31,53].

### *Method of Genetic Transformation*
To insert the genetic construct into the plant genome, biolistic transformation of the soybean cells was performed with the PV-GMGT04 vector. The plasmid vector PV-GMGT04 contains two expression cassettes. Both contain the gene encoding CP4 EPSPS. In the first cassette, expression of the *cp4epsps* gene is controlled by the cauliflower mosaic virus (CaMV) 35S promoter which contains a duplicated enhancer region. The chloroplast transit peptide (CTP) coding region from *Petunia hybrida* EPSPS is fused to the coding region of EPSPS from *Agrobacterium* sp. strain CP4, to target the CP4 EPSPS to the chloroplast, the site of aromatic amino acid biosynthesis. Plant expression of the gene fusion produces a pre-protein which is rapidly imported into the chloroplasts, where the CTP is cleaved and degraded, releasing the mature CP4 EPSPS protein [19,34].

In the second cassette, the *cp4 epsps* gene is controlled by P-FMV promoter of scropula plant mosaic virus. In both cassettes the 3′ region of the gene is from the 3′ non-translated region of the nopaline synthase (NOS) gene of the Ti-plasmid, pTiT37 from *Agrobacterium tumefaciens* strain T37. Only a portion of the DNA sequence of plasmid PV-GMGT04 was inserted into the parental variety, A5403, to produce Roundup Ready soybean line 40-3-2. Insert 1 was also responsible for the expression of the GUS marker protein and a weak expression of the glyphosate tolerance trait. Insert 2 had a strong expression of the glyphosate tolerance trait, but did not express the GUS protein. Thus, in 40-3-2, insert 1 had been lost through normal genetic segregation. This was confirmed by the fact that none of the progenies from line 40-3-2 expressed GUS, based on leaf GUS enzyme assays. The *nptII* gene of neomycin phosphotransferase II of transposon Tn5 *E. coli* was used as selectable marker gene during the transformation process, and is absent in the genome of 40-3-2 line soybean [19,57].

### Global Registration Status of Soybean Line 40-3-2
Table 5.1 shows the status of global registrations to use transgenic soybean line 40-3-2 at the time of registration in Russia [19].

### Safety Assessment of Soybean Line 40-3-2 Conducted in the Russian Federation
The studies were conducted in accordance with the requirements of the Ministry of Health of the Russian Federation authorized for risk and safety assessment of food derived from GM sources [8]. PCR analysis of the soybean

**Table 5.1** Registration Status of Transgenic Soybean Line 40-3-2 in Various Countries [19]

| Country | Approval Date | Application |
|---|---|---|
| Argentina | 1996 | Food, feed, environmental release |
| Brazil | 1998 | Food, feed, environmental release |
| UK | 1996 | Food, feed |
| Canada | 1995 | Feed, environmental release |
|  | 1996 | Food |
| Mexico | 1998 | Food, feed, environmental release |
| USA | 1994 | Food, feed, environmental release |
| Uruguay | 1997 | Food, feed, environmental release |
| Switzerland | 1996 | Food, feed |
| Japan | 1996 | Food, feed, environmental release |

*For an up-to-date registration status of transgenic crops, see http://www.biotradestatus.com/*

test and control samples was performed to confirm the identity of the transformation event and its absence in the conventional control line.

### Biochemical Composition of Soybean 40-3-2 and Soybean Protein Concentrate

Protein content in transgenic soybean did not differ from that in its conventional counterpart. The corresponding values for protein concentrate derived from soybean were similar [17,18]. Amino acid composition in the beans and soybean protein concentrate did not significantly differ in the examined transgenic and conventional varieties (Table 5.2).

The content of carbohydrates in soybeans and soybean protein concentrate did not significantly differ between transgenic and conventional soybean (Table 5.3). Similarly, the content of lipids in soybeans and soybean protein concentrate did not significantly differ between transgenic and conventional soybean (Table 5.4). Fatty acid and phospholipid compositions in transgenic and conventional soybean also were almost identical (Tables 5.5 and 5.6).

The content of vitamins in soybeans and soybean protein concentrate did not significantly differ between transgenic and conventional soybean (Table 5.7). The changes in the content of α-tocopherol in soybean protein concentrate derived from any of the examined soybean varieties remained within the range characteristic of this product (0.10–0.50 mg/100 g).

The content of minerals in GM soybean and the protein concentrate derived from it did not significantly differ from the corresponding values for conventional soybean (Table 5.8). However, according to the data of long-term studies

**Table 5.2** Protein Content (%) and Amino Acid Composition (g/100 g protein) in Soybean and Soybean Protein Concentrate

| Ingredient | Conventional Soybean | Transgenic Soybean | Protein Concentrate (Conventional Soybean) | Protein Concentrate (Transgenic Soybean) |
|---|---|---|---|---|
| Protein content | 36.2 | 38.6 | 74.3 | 68.6 |
| *Amino acids* | | | | |
| Aspartic acid | 14.74 | 14.93 | 14.52 | 13.75 |
| Threonine | 4.54 | 4.35 | 4.12 | 4.08 |
| Serine | 5.45 | 5.60 | 5.54 | 5.40 |
| Glutamic acid | 22.50 | 23.80 | 23.80 | 22.90 |
| Proline | 6.47 | 6.87 | 6.86 | 7.34 |
| Cysteine | 0.74 | 0.76 | 0.85 | 0.83 |
| Glycine | 3.52 | 3.54 | 3.51 | 3.48 |
| Alanine | 4.00 | 3.99 | 4.05 | 4.10 |
| Valine | 3.58 | 3.64 | 3.57 | 3.67 |
| Methionine | 1.86 | 1.87 | 1.90 | 1.91 |
| Isoleucine | 2.75 | 2.68 | 2.81 | 2.74 |
| Leucine | 6.54 | 6.48 | 6.81 | 6.64 |
| Tyrosine | 3.66 | 3.54 | 3.70 | 3.70 |
| Phenylalanine | 4.47 | 4.55 | 4.75 | 4.62 |
| Tryptophane | 1.15 | 1.19 | 1.17 | 1.16 |
| Histidine | 3.18 | 2.85 | 2.89 | 3.25 |
| Lysine | 7.56 | 7.27 | 7.48 | 7.01 |
| Arginine | 8.71 | 8.09 | 7.77 | 8.48 |

**Table 5.3** Carbohydrate Content (g/100 g product) in Soybean and Soybean Protein Concentrate

| Carbohydrate | Conventional Soybean | Transgenic Soybean | Protein Concentrate (Conventional Soybean) | Protein Concentrate (Transgenic Soybean) |
|---|---|---|---|---|
| Fructose | 0.18 | 0.19 | Not detected | Not detected |
| Sucrose | 3.50 | 3.70 | Not detected | Not detected |
| Cellulose | 6.56 | 5.16 | 3.69 | 3.68 |

carried out for many years in the State Research Institute of Nutrition of the Russian Academy of Medical Science (RAMS), there were differences in the content of sodium, potassium, calcium, and magnesium, although they did not surpass the range of physiological variations of soybean: 50–500 mg/kg (sodium),

**Table 5.4** Content of Lipids (g/100 g product) in Soybean and Soybean Protein Concentrate

| Ingredient | Conventional Soybean | Transgenic Soybean | Protein Concentrate (Conventional Soybean) | Protein Concentrate (Transgenic Soybean) |
|---|---|---|---|---|
| Lipids | 20.0 | 18.5 | 0.6 | 1.2 |

**Table 5.5** Comparative Content of Fatty Acids (Rel. %) in Soybean

| Fatty Acid | Conventional Soybean | Transgenic Soybean |
|---|---|---|
| Lauric 12:0 | 0.1 | 0.1 |
| Myristic 14:0 | 0.3 | 0.3 |
| Palmitic 16:0 | 10.2 | 12.5 |
| Palmitoleic 16:1 | 0.2 | 0.2 |
| Stearic 18:0 | 3.9 | 3.1 |
| Oleic 18:1 | 19.9 | 19.6 |
| Linoleic 18:2 | 55.4 | 54.8 |
| Linolenic 18:3 | 9.1 | 9.0 |
| Arachidic 20:0 | 0.1 | 0.1 |
| Eicosenoic 20:1 | 0.8 | 0.3 |

**Table 5.6** Comparative Content of Phospholipids (%) in Soybean

| Phospholipid | Conventional Soybean | Transgenic Soybean |
|---|---|---|
| Lysophosphatidylcholine | 8.0 | 8.0 |
| Phosphatidylcholine | 28.0 | 27.5 |
| Lysophosphatidylethanolamine | 3.0 | 3.5 |
| Phosphatidylserine | 5.0 | 5.0 |
| Phosphatidylethanolamine | 31.5 | 32.0 |
| Phosphatidylglycerol | 8.5 | 8.0 |
| Phosphatidic acids | 16.0 | 16.0 |

10 000–15 000 mg/kg (potassium), and 1500–2500 mg/kg (calcium). The corresponding values for soybean protein concentrate were: 6000–14 000 mg/kg (sodium), 2000–7000 mg/kg (potassium), 2000–5000 mg/kg (calcium), and 1000–3000 mg/kg (magnesium).

The contents of raffinose and stachyose in transgenic and conventional soybean was virtually identical (Table 5.9). The contents of these antinutrients in protein concentrates derived from both varieties of soybean did not surpass

## 5.1.1 Glyphosate-Tolerant 40-3-2 Soybean Line

**Table 5.7** Content of Vitamins (mg/100 g product) in Soybean and Soybean Protein Concentrate

| Ingredient | Conventional Soybean | Transgenic Soybean | Protein Concentrate (Conventional Soybean) | Protein Concentrate (Transgenic Soybean) |
|---|---|---|---|---|
| Vitamin $B_6$ | 0.52 | 0.54 | 0.48 | 0.45 |
| Vitamin $B_1$ | 0.36 | 0.27 | 0.16 | 0.09 |
| Vitamin $B_2$ | 0.15 | 0.18 | 0.09 | 0.05 |
| δ-Tocopherol | 4.51 | 3.72 | 0.32 | 0.34 |
| γ-Tocopherol | 6.00 | 6.86 | 0.74 | 0.61 |
| α-Tocopherol | 0.55 | 0.47 | 0.14 | 0.35 |
| Total tocopherols | 11.06 | 11.05 | 1.20 | 1.30 |
| Lutein + zeaxanthin | 0.07 | 0.08 | Not detected | Not detected |
| Non-identified carotenoids | 0.03 | 0.04 | Not detected | Not detected |
| Total carotenoids | 0.10 | 0.12 | Not detected | Not detected |

**Table 5.8** Mineral Composition (mg/kg product) of Soybean and Soybean Protein Concentrate

| Ingredient | Conventional Soybean | Transgenic Soybean | Protein Concentrate (Conventional Soybean) | Protein Concentrate (Transgenic Soybean) |
|---|---|---|---|---|
| Sodium | 389 | 100 | 8718 | 12329 |
| Potassium | 12721 | 14064 | 2571 | 3711 |
| Calcium | 1743 | 2196 | 4654 | 2257 |
| Magnesium | 1525 | 1786 | 2727 | 1627 |
| Iron | 48.6 | 68.9 | 99.3 | 95.2 |
| Copper | 9.62 | 7.64 | 9.50 | 7.64 |

**Table 5.9** Antinutrients (g/100 g product) in Soybean and Soybean Protein Concentrate

| Ingredient | Conventional Soybean | Transgenic Soybean | Protein Concentrate (Conventional Soybean) | Protein Concentrate (Transgenic Soybean) |
|---|---|---|---|---|
| Raffinose | 1.79 | 1.92 | Not detected | Not detected |
| Stachyose | 7.95 | 6.77 | Not detected | 0.41 |

**Table 5.10 Human and Animal Safety Parameters of Soybean**

| Ingredient | Conventional Soybean | Transgenic Soybean | Protein Concentrate (Conventional Soybean) | Protein Concentrate (Transgenic Soybean) |
|---|---|---|---|---|
| Aflatoxin B1, mg/kg | Not detected | Not detected | Not detected | Not detected |
| Lead, mg/kg | 0.072 | <0.001 | 0.091 | 0.017 |
| Cadmium, mg/kg | 0.113 | 0.049 | 0.036 | 0.051 |

the acceptable levels of the products for children's diet and dietary nutrition levels established by corresponding regulations of the Russian Federation [5]

The content of cadmium, lead, and mycotoxins in conventional or transgenic soybean and in the protein concentrates derived from both varieties did not surpass the acceptable limits of the regulations valid in Russia [5] (Table 5.10).

Thus, the data shown in Tables 5.2 to 5.10 attest to the compositional equivalence of glyphosate-tolerant transgenic soybean line 40-3-2 and corresponding protein concentrate to their conventional counterparts. The variations of the examined parameters (relative or absolute concentrations) remained within the range characteristic of soybean [4,17,18,22,35,46,56].

The safety parameters (presence of mycotoxins and toxic contaminants) of transgenic soybean line 40-3-2 and conventional soybean, as well as the corresponding protein concentrates, comply with the requirements of the regulations valid in the Russian Federation [5].

The content of antinutrients (stachyose and raffinose) in the protein concentrate derived from transgenic soybean line 40-3-2 or conventional soybean did not surpass the acceptable levels established in the Russian Federation for children's diet and dietary nutrition [5].

### *Toxicological Assessment of Soybean 40-3-2*

The chronic experiment (150 days) was carried out on male Wistar rats ($n = 60$) with an initial body weight of 80–100 g. The rats were randomized into two groups. All animals were maintained on isocaloric diets (203.7 kcal/100 g feed) with identical content of protein (10.8 g/100 g feed), fat (9.3 g/100 g feed), and carbohydrates (22.2 g/100 g feed). In the control group, the daily diet was supplemented with protein concentrate derived from the conventional soybean (1.25 g per animal). The test rats were provided with the same amount of protein concentrate prepared from transgenic soybean line 40-3-2 (Table 5.11).

**Table 5.11** Standard Rat Diet with Soybean Protein Concentrate

| Ingredient | Weight, g |
|---|---|
| Soybean protein concentrate | 3.10 |
| Grain mix | 40.0 |
| Bread baked of second-grade flour | 9.23 |
| Curd | 4.71 |
| Fish flour | 1.15 |
| Carrot | 20.5 |
| Greens | 20.5 |
| Cod-liver oil | 0.23 |
| Yeast | 0.23 |
| NaCl | 0.35 |
| Total | 100 |

In this study, the diet was given in two stages: in the morning, the rats were fed millet porridge supplemented with soybean protein concentrate, and in 6 hours they were given the remaining dietary ingredients. The amount of the daily diet was 40.5 g per animal. The actual feed intake was recorded throughout the experiment.

Biochemical, hematological, and morphological studies were conducted in accordance with the requirements of the Ministry of Health of the Russian Federation authorized for risk and safety assessment of food derived from GM sources [8]. Samples were collected on days 30 and 150 of the experiment.

### Assessment of Proximate Parameters

During the entire duration of the experiment, the daily portion of millet porridge containing 1.25 g protein concentrate derived from transgenic soybean line 40-3-2 (test group) or the same amount of protein concentrate made of conventional soybean (control group) were completely consumed. The difference in body weight of the rats from control and test groups was insignificant (Table 5.12; Figure 5.1).

The absolute and relative weights of internal organs were determined on days 30 and 150 after the start of the experiment. The values obtained did not significantly differ for rats from the control and test groups (Table 5.13).

### Assessment of Biochemical Parameters

The content of total protein, glucose, activity of alkaline phosphatase, alanine aminotransferase, and aspartate aminotransferase in blood serum, pH and the relative density of urine, urinary concentration of creatinine and its urinary

**Table 5.12** Body Weight of Rats (g) Fed Diet with Protein Concentrate Derived from Conventional Soybean or Transgenic Soybean Line 40-3-2 ($M \pm m$; $n = 8$)

| Duration of the Experiment, Weeks | Control | Test |
|---|---|---|
| 0 | 72.5 ± 2.5 | 74.4 ± 1.6 |
| 1 | 102.0 ± 2.0 | 108.0 ± 3.1 |
| 2 | 128.4 ± 3.2 | 133.5 ± 3.8 |
| 3 | 150.5 ± 3.9 | 156.7 ± 2.9 |
| 4 | 178.3 ± 10.2 | 179.2 ± 11.1 |
| 8 | 235.5 ± 8.6 | 233.1 ± 11.6 |
| 12 | 296.5 ± 16.3 | 301.6 ± 13.9 |
| 16 | 347.5 ± 19.0 | 352.5 ± 15.7 |
| 20 | 380.1 ± 28.2 | 375.7 ± 22.4 |

*Note: Here and in Tables 5.13–5.23 the differences are not significant ($p > 0.05$).*

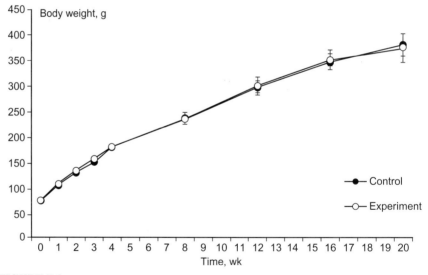

**FIGURE 5.1**

Comparative dynamics of body weight of rats fed a diet containing protein concentrate derived from transgenic (test) or conventional (control) soybean.

excretion did not significantly differ in control and test rats on Day 30 and on Day 150 after the onset of the experiments (Tables 5.14 and 5.15). Insignificant moderation of activity of alkaline phosphatase in blood serum of rats fed diet with protein concentrate derived from transgenic soybean did not surpass the physiological range characteristic of the rats of corresponding age.

### 5.1.1 Glyphosate-Tolerant 40-3-2 Soybean Line

**Table 5.13** Absolute and Relative Weight of Internal Organs of Rats Fed Diet with Protein Concentrate Derived from GM Soybean Line 40-3-2 or Conventional Soybean ($M \pm m$, $n = 6–8$)

| Organ | | 30 Days | | 150 Days | |
|---|---|---|---|---|---|
| | | Control | Test | Control | Test |
| Kidneys | Abs.[a], g | 1.45 ± 0.08 | 1.42 ± 0.11 | 1.99 ± 0.23 | 2.37 ± 0.20 |
| | Rel.[b], g/100 g | 0.81 ± 0.01 | 0.79 ± 0.03 | 0.60 ± 0.08 | 0.54 ± 0.08 |
| Liver | Abs., g | 8.52 ± 0.57 | 8.61 ± 0.62 | 12.40 ± 0.88 | 13.50 ± 0.62 |
| | Rel., g/100 g | 4.77 ± 0.10 | 4.80 ± 0.14 | 3.20 ± 0.32 | 3.60 ± 0.45 |
| Spleen | Abs., g | 1.21 ± 0.08 | 1.19 ± 0.11 | 1.70 ± 0.32 | 1.67 ± 0.30 |
| | Rel., g/100 g | 0.68 ± 0.05 | 0.66 ± 0.03 | 0.44 ± 0.11 | 0.45 ± 0.12 |
| Heart | Abs., g | 0.73 ± 0.02 | 0.73 ± 0.05 | 1.06 ± 0.11 | 1.21 ± 0.21 |
| | Rel., g/100 g | 0.42 ± 0.02 | 0.40 ± 0.02 | 0.33 ± 0.09 | 0.32 ± 0.08 |
| Testicles | Abs., g | 1.96 ± 0.21 | 2.05 ± 0.12 | 2.92 ± 0.34 | 3.18 ± 0.41 |
| | Rel., g/100 g | 1.08 ± 0.06 | 1.15 ± 0.05 | 0.78 ± 0.17 | 0.75 ± 0.16 |
| Hypophysis | Abs., mg | 5.40 ± 1.27 | 6.50 ± 0.22 | 10.50 ± 0.50 | 10.80 ± 0.65 |
| | Rel., mg/100 g | 3.12 ± 0.75 | 3.67 ± 0.17 | 3.20 ± 0.50 | 3.00 ± 0.43 |
| Adrenal glands | Abs., mg | 22.0 ± 1.57 | 20.8 ± 1.25 | 28.0 ± 1.2 | 28.5 ± 1.0 |
| | Rel., mg/100 g | 12.30 ± 0.51 | 12.10 ± 1.50 | 8.20 ± 0.97 | 7.10 ± 0.98 |
| Seminal vesicles | Abs., mg | 72.33 ± 23.91 | 86.83 ± 30.60 | 532.4 ± 28.7 | 657.1 ± 37.0 |
| | Rel., mg/100 g | 38.20 ± 11.03 | 47.27 ± 15.61 | 164.80 ± 14.00 | 162.00 ± 12.10 |
| Prostate | Abs., mg | 59.00 ± 16.57 | 53.67 ± 20.15 | 349.0 ± 30.8 | 427.0 ± 29.6 |
| | Rel., mg/100 g | 31.30 ± 7.47 | 29.42 ± 10.31 | 119.9 ± 18.5 | 128.9 ± 18.7 |

[a]Absolute weight of internal organs (g).
[b]Relative weight of internal organs (g/100 g body weight).

**Table 5.14** Biochemical Parameters of Blood Serum of Rats Fed Diet with Protein Concentrate Derived from GM Soybean Line 40-3-2 or Conventional Soybean ($M \pm m$, $n = 6–8$)

| Parameter | 30 Days | | 150 Days | |
|---|---|---|---|---|
| | Control | Test | Control | Test |
| Total protein, g/L | 65.2 ± 2.9 | 64.8 ± 3.3 | 75.2 ± 4.9 | 76.8 ± 2.2 |
| Glucose, mmol/L | 9.0 ± 0.5 | 9.7 ± 0.5 | 6.9 ± 0.4 | 6.9 ± 0.3 |
| Alanine aminotransferase, μM/min/L | 66.1 ± 6.4 | 56.0 ± 4.8 | 61.8 ± 3.5 | 65.8 ± 6.0 |
| Aspartate aminotransferase, μM/min/L | 67.5 ± 4.4 | 61.0 ± 3.4 | 66.0 ± 3.8 | 47.7 ± 4.7 |
| Alkaline phosphatase, IU/L | 722.2 ± 65.3 | 822.5 ± 36.5 | 522.2 ± 45.7 | 529.5 ± 38.3 |

**Table 5.15** Urinary Biochemical Parameters of Rats Fed Diet with Protein Concentrate Derived from Conventional Soybean or Transgenic Soybean Line 40-3-2 ($M \pm m$, $n = 6-8$)

| Parameter | 30 Days | | 150 Days | |
|---|---|---|---|---|
| | Control | Test | Control | Test |
| pH | 6.9 | 7.0 | 7.0 | 7.0 |
| Daily diuresis, mL | 11.5 ± 1.2 | 9.5 ± 1.4 | 6.1 ± 0.5 | 6.5 ± 1.0 |
| Relative density, g/mL | 0.99 ± 0.01 | 0.98 ± 0.01 | 1.02 ± 0.01 | 1.01 ± 0.01 |
| Creatinine, mg/mL | 0.26 ± 0.02 | 0.32 ± 0.14 | 1.53 ± 0.13 | 1.33 ± 0.17 |
| Creatinine, mg/day | 3.02 ± 0.44 | 3.07 ± 0.38 | 9.05 ± 0.51 | 8.08 ± 1.04 |

**Table 5.16** Activity of Enzymes Involved in Metabolism of Xenobiotics in Liver Microsomes of Rats Fed Diet with Protein Concentrate Derived from Conventional Soybean or Transgenic Soybean Line 40-3-2 ($M \pm m$, $n = 6-8$)

| Parameter | 30 Days | | 150 Days | |
|---|---|---|---|---|
| | Control | Test | Control | Test |
| Cytochrome P450, nM/mg | 0.72 ± 0.06 | 0.71 ± 0.02 | 0.58 ± 0.05 | 0.61 ± 0.03 |
| Cytochrome b5, nM/mg | 0.66 ± 0.05 | 0.66 ± 0.02 | 0.56 ± 0.09 | 0.67 ± 0.07 |
| 7-ethoxycoumarin-O-deethylation, nM/min/mg | 3.19 ± 0.52 | 3.52 ± 0.25 | 1.42 ± 0.15 | 1.33 ± 0.23 |
| Aminopyrine N-demethylation, nM/min/mg | 10.45 ± 0.44 | 10.13 ± 0.38 | 10.57 ± 0.37 | 11.13 ± 0.18 |
| Acetylesterase, µM/min/mg | 7.77 ± 0.24 | 7.74 ± 0.30 | 5.58 ± 0.33 | 5.03 ± 0.60 |
| Epoxide hydrolase, nM/min/mg | 5.46 ± 0.11 | 5.54 ± 1.40 | 5.05 ± 0.53 | 4.87 ± 0.47 |
| UDP- glucuronosil transferase, nM/min/mg | 46.2 ± 3.0 | 44.8 ± 4.3 | 16.1 ± 2.1 | 15.9 ± 1.7 |

### Assessment of Sensitive Biomarkers

The studies showed that long-term (150 days) addition of protein concentrate derived from transgenic soybean line 40-3-2 to the rat diet produced no significant changes in activity of enzymes involved in phase I and phase II xenobiotic degradation located in the membranes of endoplasmic reticulum, and did not affect activity of lysosomal marker enzymes (Tables 5.16 and 5.17).

The above-mentioned data unequivocally support the conclusion that a long-term (150 days) exposure of rats to a diet supplemented with protein concentrate derived from transgenic soybean line 40-3-2 is not accompanied by any significant changes (surpassing the physiological boundaries) in the activity of enzymatic systems involved in the process of metabolism and detoxification of endogenous and exogenous chemical agents.

**Table 5.17** Total and Non-sedimentable Activity of Lysosomal Enzymes in Liver of Rats Fed Diet with Protein Concentrate Derived from Conventional Soybean or Transgenic Soybean Line 40-3-2 ($M \pm m$, $n = 6$–$8$)

| Parameters | 30 days | | 150 days | |
|---|---|---|---|---|
| | Control | Test | Control | Test |
| *Total activity, µM/min/g tissue* | | | | |
| Arylsulfatase A, B | 2.48 ± 0.04 | 2.43 ± 0.04 | 2.56 ± 0.11 | 2.62 ± 0.12 |
| β-Galactosidase | 2.16 ± 0.04 | 2.17 ± 0.04 | 2.79 ± 0.07 | 2.65 ± 0.04 |
| β-Glucuronidase | 2.57 ± 0.07 | 2.49 ± 0.05 | 2.38 ± 0.09 | 2.38 ± 0.09 |
| *Non-sedimentable activity, % total activity* | | | | |
| Arylsulfatase A, B | 3.23 ± 0.11 | 3.22 ± 0.12 | 3.53 ± 0.09 | 3.30 ± 0.12 |
| β-Galactosidase | 5.26 ± 0.14 | 5.18 ± 0.13 | 5.80 ± 0.49 | 5.88 ± 0.28 |
| β-Glucuronidase | 5.86 ± 0.21 | 6.56 ± 0.29 | 4.90 ± 0.11 | 4.91 ± 0.15 |

**Table 5.18** Hematological Parameters of Peripheral Blood Drawn from Rats Fed Diet with Protein Concentrate Derived from Conventional Soybean or Transgenic Soybean Line 40-3-2 ($M \pm m$, $n = 6$–$8$)

| Parameter | 30 Days | | 150 Days | |
|---|---|---|---|---|
| | Control | Test | Control | Test |
| Hemoglobin concentration, g/L | 137.5 ± 4.52 | 130.2 ± 3.52 | 156.4 ± 8.97 | 152.4 ± 7.66 |
| Total erythrocyte count, ×$10^{12}$/L | 5.96 ± 0.02 | 5.87 ± 0.04 | 5.55 ± 0.23 | 5.80 ± 0.34 |
| Hematocrit, vol.% | 50.00 ± 0.00 | 49.60 ± 0.19 | 48.80 ± 0.38 | 49.20 ± 0.57 |
| MCH, pg | 23.09 ± 0.77 | 23.70 ± 0.74 | 28.22 ± 1.13 | 26.30 ± 0.76 |
| MCHC, % | 27.50 ± 0.90 | 28.07 ± 0.82 | 32.00 ± 1.57 | 30.90 ± 4.00 |
| MCV, µm$^3$ | 83.89 ± 0.30 | 84.46 ± 0.55 | 88.45 ± 2.35 | 85.90 ± 4.00 |
| Total leukocyte count, ×$10^9$/L | 15.35 ± 1.54 | 13.43 ± 0.96 | 12.38 ± 0.64 | 12.10 ± 1.02 |

MCH, mean cell hemoglobin; MCHC, mean cell hemoglobin concentration; MCV, mean cell volume.

### Hematological Assessments

The hematological studies showed that addition of the protein concentrate derived from transgenic soybean line 40-3-2 to the rat diet did not induce significant changes in concentration of hemoglobin, hematocrit, total erythrocyte count, MCH, MCHC, MCV, total leukocyte count, absolute and relative count of eosinophils, neutrophils, and lymphocytes relative to the control values obtained in 30 and 150 days after the onset of the experiments (Tables 5.18 and 5.19).

Table 5.19 Levels of Leucocytes Parameters of Rats Fed Diet with Protein Concentrate Derived from Conventional or Transgenic Soybean Line 40-3-2 ($M \pm m$, $n = 6–8$)

| Parameter | 30 Days | | 150 Days | |
|---|---|---|---|---|
| | Control | Test | Control | Test |
| *Segmentonuclear neutrophils* | | | | |
| rel., % | 12.80 ± 1.90 | 10.40 ± 1.15 | 13.80 ± 1.34 | 14.00 ± 2.30 |
| abs., ×10$^9$/L | 2.04 ± 0.45 | 1.47 ± 0.14 | 1.69 ± 0.12 | 1.65 ± 3.20 |
| *Eosinophils* | | | | |
| rel., % | 0.80 ± 0.38 | 1.40 ± 0.38 | 1.20 ± 0.38 | 0.80 ± 0.38 |
| abs., ×10$^9$/L | 0.126 ± 0.07 | 0.172 ± 0.05 | 0.144 ± 0.05 | 0.110 ± 0.06 |
| *Lymphocytes* | | | | |
| rel., % | 86.40 ± 1.50 | 88.20 ± 1.15 | 85.00 ± 1.53 | 85.20 ± 2.10 |
| abs., ×10$^9$/L | 13.18 ± 1.13 | 11.19 ± 0.87 | 10.52 ± 0.72 | 10.36 ± 0.99 |

### Morphological Assessments

Post-mortem dissection of rats performed on Day 30 or Day 150 after the onset of the nutritional experiment revealed no pathological alterations in the internal organs that could be related to the diet with protein concentrate derived from transgenic soybean line 40-3-2. The results of microscopic examinations are shown in Table 5.20. The histological assessments of internal organs revealed no differences between the control and test groups.

Thus, the results of the 150-day toxicological experiment with addition of protein concentrate derived from transgenic glyphosate-tolerant soybean line 40-3-2 to the diet of male Wistar rats revealed no adverse effects on the animals based on biochemical, hematological, and morphological assessments.

### Assessment of Dietary and Nutritional Value of Protein Derived from Transgenic Soybean Line 40-3-2

Experiments were carried out on growing male Wistar rats ($n = 36$) with an initial weight of 65 g. For 28 days, all rats were fed an isocaloric diet (420 kcal/100 g dry feed) containing soybean protein or casein (8–9% mass), sunflower-seed oil (11.5% mass), maize starch (70% mass), salt mix (4% mass), and a mix of fat- and water-soluble vitamins. The rats were randomized into three groups: control group (rats fed 9% protein concentrate derived from conventional soybean), test group (rats fed 9.2% protein concentrate derived from transgenic soybean line 40-3-2), and the background control group (rats fed 8.2% casein as a single source of protein).

**Table 5.20** Microscopic Assessment of Internal Organs in Rats (Combined Data Obtained on Experimental Days 30 and 150)

| Organ | Control | Test |
| --- | --- | --- |
| Liver | Clear trabecular structure; no alterations in hepatocytes and the portal ducts | No differences from control |
| Kidneys | Usual appearance of cortical and medullar substance; no alterations in glomeruli or pelvis epithelium | No differences from control |
| Lung | Alveolar space is air-filled; no alterations in bronchi and blood vessels | No differences from control |
| Spleen | Large folliculi with wide clear marginal zones and reactive centers; splenic pulp is moderately plethoric | No differences from control |
| Small intestine | Preserved villous epithelium; usual infiltration in villous stroma | No differences from control |
| Testicle | Usual size and appearance of seminiferous tubules; clearly definable spermiogenesis | No differences from control |

During the entire experiment, the rats were kept in metabolic cages with feed and water *ad libitum*. The rats were weighed every other day, and the amount of feed consumed by each animal was recorded. During the last 3 days of the experiment, referred to as "metabolism assessment period", the amount of consumed feed and weight of excrement was recorded.

The content of total nitrogen in the feed and excrement was determined with the Kjeldahl semi-micromethod using a Kjeltec Auto 1030 Analyzer (Tecator, Sweden). Statistical analysis of the data was carried out with the routing software.

Biological and nutritional values of the proteins were calculated by standard methods and due account for endogenous nitrogen loss. To obtain the correct values, these parameters were calculated with the use of protein efficiency ratio (PER), Net Protein Efficiency Ratio (NPER), and nutritional value. The resulting absolute values were compared with control parameters and expressed as a percentage (relative biological value) [8].

Daily feed intake, daily protein intake, daily body weight gain, and the indicators of biological value (PER and NPER) in rats fed diet with protein concentrate derived from GM soybean line 40-3-2 did not significantly differ from the corresponding values for the rats fed diet with protein concentrate prepared from conventional soybean (Table 5.21). In the background

**Table 5.21** Nutritional Parameters of the Protein Concentrate Derived from Conventional Soybean and Transgenic Soybean Line 40-3-2 ($M \pm m$, $n = 12$)

| Parameter | Control | Test |
|---|---|---|
| Daily feed intake, g/day | 11.7 ± 0.38 | 12.2 ± 0.29 |
| Daily protein intake, g/day | 1.04 ± 0.03 | 1.12 ± 0.03 |
| Daily body weight gain, g/day | 2.09 ± 0.13 | 2.17 ± 0.10 |
| PER | 1.71 ± 0.11 | 1.93 ± 0.10 |
| NPER | 2.44 ± 0.13 | 2.51 ± 0.10 |

NPER, net protein efficiency ratio; PER, protein efficiency ratio.

group of rats fed diet with casein, PER and NPER (calculated for casein) were 2.28 ± 0.07 and 3.08 ± 0.07, correspondingly. The relative biological value of the protein of transgenic soybean line 40-3-2 was 81.4%, while the corresponding value of the conventional soybean was 79.2%.

True digestibility of proteins in the concentrates derived from GM soybean line 40-3-2 and conventional soybean did not significantly differ, being 91.30 ± 0.58 and 90.70 ± 0.64%, correspondingly ($n = 12$). However, in both cases, digestibility parameters for these proteins were smaller than that of casein.

The experiments were terminated with an 18 h fasting period; thereafter the rats were decapitated under ether anesthesia. Visual inspection of the internal organs in the control and test rats did not reveal any structural alterations in comparison with the rats fed diet with casein as a single source of the protein.

The weight of internal organs of the rats, kept for 29 days on the diet with the proteins from transgenic soybean, did not differ from the corresponding values for the control rats (Table 5.22).

The data on the content of essential amino acids in the proteins of control and test samples, as well as the TPD values of these proteins (0.913 and 0.907, correspondingly) made it possible to calculate their biological value by the method recommended for *in vitro* studies (Table 5.23) [8].

The calculations showed that the proteins of both soybean varieties are full-value; they are not limited by any essential amino acid or by the value of amino acid score (1.0 and 0.98 for control and test soybean samples, correspondingly).

These data were used to calculate the biological value (BV) of soybean proteins: 0.89 (GM soybean) and 0.91 (conventional soybean).

### 5.1.1 Glyphosate-Tolerant 40-3-2 Soybean Line

**Table 5.22** Weight of Internal Organs of Rats Fed Diet with Protein Concentrate Derived from GM Soybean Line 40-3-2 or Conventional Soybean ($M \pm m$, $n = 12$)

| Weight (g) | Control | Test |
|---|---|---|
| Body weight | 118.3 ± 4.9 | 126.8 ± 3.3 |
| Liver | 4.64 ± 0.18 | 4.75 ± 0.14 |
| Kidneys | 1.12 ± 0.03 | 1.11 ± 0.07 |
| Heart | 0.56 ± 0.02 | 0.55 ± 0.01 |
| Spleen | 0.61 ± 0.05 | 0.60 ± 0.03 |
| Testicles | 1.48 ± 0.11 | 1.43 ± 0.10 |

**Table 5.23** Amino Acid Score of Soybean Protein Concentrate Corrected for Protein Digestibility

| | | Soybean | | | |
|---|---|---|---|---|---|
| | | Control | | Test | |
| Essential Amino Acids | Reference Scale | Content, g/100 g protein | Score | Content, g/100 g protein | Score |
| Valine | 3.5 | 3.57 | 1.02 | 3.67 | 1.05 |
| Threonine | 3.4 | 4.12 | 1.21 | 4.08 | 1.20 |
| Isoleucine | 2.8 | 2.81 | 1.00 | 2.74 | 0.98 |
| Leucine | 6.6 | 6.81 | 1.03 | 6.64 | 1.00 |
| Lysine | 5.8 | 7.48 | 1.29 | 7.01 | 1.21 |
| Methionine + Cysteine | 2.5 | 2.75 | 1.10 | 2.74 | 1.09 |
| Phenylalanine + Tyrosine | 6.3 | 8.45 | 1.34 | 8.32 | 1.32 |
| Tryptophan | 1.1 | 1.17 | 1.06 | 1.16 | 1.05 |

The revealed similarity between BV of the proteins in conventional soybean and transgenic soybean line 40-3-2 attests to equivalence of both soybean varieties by this criterion.

### Assessment of Potential Impact of Transgenic Soybean Line 40-3-2 on Immune System in Studies on Mice
#### Potential Effect on Humoral Component of Immune System

The immunomodulating effect of GM soybean on the humoral component of the immune system was examined by determining the level of hemagglutination to sheep erythrocytes (SE) in mice lines C57Bl/6 (low sensitivity to SE) and CBA (high sensitivity to SE).

Soybean protein concentrate was fed to mice for 21 days. The control and test mice were fed a diet with conventional and transgenic soybean line 40-3-2, correspondingly (Table 5.11). On Day 21 the mice of both groups were intraperitoneally injected with 0.5 mL sheep erythrocytes (SE) (10 million cells). Blood was drawn on day 7, 14, and 21 after the onset of the experiment. The protein concentrates from conventional and transgenic soybean were fed to the mice of control and test groups during the entire period of the experiment. Blood serum was titrated in reaction of hemagglutination by the routine method.

All mice demonstrated the presence of antibodies against SE. At any term of the experiment, the antibody titers were 1:64 in C57Bl/6 mice and 1:128 in CBA mice. Thus, the control and test mice had identical titers of antibodies raised against SE. These data support the conclusion that transgenic soybean line 40-3-2 produces no effect on the humoral component of the immune system.

### Potential Effect on Cellular Component of Immune System

The immunomodulating effect of transgenic soybean was assessed with delayed hypersensitivity reaction to sheep erythrocytes (SE). Both lines of mice were used in this test. The soybean protein concentrate was added to the diet for 21 days; thereafter, SE was injected subcutaneously (1 million cells per mouse). On post-injection Day 5, SE (0.02 mL, $10^9$ cells) was injected into the finger-pad of the right hindleg of control and test mice. The left hindleg was injected with 0.02 mL physiological saline solution. Local inflammatory reaction was assessed 18 h after the injections by comparison of the weights of both injected paws.

In CBA mice fed diet with protein concentrate derived from transgenic soybean, the reaction index (RI) was $23 \pm 13$. In CBA mice fed the conventional protein concentrate, RI was $41 \pm 15$; in the control (fed a soy-free diet) CBA mice, RI was $30 \pm 11$.

In C57Bl/6 mice, a similar trend was observed with corresponding RI parameters of $48 \pm 18$ (40-3-2), $59 \pm 24$ (conventional soybean), and $51 \pm 21$ (control). These data support the conclusion that transgenic soybean line 40-3-2 produces no effect on the cellular component of the immune system.

### Assessment of a Potential Sensibilization Effect of Transgenic Soybean

Assessment of possible sensibilization action of the transgenic soybean on the immune response to endogenous metabolic products was carried out in the experiment of mouse sensitivity to histamine. For 21 days, the control and test mice were fed diets with protein concentrate derived from conventional and transgenic soybean (Table 5.11). Then the mice of both groups were injected intraperitoneally with 2.5 mg histamine hydrochloride dissolved in 0.5 mL physiological solution. After 24 h post-injection, all mice were alive, which attests to the absence of sensitizing ingredient in transgenic soybean line 40-3-2.

**Potential Effect of Soybean on Susceptibility of Mice to *Salmonella typhimurium***

The effect of transgenic soybean on susceptibility of mice to infection by salmonella of murine typhus was examined in experiments on mice injected intraperitoneally with various doses of Strain 415 *Salmonella typhimurium*. Four weeks prior to infection, the diet of control and test mice was supplemented with protein concentrate derived from conventional or transgenic soybean, respectively. The injected doses ranged from 10 to $10^5$ microbial cells per mouse and varied on a 10-fold basis. The post-injection observation period was 14 days.

In both groups, the death of mice started on post-injection Day 3, and all mice infected with $10^3$–$10^5$ microbial cells died by post-injection Day 6. The smaller doses of the virulent culture ($10$–$10^3$ microbial cells per animal) did not produce 100% death in the first post-injection days, so the loss of mice in both groups was observed during the entire observation period.

The lifetime of the mice in the test group was somewhat longer than that of the control mice: the test mice infected with $10^5$ or $10^4$ microbial cells lived 4.2 and 6.2 days as compared with 1.2 and 2.2 days of the control mice, correspondingly. The smaller doses did not reveal any difference in the lifetime of mice in both groups. The values of $LD_{50}$ were 256 and 175 bacterial cells per mouse in the test and control groups, respectively.

These data showed that *Salmonella typhimurium* produced typical infection both in control mice fed diet with conventional soybean protein concentrate and in the test mice fed diet with transgenic protein concentrate. According to the difference in the time to death, the test group took longer to die than the controls, although the differences in $LD_{50}$ values remained within the experimental error.

Thus, introduction of protein concentrate derived from transgenic soybean line 40-3-2 into mouse diet produced no effect on the humoral and cellular components of the immune system, did not sensitize the mouse organism, and did not disturb the natural resistance against typical infection such as murine typhus. Taken together, these data support the conclusion that transgenic soybean line 40-3-2 has no immunomodulating properties.

### Assessment of Potential Impact of Transgenic Soybean 40-3-2 on Immune System in Studies on Rats

The study was carried out on male Wistar rats ($n = 49$) weighing $180 \pm 10$ g. After a 7-day adaptation period to standard vivarium diet, in the following 28 days the rats were fed diet supplemented with protein concentrate (3.3 g/day/rat) derived from conventional soybean (control group) or from transgenic soybean line 40-3-2 (test group), as presented in Table 5.11.

### Model of Systemic Anaphylaxis

The model of systemic anaphylaxis was developed according to the standard protocols as described in [8].

On experimental days 1, 3, and 5, the rats were sensitized intraperitoneally with 100 µg ovalbumin from hens' eggs (OVA, the preparation was re-crystallized five times) absorbed by 10 mg aluminum hydroxide. On Day 21, another portion of 10 µg OVA was administered under the same conditions to induce the secondary immune response. After termination of feeding animals with the diets on experimental Day 29, blood (0.2 mL) was drawn from the tail vein to assess the response of antibodies. Then the booster dose of OVA (30 mg/kg in 0.5 mL isotonic apyrogenic 0.15 M NaCl saline) was injected intravenously. During the following 24 h, the development of symptoms of active anaphylactic shock was observed.

Severity of anaphylactic shock was scored as follows: +(1), shiver, chill, dyspnea; ++(2), asthenia, ataxia, peripheral cyanosis; +++(3), convulsions, paralysis; ++++(4) fatal outcome.

The anaphylactic index (AI) was calculated according to [8] as the mean of anaphylactic severity scores in a group in 24 h after injection of the booster dose.

Intensity of humoral immune response was assessed according to concentration of circulating specific immunoglobulin antibodies (the sum of $IgG_1$ and $IgG_4$ fractions) by the method of indirect solid-phase enzyme-linked immunosorbent assay (standard ELISA) on polystyrene.

### Mathematical Processing of Experimental Data

Concentration of IgG antibodies in rat blood serum was determined from the standard plot by linear interpolation in semilogarithmic coordinates using Excel 5.0 software (Figure 5.2). Significance of differences between the mean values of animal's body weight and the titers of antibodies against OVA in two groups was determined with the two-sided Student's $t$-test and Fisher's $F$-test for residual variance. The examined parameters were optical density, concentration of antibodies, and common logarithm of this concentration. Initially, the normal distribution of data was examined using histograms plotted by Winstat

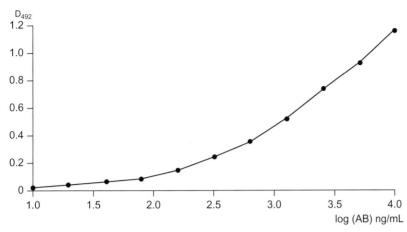

**FIGURE 5.2**
Standard plot used to assess concentration of antibodies.

**Table 5.24** Severity of Active Anaphylactic Shock in Control and Test Rats

| Group of Rats | AI | | Severe Reactions, % | Mortality, % |
|---|---|---|---|---|
| | 3 h after Injection | 24 h after Injection | | |
| Control (n = 25) | 1.88 | 2.64 | 56.00 | 52.00 |
| Test (n = 24) | 1.67 | 2.42 | 45.80 | 45.80 |

4.3 software that was used throughout this study. Significance of the difference between the fraction parameters (mortality percentage and AI) was proved with the U-test (Fisher's angular transformation) [3].

### Results
During the entire experiment the rats of both groups fed diet with protein concentrate derived from conventional and transgenic soybean line 40-3-2 grew normally, which indicates nutritional adequacy of the diets used. On experimental Day 29, the body weights of test and control rats were $248 \pm 7$ g and $242 \pm 6$ g, respectively ($p > 0.1$).

Table 5.24 shows the data on severity of active anaphylactic shock in control and test rats. All parameters (mortality in a group, severe anaphylactic reactions, AI in 3 h and 24 h) showed that the differences between the rats fed diets with protein concentrate derived from conventional soybean (control

**Table 5.25** Parameters of Humoral Immune Response in Control and Test Rats ($M \pm m$)

| Group of Rats | $D_{492}$ | Concentration of Antibodies, mg/mL | Logarithm of Antibody Concentration |
|---|---|---|---|
| Control ($n = 25$) | $0.563 \pm 0.022$ | $4.39 \pm 0.74$ | $0.491 \pm 0.076$ |
| Test ($n = 24$) | $0.491 \pm 0.028$ | $2.76 \pm 0.44$ | $0.278 \pm 0.086$ |

group) or GM soybean line 40-3-2 (test group) were insignificant ($p > 0.05$). There was only an insignificant trend to moderation of anaphylactic reaction in the test group.

Table 5.25 shows the mean values of parameters $D_{492}$, concentration of antibodies, and common logarithm of this concentration in the control and test groups. The parameters of antibody concentration and its common logarithm did not significantly differ between groups ($p > 0.05$). The scattering of antibody concentration was significantly greater in the control group than in the test. However, the difference in $D_{492}$ value between groups was small but significant ($p < 0.05$): in the test group, $D_{492}$ was smaller than that in the control group, although standard deviation of this value did not differ between groups.

Thus, the intensity of humoral immune response in the rats fed diet with protein concentrate derived from transgenic soybean line 40-3-2 demonstrated a declining trend in comparison with the control group. Therefore, the degree of sensitization by ovalbumin in these rats did not increase in comparison with the rats fed diet with protein concentrate derived from conventional soybean.

The studies showed that the protein concentrate prepared from transgenic soybean line 40-3-2 did not elevate allergic reactivity and sensitization towards the model allergen in test rats in comparison with the control rats fed conventional soybean.

### Assessment of Potential Genotoxicity of Transgenic Soybean

Genotoxicity studies were carried out on C57Bl/6 and CBA mice sensitive to mutagenesis. For 35 days, the mice weighing 16–18 g were fed diet with transgenic soybean (test group) or its conventional counterpart (control group) with daily feed intake of 1.5 g/day/animal (Table 5.11).

These studies examined chromosomal aberrations in the cells of bone marrow and the dominant lethal mutations in the gametes of control and test mice. The cytogenetic analysis was carried out by metaphasic method [8].

### 5.1.1 Glyphosate-Tolerant 40-3-2 Soybean Line

**Table 5.26** Bone Marrow Cytogenetic Parameters in Mice Fed a 35-Day Diet with Protein Concentrate Derived from Transgenic Soybean Line 40-3-2 (Test) or Conventional Soybean (Control)

| Number of Cells, % | Control | Test |
| --- | --- | --- |
| with chromosomal aberrations | 1.08 ± 0.53 | 0.77 ± 0.44 |
| with gaps | 1.89 ± 0.70 | 1.55 ± 0.62 |
| with polyploid chromosome set | 1.89 ± 0.70 | 1.81 ± 0.67 |

*Note: The numbers of analyzed metaphases were: 386 (test) and 370 (control). The differences were insignificant ($p > 0.05$).*

The mice of both groups were sacrificed in 24 h after the last feeding. Two hours prior to termination of the experiment, the mice were intraperitoneally injected with colchicine to accumulate the cells with metaphases. Bone marrow was isolated from both femoral bones. After hypotenization of the cells in a thermostat with 0.5% KCl saline and fixation in ethanol + acetic acid mixture, the cytogenetic preparations were stained with azure-eosin and examined under an MBI-6 microscope. A total of 70–80 cells at the metaphasic stage of nuclear division were taken for analysis from each mouse from the group of 2-month male C57Bl/6 mice weighing 20–22 g. Genetic alterations in gametes were examined by assessing dominant lethal mutations in C57Bl/6 male mice [8].

After a 35-day soybean diet, the test and control male mice were caged with virgin CBA female mice in 1:2 ratio. The mating period of 3 weeks was sufficient to assess the effect of soybean diet on sex cells (spermatids and spermatozoa) during the postmeiotic period. Gravid females were isolated and sacrificed on gestation days 15–17 by cervical dislocation. Numbers of corpus lutea and live and dead embryos were recorded. These data were used to calculate the mutagenic parameters: pre-implantation, post-implantation, and inducible mortality (Tables 5.26 and 5.27).

Among various structural chromosomal abnormalities in animals of both groups, there were single segments, one circular chromosome (in a test mouse), and gaps. The number of cells with such chromosomal abnormalities did not significantly differ in control and test mice. In general, they can appear spontaneously and are not considered as a persistent disorder, because these aberrations usually disappear in the next divisions of the cells. The numbers of cells with polyploid chromosome set were similar in test and control mice: 1.81 ± 0.67 and 1.89 ± 0.70%, correspondingly (Table 5.26).

To examine the dominant lethal mutations, 90 test and 78 control females were dissected to analyze 473 (test) and 447 (control) embryos, 502 test and 475 control implantation sites, and 537 (test) and 524 (control) corpus lutea.

**Table 5.27** Dominant Lethal Mutations in Sex Cells of Mice Fed Diet with Protein Concentrate Derived from Transgenic Soybean Line 40-3-2 (Test) or Conventional Soybean (Control)

| Parameter, % | Week 1, Mature Spermatozoa | | Week 2, Late Spermatids | | Week 3, Early Spermatids | |
|---|---|---|---|---|---|---|
| | Control | Test | Control | Test | Control | Test |
| Pre-implantation mortality | 9.28 | 6.60 | 5.35 | 7.54 | 6.93 | 8.69 |
| Post-implantation mortality | 7.22 | 7.73 | 7.34 | 6.70 | 6.77 | 7.82 |
| Survival rate | 84.15 | 86.10 | 86.90 | 86.44 | 84.97 | 85.93 |
| Induced mortality | 0 | 0 | 0 | 0 | 0 | 0 |

At the stages of early and late spermatids or mature spermatozoa, the pre-implantation mortality in the test group was smaller than in the control. At these stages, the post-implantation mortality in the test group (the most significant index of mutagenic activity of the examined agent) did not surpass that in the control group (Table 5.27).

Induced mortality at the stages of early and late spermatids or mature spermatozoa was absent, indicating absence of the negative effect of the protein concentrate derived from transgenic soybean line 40-3-2 on spermiogenesis in mice.

The above data support the conclusion that glyphosate-tolerant soybean line 40-3-2 produced no mutagenic effect in the described experiments.

### Assessment of Potential Effect of Transgenic Soybean 40-3-2 on Reproduction

The experiments were performed on male and female Wistar rats as described in standard protocols [8]. The diet of control rats comprised the protein concentrate derived from conventional soybean, while that of test rats included similar concentrate prepared from the transgenic soybean line 40-3-2 (1.25 g/rat/day). The composition of this diet is shown in Table 5.11. The weight of the rats during mating was 280–350 g. The males and females were given soybean protein concentrate during the entire term of the experiment, which included a 20-day preliminary diet, the mating period, the total term of gestation, and the entire period of lactation. The progeny was fed diet with soybean protein concentrate *ad libitum* for 1 month after birth. To examine the potential embryotoxic effect of the soybean, 8 pregnant rats were sacrificed on Day 20 of gestation. The fetuses were extracted from the uteri and visually inspected. The corpus lutea and resorbed or dead embryos were counted. The overall pre- and post-implantation mortalities of the embryos were

**Table 5.28** Parameters of Prenatal Development of Fetuses Whose Parents were Fed Protein Concentrate Derived from Conventional Soybean (Control) or Transgenic Soybean Line 40-3-2 (Test) ($M \pm m$, $n = 8$)

| Parameter | Generation | Control | Test |
|---|---|---|---|
| Overall embryonic mortality, % | 1 | 17.4 ± 2.8 | 14.6 ± 1.9 |
|  | 2 | 16.0 ± 1.5 | 14.8 ± 2.0 |
| Pre-implantation mortality of embryos, % | 1 | 8.9 ± 2.0 | 8.0 ± 3.1 |
|  | 2 | 8.7 ± 1.5 | 7.9 ± 2.6 |
| Post-implantation mortality of embryos, % | 1 | 8.4 ± 1.7 | 6.5 ± 2.5 |
|  | 2 | 7.3 ± 1.6 | 6.9 ± 1.8 |
| Craniocaudal size, mm | 1 | 28.9 ± 2.7 | 28.7 ± 5.1 |
|  | 2 | 29.5 ± 2.6 | 28.8 ± 2.8 |
| Fetus weight, g | 1 | 2.3 ± 0.6 | 2.4 ± 0.4 |
|  | 2 | 2.4 ± 0.3 | 2.3 ± 0.3 |

*Note: Here and in Tables 5.29–5.30 the differences are not significant ($p > 0.05$).*

calculated. The craniocaudal size and weight of the fetuses were measured, and their sagittal histological sections were examined.

To study postnatal development, the progeny of 10 female rats in each group were examined: the number of pups delivered by one female, the body weight of these pups (measured every week), the terms of total unfolding of the external ears, eye opening, fur development, eruption of incisors, and survival of progeny in 30 days. During the experiment, general condition of the males, females, and progeny was satisfactory in both groups (Table 5.28).

Comparison of the parameters of prenatal development of the progeny revealed no significant differences between the control and test groups of rats in terms of total embryonic mortality, the pre- and post-implantation embryonic mortality, and the size and weight of the fetuses. All these parameters varied within the physiological boundaries characteristic of Wistar rats. Examination of a series of sagittal sections revealed no abnormalities in the fetal development.

Table 5.29 shows that the number of rat pups delivered by one female did not significantly differ between test and control groups in either the first or second generation.

During the entire growth period, the body weight of the infant rats in either the first or second generation of the test group did not significantly differ from that of the control group.

Survival of the newborn rats on Day 30 did not differ between test and control groups of rats in generations 1 and 2 (Table 5.30).

**Table 5.29** Number of Rat Pups Delivered by One Female Fed Diet with Protein Concentrate Derived from Conventional Soybean (Control) or Transgenic Soybean Line 40-3-2 (Test) ($M \pm m$, $n = 8$)

| Rat Pups | Control | Test |
| --- | --- | --- |
| Generation 1 | 7.25 ± 0.70 | 7.66 ± 0.95 |
| Generation 2 | 10.10 ± 1.00 | 8.50 ± 0.90 |

**Table 5.30** Survival of Progeny (%) on Day 30 of Rats Fed Diet with Protein Concentrate Derived from Conventional Soybean (Control) or Transgenic Soybean Line 40-3-2 (Test)

| Rat Pups | Control | Test |
| --- | --- | --- |
| Generation 1 | 100 | 100 |
| Generation 2 | 92.8 | 95.5 |

Thus, the protein concentrate derived from transgenic soybean line 40-3-2 (1.25 g/rat/day) produced no embryotoxic, gonadotoxic, or teratogenic effects in two generations of rats. Moreover, it induced no adverse effects on the postnatal development of two rat generations.

### Assessment of Technological Parameters

The assessment of technological parameters was carried out in Moscow State University of Applied Biotechnology (Ministry of Science and Education of Russian Federation). Comparative analysis of functional properties of the proteins derived from transgenic soybean line 40-3-2 and its conventional counterpart was carried out with the following parameters: solubility, emulsion stability, critical concentration of gel formation, as well as water binding and lipid-retention capacities. There were no differences between the proteins obtained from conventional and transgenic soybean, which substantiates the conclusion of possibility to use the transgenic soybean line 40-3-2 in the food industry.

### Conclusions

The result of the complex safety assessment of glyphosate-tolerant transgenic soybean line 40-3-2 is the established absence of any toxic, genotoxic, immune system modulating, or allergenic effects as indicated by all examined parameters. Analysis of chemical composition of transgenic soybean line 40-3-2 and the protein concentrate derived from it, showed its identity to the composition of their conventional counterparts. Based on the results of the studies, the State Sanitation Service of the Russian Federation

(Department of State Sanitation and Epidemiological Inspectorate) granted the Registration Certificate, which allows use of the transgenic soybean line 40-3-2 in the food industry and allows it to be placed on the market without restrictions.

## 5.1.2 GLUFOSINATE-TOLERANT SOYBEAN LINE A2704-12

### Molecular Characteristics of Soybean Line A2704-12

*Recipient Organism*

Family legumes (*Leguminosae*), the genus *Glycine*, species *max*, has a very long history of cultivation and use as human food (about 4000 years). Soybean variety *Glycine max* A2704 is a fast-ripening soybean variety, which combines a high crop capacity, resistance to lodging, and good germination.

*Donor Organism*

The donor of the *pat* gene responsible for tolerance to ammonium glufosinate, *Streptomyces viridochromogenes* strain Tu 494 [12], is a gram-positive spore-forming soil bacterium, which produces bialafos (phosphinothricin), a tripeptide composed of two molecules of L-alanine and an analog of L-glutamine acid. The ammonium salts of phosphinothricin (common name is ammonium glufosinate) are used to produce herbicides that inhibit the enzyme glutamine synthase, which plays an important role in nitrogen metabolism in plants. Inhibition of this enzyme results in an accumulation of ammonia and subsequent death of the plant cells.

The *pat* gene encodes the synthesis of phosphinothricin acetyltransferase (PAT), which acetylates the free $NH_2$-group of ammonium glufosinate and therefore prevents accumulation of ammonia [32,47].

The natural *pat* gene contains many G:C nucleotides, which is not typical for plants. To enhance expression of PAT protein in the plant, a modified nucleotide sequence was synthesized without changing the amino acid sequence in the protein.

*Method of Genetic Transformation*

The plasmid DNA was inserted into the plant genome by biolistic transformation. The *pat* gene is stably integrated in the genome of soybean line A2704 as part of the expression vector pB2/35SAcK, which has the following basic genetic elements: synthetic *pat* gene providing tolerance to ammonium glufosinate; promoter 35S of the cauliflower mosaic virus (CaMV) to stimulate expression of the *pat* gene in the plant tissue; terminator 35S of CaMV signaling termination of the transcription; and the *bla* gene isolated from *E. coli*, which encodes synthesis of β-lactamase to provide tolerance to

**Table 5.31** Registration Status of Transgenic Soybean Line A2704-12 in Various Countries

| Country | Date of Approval | Application |
| --- | --- | --- |
| Canada | 1999 | Environmental release |
|  | 2000 | Food, feed |
| USA | 1996 | Environmental release |
|  | 1998 | Food, feed |
| South Africa | 2001 | Food |
| Japan | 1999 | Environmental release |
|  | 2002 | Food |

*Note: The current registration status of the transgenic crop is on http://www.biotradestatus.com/*

some antibiotics. The *bla* gene is used as a marker and is not expressed in the A2704-12 soybean plant [29,36].

## Global Registration Status of Soybean Line A2704-12

Table 5.31 shows the global registration status of the transgenic soybean line A2704-12 at the time of registration in Russia [19].

## Safety Assessment of Soybean Line A2704-12 Conducted in the Russian Federation

The studies were conducted in accordance with the requirements of the Ministry of Health of the Russian Federation authorized for risk and safety assessment of food derived from GM sources [8]. PCR analysis of the soybean test and control samples was performed to confirm the identity of the transformation event and its absence in the conventional control line.

### *Biochemical Composition of Transgenic Soybean and Soybean Protein Concentrate*

Tables 5.32–5.37 show the biochemical composition of the beans from conventional soybean and transgenic soybean line A2704-12 and soy protein concentrate derived from conventional and transgenic soybean line A2704-12.

Protein content and amino acid composition in transgenic soybean line A2704-12 and soy protein concentrate derived from transgenic soybean line A2704-12 did not differ from the corresponding values for the conventional counterpart (Table 5.32).

The content of carbohydrates in the beans of transgenic soybean line A2704-12 and soy protein concentrate derived from transgenic soybean line A2704-12 did not significantly differ from that in conventional soybean

**Table 5.32** Protein Content (%) and Amino Acid Composition (g/100 g protein) in Soybean and Soybean Protein Concentrate

| Ingredient | Conventional Soybean | Transgenic Soybean A2704-12 | Protein Concentrate (Conventional Soybean) | Protein Concentrate (Transgenic Soybean A2704-12) |
| --- | --- | --- | --- | --- |
| Protein | 37.50 | 37.78 | 66.64 | 67.66 |
| *Amino acids* | | | | |
| Lysine | 6.44 | 6.34 | 6.52 | 6.48 |
| Histidine | 2.96 | 2.97 | 2.79 | 2.98 |
| Arginine | 6.96 | 7.07 | 7.82 | 7.70 |
| Aspartic acid | 10.87 | 10.52 | 10.42 | 10.52 |
| Threonine | 3.66 | 3.74 | 3.52 | 3.51 |
| Serine | 4.58 | 4.96 | 4.81 | 4.71 |
| Glutamic acid | 17.77 | 17.33 | 18.20 | 17.56 |
| Proline | 4.53 | 4.56 | 4.83 | 4.50 |
| Glycine | 4.53 | 4.45 | 4.43 | 4.41 |
| Alanine | 3.76 | 3.84 | 3.55 | 3.57 |
| Cysteine | 1.31 | 1.33 | 1.31 | 1.28 |
| Valine | 4.80 | 4.64 | 4.84 | 4.79 |
| Methionine | 1.34 | 1.38 | 1.36 | 1.34 |
| Isoleucine | 4.52 | 4.20 | 4.47 | 4.38 |
| Leucine | 6.82 | 6.77 | 6.95 | 6.94 |
| Tyrosine | 3.65 | 3.78 | 3.20 | 3.33 |
| Phenylalanine | 4.73 | 4.96 | 5.10 | 5.15 |

**Table 5.33** Carbohydrates Content (g/100 g product) in Soybean and Soybean Protein Concentrate

| Carbohydrate | Conventional Soybean | Transgenic Soybean A2704-12 | Protein Concentrate (Conventional Soybean) | Protein Concentrate (Transgenic Soybean A2704-12) |
| --- | --- | --- | --- | --- |
| Fructose | 0.18 | 0.18 | – | – |
| Sucrose | 2.12 | 2.60 | 0.35 | 1.12 |
| Starch | 2.90 | 2.90 | 1.10 | 1.10 |
| Cellulose | 3.56 | 3.72 | 3.83 | 3.73 |
| Raffinose | 0.41 | 1.30 | 0.02 | 0.02 |
| Stachyose | 2.10 | 3.10 | 0.14 | 0.11 |

**Table 5.34** Lipids and Fatty Acids Content in Soybean and Soybean Protein Concentrate

| Ingredient | Conventional Soybean | Transgenic Soybean A2704-12 | Protein Concentrate (Conventional Soybean) | Protein Concentrate (Transgenic Soybean A2704-12) |
|---|---|---|---|---|
| Lipids, % | 18.5 | 18.7 | 0.25 | 0.25 |
| *Fatty acids, rel. %* | | | | |
| Lauric 12:0 | 0.01 | 0.01 | – | – |
| Myristic 14:0 | 0.10 | 0.08 | 0.38 | 0.38 |
| Pentadecanoic 15:0 | 0.02 | 0.02 | 0.11 | 0.17 |
| Palmitic 16:0 | 11.95 | 9.80 | 29.93 | 35.15 |
| Palmitoleic 16:1 | 0.09 | 0.10 | 1.32 | 1.04 |
| Margaric (heptadecanoic) 17:0 | 0.08 | 0.08 | 0.18 | 0.25 |
| Heptadecenoic 17:1 | 0.04 | 0.05 | – | – |
| Stearic 18:0 | 3.49 | 3.78 | 5.57 | 8.01 |
| Cis-9-Oleic 18:1 | 17.47 | 22.10 | 9.41 | 11.78 |
| trans-11-Vaccenic 18:1 | 0.64 | 1.08 | 2.33 | 2.08 |
| Linoleic 18:2 | 56.68 | 55.65 | 44.52 | 37.10 |
| γ-Linolenic 18:3 | 0.02 | 0.05 | 2.28 | 1.07 |
| α-Linolenic 18:3 | 8.50 | 7.56 | 3.38 | 2.04 |
| Arachidic 20:0 | 0.31 | 0.28 | 0.16 | 0.39 |
| Gondoic 20:1 | 0.17 | 0.14 | 0.13 | 0.33 |
| Behenic 22:0 | 0.03 | 0.02 | – | – |
| Erucic 22:1 | 0.35 | 0.23 | 0.27 | 0.18 |

**Table 5.35** Vitamins Content (mg/100 g product) in Soybean and Soybean Protein Concentrate

| Ingredient | Conventional Soybean | Transgenic Soybean A2704-12 | Protein Concentrate (Conventional Soybean) | Protein Concentrate (Transgenic Soybean A2704-12) |
|---|---|---|---|---|
| Vitamin $B_1$ | 0.69 | 0.79 | – | – |
| Vitamin $B_2$ | 0.22 | 0.23 | – | – |
| Vitamin $B_6$ | 0.73 | 0.68 | 0.48 | 0.54 |
| Vitamin E (α-tocopherol) | 1.9 | 1.7 | – | – |
| γ-Tocopherol | – | – | 0.29 | 0.32 |
| Carotenoids (lutein) | 0.41 | 0.40 | Traces | Traces |

**Table 5.36** Mineral Content in Soybean and Soybean Protein Concentrate

| Ingredient | Conventional Soybean | Transgenic Soybean A2704-12 | Protein Concentrate (Conventional Soybean) | Protein Concentrate (Transgenic Soybean A2704-12) |
|---|---|---|---|---|
| Sodium, mg/kg | 67.8 | 152 | 10 687 | 6680 |
| Calcium, mg/kg | 2224 | 2080 | 2131 | 2536 |
| Magnesium, mg/kg | 3227 | 3377 | 1558 | 1549 |
| Iron, mg/kg | 76.1 | 71.0 | 139 | 135 |
| Potassium, mg/kg | 18 663 | 17 923 | 4539 | 4604 |
| Zinc, mg/kg | 43.1 | 40.9 | 24.9 | 27.3 |
| Copper, mg/kg | 10.45 | 11.55 | 10.9 | 8.43 |
| Selenium, µg/kg | 178 | 134 | 203 | 213 |

**Table 5.37** Analysis of Toxic Elements in Soybean and Soybean Protein Concentrate

| Ingredient | Conventional Soybean | Transgenic Soybean A2704-12 | Protein Concentrate (Conventional Soybean) | Protein Concentrate (Transgenic Soybean A2704-12) |
|---|---|---|---|---|
| Aflatoxin $B_1$, mg/kg | Not detected | Not detected | Not detected | Not detected |
| Cadmium, mg/kg | 0.083 | 0.068 | 0.039 | 0.033 |
| Lead, mg/kg | <0.001 | 0.001 | 0.052 | 0.096 |

(Table 5.33). The revealed changes in the content of raffinose and stachyose were within the mean statistical variations characteristic of these parameters, which are 0–3.0 (raffinose) and 0–5.0 (stachyose) according to the data of the State Research Institute of Nutrition, RAMS.

The content of lipids and fatty acid compositions in the beans and soy protein concentrate of both soybean varieties also did not differ significantly (Table 5.34).

The vitamin composition of conventional and transgenic soybean line A2704-12 and soy protein concentrate derived from conventional and transgenic soybean line A2704-12 were virtually identical (Table 5.35).

The contents of minerals in conventional and transgenic soybean line A2704-12 and soy protein concentrate derived from conventional and transgenic soybean line A2704-12 did not significantly differ (Table 5.36). The revealed changes in the content of sodium remained within the mean statistical variations characteristic of soybean (25–500 mg/kg) according to the data of the State Research Institute of Nutrition, RAMS. The content of heavy metals (cadmium, lead, copper, and zinc) and mycotoxins in transgenic soybean line A2704-12 did not surpass the levels acceptable under the regulations valid in Russia [5] (Table 5.37).

Thus, the above-mentioned data showed that the biochemical composition of glufosinate-tolerant transgenic soybean line A 2704-12 and its protein concentrate did not significantly differ from that of conventional soybean. The revealed variations of the examined parameters (relative and absolute concentrations) remained within the range characteristic of soybean [4,17,18,22,35,37,42].

The safety parameters (mycotoxins and toxic contaminants) of conventional and transgenic soybean line A2704-12, as well as the corresponding soy protein concentrates, met the requirements of the regulations valid in the Russian Federation [5]. The content of antinutrients (stachyose and raffinose) in the protein concentrate derived from GM soybean line A2704-12 or conventional soybean did not surpass the acceptable levels established in the Russian Federation for children's diet and medical dietary nutrition [4].

### *Toxicological Assessment of Soybean A2704-12*

The chronic experiment (180 day) was carried out on male Wistar rats ($n = 60$) with an initial body weight of 70–80 g. After admission to the vivarium of the State Research Institute of Nutrition, the rats were placed in quarantine for 10 days. At the onset of feeding experimental diet, the body weight of rats was 85–95 g. In the test group, the rats were fed a standard semi-synthetic diet supplemented with soy protein concentrate derived from transgenic soybean line A2704-12. In the control group, the diet was supplemented with the protein concentrate derived from conventional soybean (Tables 5.38 to 5.41).

Samples were obtained on days 30 and 150 of the experiment. During the entire length of the experiment, the general condition of the rats was similar in the control and test groups. No mortality was observed in either group.

The body weight of the rats fed protein concentrate derived from transgenic soybean line A2704-12 did not statistically differ from that of the control rats fed diet with protein concentrate derived from conventional soybean (Figure 5.3; Table 5.42). The absolute and relative weights of internal organs of test rats did not significantly differ from the corresponding values for the control rats (Table 5.43).

### Assessment of Biochemical Parameters

During the entire experiment, the content of total protein, glucose, activity of aspartate aminotransferase in blood serum of rats fed diet with protein

### Table 5.38 Standard Semi-Synthetic Rat Diet with Soy Protein Concentrate

| Ingredient | Mass, g/100 g feed |
| --- | --- |
| Soybean protein concentrate | 22.5 |
| Starch | 57.4 |
| Vegetable oil | 5.0 |
| Lard | 5.0 |
| Salt mix[a] | 4.0 |
| Liposoluble vitamins[b] | 1.0 |
| Vitamin mix[c] | 0.1 |
| Bran | 5.0 |
| Total | 100.0 |

[a]See Table 5.39.
[b]see Table 5.40.
[c]see Table 5.41.

### Table 5.39 Composition of Salt Mix

| Ingredient | Chemical Formula | Quantity, g |
| --- | --- | --- |
| Sodium chloride | $NaCl$ | 139.3 |
| Monopotassium phosphate | $KH_2PO_4$ | 388.8 |
| Magnesium sulfate | $MgSO_4$ | 57.4 |
| Calcium carbonate | $CaCO_3$ | 380.4 |
| Ferrous sulfate | $FeSO_4 \times 7H_2O$ | 26.4 |
| Potassium iodide | $KI$ | 0.77 |
| Manganese sulfate | $MnSO_4 \times 7H_2O$ | 4.55 |
| Zinc sulfate | $ZnSO_4 \times 7H_2O$ | 0.53 |
| Copper sulfate | $CuSO_4 \times 5H_2O$ | 0.48 |
| Cobalt chloride | $CoCl_2 \times 6H_2O$ | 0.024 |
| Sodium fluoride | $NaF$ | 0.50 |
| Potassium alum | $K_2SO_4Al_2(SO_4)_3 \times 24H_2O$ | 0.11 |
| Total | | 1000 |

### Table 5.40 Composition of Liposoluble Vitamins

| Ingredient | Per 0.1 mL |
| --- | --- |
| Tocopherol, IU | 5 |
| Retinol, IU | 800 |
| Ergocalciferol, IU | 70 |
| Sunflower-seed oil, mL | up to 0.1 mL |

**Table 5.41** Composition of Vitamin Mix

| Ingredient | Mass, mg/g Feed |
|---|---|
| $B_1$ | 0.40 |
| $B_2$ | 0.60 |
| $B_6$ | 0.60 |
| Calcium pantothenate | 1.50 |
| Nicotinic acid | 3.00 |
| Folic acid | 0.20 |
| $B_{12}$ | 0.003 |
| Menadione | 0.10 |
| L-methionine | 50.00 |
| Glucose (sucrose, fructose) | up to 1 g |

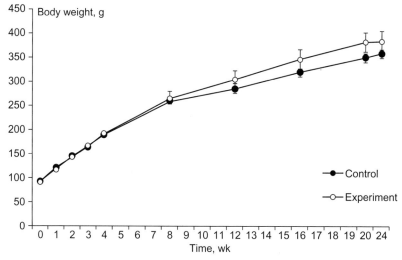

**FIGURE 5.3**

Comparative dynamics of body weight of rats fed the diet containing protein concentrate derived from transgenic soybean line A2704-12 (test) or its conventional counterpart (control).

concentrate made of transgenic soybean line A2704-12 (test group) did not statistically differ from the corresponding parameters in control rats fed diet with conventional soy protein concentrate. Although activity of alanine aminotransferase in test rats was significantly lower than that in the control rats (Table 5.44), this decrease remained within the physiological range 0.40–0.80 μcat/L. Within 180 days after the onset of the experiment, the activity of alkaline phosphatase in the serum of test rats was significantly lower than

## 5.1.2 Glufosinate-Tolerant Soybean Line A2704-12

**Table 5.42** Body Weight of Rats (g) Fed Diet with Protein Concentrate Derived from Conventional Soybean or Transgenic Soybean Line A2704-12 ($M \pm m$; $n = 6$–$8$)

| Duration of the Experiment, Weeks | Control | Test |
|---|---|---|
| 0 | 93.2 ± 1.1 | 93.2 ± 5.0 |
| 1 | 121.2 ± 2.2 | 117.6 ± 5.5 |
| 2 | 144.4 ± 2.8 | 144.4 ± 5.2 |
| 3 | 166.8 ± 2.0 | 167.6 ± 4.6 |
| 4 | 190.0 ± 5.0 | 193.7 ± 2.8 |
| 8 | 259.5 ± 10.5 | 266.5 ± 15.9 |
| 12 | 285.5 ± 7.8 | 305.0 ± 18.7 |
| 16 | 320.5 ± 8.6 | 347.0 ± 19.1 |
| 20 | 350.0 ± 9.0 | 382.5 ± 20.3 |
| 24 | 357.5 ± 9.0 | 384.5 ± 21.4 |

*Note: Here and in Table 5.43 the differences are not significant ($p > 0.05$).*

**Table 5.43** Absolute and Relative Weight of Internal Organs of Rats Fed Diet with Protein Concentrate Derived from GM Soybean Line A 2704-12 or Conventional Soybean ($M \pm m$, $n = 6$–$8$)

| Organ | | 30 Days | | 180 Days | |
|---|---|---|---|---|---|
| | | Control | Test | Control | Test |
| Kidneys | Abs.[a], g | 1.70 ± 0.06 | 1.65 ± 0.06 | 2.23 ± 0.10 | 2.05 ± 0.03 |
| | Rel.[b], g/100g | 0.56 ± 0.04 | 0.61 ± 0.01 | 0.56 ± 0.02 | 0.50 ± 0.01 |
| Liver | Abs., g | 9.80 ± 0.18 | 9.66 ± 0.46 | 11.21 ± 0.31 | 10.88 ± 0.49 |
| | Rel., g/100g | 3.54 ± 0.06 | 3.55 ± 0.04 | 2.80 ± 0.07 | 2.68 ± 0.09 |
| Spleen | Abs., g | 1.21 ± 0.06 | 1.14 ± 0.07 | 1.18 ± 0.09 | 1.23 ± 0.09 |
| | Rel., g/100g | 0.43 ± 0.02 | 0.42 ± 0.02 | 0.30 ± 0.01 | 0.31 ± 0.03 |
| Heart | Abs., g | 0.91 ± 0.03 | 0.89 ± 0.05 | 1.15 ± 0.05 | 1.12 ± 0.03 |
| | Rel., g/100g | 0.33 ± 0.01 | 0.33 ± 0.01 | 0.29 ± 0.001 | 0.28 ± 0.005 |
| Testicles | Abs., g | 2.70 ± 0.09 | 2.62 ± 0.10 | 2.80 ± 0.06 | 2.88 ± 0.07 |
| | Rel., g/100g | 0.97 ± 0.02 | 0.97 ± 0.02 | 0.70 ± 0.02 | 0.71 ± 0.02 |
| Hypophysis | Abs., mg | 7.83 ± 0.70 | 7.33 ± 0.99 | 9.50 ± 0.56 | 9.67 ± 0.58 |
| | Rel., mg/100g | 2.85 ± 0.28 | 2.73 ± 0.36 | 2.39 ± 0.16 | 2.40 ± 0.19 |
| Adrenal glands | Abs., mg | 22.00 ± 2.57 | 18.67 ± 1.09 | 20.3 ± 1.39 | 22.33 ± 2.52 |
| | Rel., mg/100g | 7.91 ± 0.86 | 6.87 ± 0.27 | 5.09 ± 0.36 | 5.47 ± 0.29 |
| Seminal vesicles | Abs., mg | 386.7 ± 35.65 | 423.3 ± 35.28 | 665.0 ± 44.5 | 730.00 ± 33.2 |
| | Rel., mg/100g | 139.24 ± 12.48 | 155.96 ± 11.62 | 165.9 ± 12.78 | 179.6 ± 8.30 |
| Prostate | Abs., mg | 215.7 ± 33.63 | 221.7 ± 22.57 | 496.67 ± 35.09 | 518.33 ± 65.06 |
| | Rel., mg/100g | 78.55 ± 12.70 | 81.54 ± 7.54 | 124.2 ± 9.51 | 129.6 ± 18.49 |

[a]Absolute weight of internal organs, g.
[b]Relative weight of internal organs (g/100g body weight).

**Table 5.44** Biochemical Parameters of Blood Serum of Rats Fed Diet with Protein Concentrate Derived from Conventional Soybean or Transgenic Soybean Line A2704-12 ($M \pm m$, $n = 6$–$8$)

| Parameter | 30 Days | | 180 Days | |
|---|---|---|---|---|
| | Control | Test | Control | Test |
| Total protein, g/L | 88.30 ± 6.74 | 95.50 ± 6.77 | 80.20 ± 8.34 | 81.20 ± 7.00 |
| Glucose, mmol/L | 6.21 ± 0.21 | 6.55 ± 0.75 | 8.96 ± 1.28 | 8.32 ± 0.46 |
| Alanine aminotransferase, μcat/L | 0.72 ± 0.02 | 0.59 ± 0.03* | 1.05 ± 0.06 | 1.08 ± 0.06 |
| Aspartate aminotransferase, μcat/L | 0.96 ± 0.04 | 0.84 ± 0.02 | 0.78 ± 0.05 | 0.83 ± 0.03 |
| Alkaline phosphatase, μcat/L | 13.50 ± 1.30 | 10.97 ± 1.20 | 11.14 ± 0.63 | 8.98 ± 0.60* |

*$p < 0.05$ in comparison with control.

**Table 5.45** Biochemical Parameters of Urine of Rats Fed Diet with Protein Concentrate Derived from Conventional Soybean or Transgenic Soybean Line A2704-12 ($M \pm m$, $n = 6$–$8$)

| Parameter | 30 Days | | 180 Days | |
|---|---|---|---|---|
| | Control | Test | Control | Test |
| pH | 7.0 | 7.0 | 7.0 | 7.0 |
| Daily diuresis, mL | 4.2 ± 0.6 | 4.0 ± 0.5 | 4.1 ± 0.4 | 4.9 ± 0.3 |
| Relative density, g/mL | 1.07 ± 0.01 | 1.08 ± 0.01 | 1.12 ± 0.01 | 1.10 ± 0.01 |
| Creatinine, mg/mL | 1.93 ± 0.21 | 1.56 ± 0.10 | 2.57 ± 0.26 | 2.72 ± 0.20 |
| Creatinine, mg/day | 7.56 ± 0.69 | 5.71 ± 0.95 | 9.62 ± 1.92 | 13.34 ± 1.07 |

Note: Here and in Tables 5.46 and 5.47 the differences are not significant ($p > 0.05$).

that in the control rats. These changes also remained within the range of physiological variation measured with this method (6.0–15.0 μcat/L).

The content and daily urinary excretion of creatinine, the pH and the relative density of urine did not significantly differ in rats fed diet with protein concentrate made of transgenic soybean line A2704-12 from the corresponding parameters in control rats fed diet with an equivalent amount of conventional soy protein concentrate derived from soybean line A2704-12 (Table 5.45).

### Assessment of Sensitive Biomarkers

Comparison of activity of the enzymes involved in degradation of xenobiotics in liver of rats, fed for 180 days diet with soy protein concentrate derived from transgenic soybean line A 2704-12 (test group) or equivalent amount of soy protein concentrate derived from conventional soybean (control group) revealed no significant differences (Table 5.46).

**Table 5.46** Activity of Enzymes Involved in Metabolism of Xenobiotics and Protein Content in Liver Microsomes in Rats Fed Diet with Protein Concentrate Derived from Conventional Soybean or Transgenic Soybean Line A2704-12 ($M \pm m$, $n = 6–8$)

| Parameter | 30 Days | | 180 Days | |
|---|---|---|---|---|
| | Control | Test | Control | Test |
| Cytochrome P450, nM/mg protein | 0.83 ± 0.02 | 0.90 ± 0.03 | 0.79 ± 0.03 | 0.81 ± 0.06 |
| Cytochrome $b_5$, nM/mg protein | 0.75 ± 0.02 | 0.74 ± 0.02 | 0.69 ± 0.02 | 0.71 ± 0.02 |
| Aminopyrine N-demethylation, nM/min/mg protein | 8.00 ± 0.17 | 7.95 ± 0.10 | 10.2 ± 0.7 | 10.7 ± 0.5 |
| Benzpyrene hydroxylation, Fl-units/min/mg protein | 13.1 ± 0.5 | 10.9 ± 1.3 | 8.20 ± 0.23 | 8.57 ± 0.38 |
| Epoxide hydrolase, nM/min/mg protein | 6.47 ± 0.23 | 6.21 ± 0.25 | 6.08 ± 0.23 | 5.39 ± 0.22 |
| UDP-glucuronosil transferase, nM/min/mg protein | 40.8 ± 1.7 | 36.8 ± 1.4 | 37.2 ± 1.7 | 31.3 ± 2.3 |
| CDNB-glutathione transferase, μM/min/mg protein | 1.23 ± 0.05 | 1.15 ± 0.02 | 1.68 ± 0.08 | 1.55 ± 0.05 |
| Microsomal protein, mg/g | 16.1 ± 0.5 | 17.1 ± 0.4 | 14.9 ± 0.3 | 14.3 ± 0.4 |
| Cytosolic protein, mg/g | 79.0 ± 1.2 | 80.1 ± 1.0 | 92.1 ± 1.5 | 89.4 ± 1.5 |

**Table 5.47** Total and Non-sedimentable Activity of Hepatic Lysosomal Enzymes of Rats Fed Diet with Protein Concentrate Derived from Conventional Soybean or Transgenic Soybean Line A 2704-12 ($M \pm m$, $n = 6–8$)

| Parameters | 30 Days | | 180 Days | |
|---|---|---|---|---|
| | Control | Test | Control | Test |
| *Total activity, μM/min/g tissue* | | | | |
| Arylsulfatase A, B | 2.20 ± 0.03 | 2.19 ± 0.03 | 2.04 ± 0.03 | 2.02 ± 0.03 |
| β-Galactosidase | 2.47 ± 0.05 | 2.50 ± 0.07 | 2.23 ± 0.09 | 2.31 ± 0.10 |
| β-Glucuronidase | 2.26 ± 0.05 | 2.35 ± 0.08 | 2.37 ± 0.08 | 2.14 ± 0.06 |
| *Non-sedimentable activity, % total activity* | | | | |
| Arylsulfatase A, B | 4.10 ± 0.08 | 4.07 ± 0.07 | 3.38 ± 0.09 | 3.36 ± 0.09 |
| β-Galactosidase | 5.57 ± 0.22 | 5.56 ± 0.26 | 5.30 ± 0.22 | 5.55 ± 0.35 |
| β-Glucuronidase | 6.36 ± 0.28 | 6.46 ± 0.21 | 4.54 ± 0.12 | 4.15 ± 0.12 |

The total and non-sedimentable activity of hepatic lysosomal enzymes of the rats fed a 180-day diet with soy protein concentrate derived from transgenic soybean line A2704-12 did not significantly differ from the corresponding values for the control rats fed diet with the equivalent amount of conventional soy protein concentrate (Table 5.47).

Table 5.48 Content of LPO Products in Blood and Liver of Rats Fed Diet with Soy Protein Concentrate Derived from Conventional Soybean (Control) or Transgenic Soybean Line A2704-12 (Test) ($M \pm m$, $n = 6$–8)

| Parameter | 30 Days | | 180 Days | |
|---|---|---|---|---|
| | Control | Test | Control | Test |
| Erythrocytes | | | | |
| DC, nM/mL | 4.28 ± 0.4 | 4.12 ± 0.2 | 5.31 ± 0.5 | 4.51 ± 0.3 |
| MDA, nM/mL | 2.23 ± 0.3 | 2.08 ± 0.3 | 2.60 ± 0.2 | 2.73 ± 0.2 |
| Blood serum | | | | |
| DC, nM/mL | 2.96 ± 0.1 | 3.07 ± 0.4 | 2.76 ± 0.2 | 2.85 ± 0.2 |
| MDA, nM/mL | 5.49 ± 0.5 | 5.01 ± 0.4 | 6.11 ± 0.2 | 5.63 ± 0.2 |
| Liver | | | | |
| DC, Unit | 0.99 ± 0.03 | 1.03 ± 0.02 | 1.03 ± 0.02 | 1.00 ± 0.02 |
| MDA, nM/g | 273.7 ± 2.8 | 307.5 ± 7.9*. | 459.2 ± 15.5 | 482.2 ± 23.1 |

*$p < 0.05$ in comparison with control.

The studies showed that long-term addition of protein concentrate derived from transgenic soybean line A2704-12 to the rat diet produced no significant changes in activity of enzymes involved in phase I and phase II xenobiotic degradation and did not affect activity of lysosomal marker enzymes. Parameters characterizing LPO level and activity of the enzymes of antioxidant protection system were also examined in rats of both groups.

In 30 days, the content of diene conjugates (DC) and MDA in the blood of control and test rats did not significantly differ (Table 5.48). The content of MDA in the liver of test rats was 12% higher ($p < 0.05$) than in the control rats, but the difference was within the physiological variations.

On experimental Day 180, the content of LPO products in the blood and liver did not significantly differ in control and test rats.

In 30 days study, activity of the enzymes of the erythrocytic antioxidant protection system did not significantly differ in control and test rats, except for a significant 10% elevation of glutathione peroxidase activity in the test rats relative to the control value (Table 5.49). In the norm, physiological variation of activity of the antioxidant protection enzymes is about 15–20%, therefore the difference was within the physiological variations.

On experimental Day 180, activity of the antioxidant protection enzymes did not significantly differ in control and test rats.

The above-mentioned data shows that the transgenic soybean line A2704-12 produces no effect on the antioxidant status of the rats and does not contain ingredients with pro-oxidant action.

**Table 5.49** Activity of Enzymes of Erythrocytic Antioxidant Protection System in Rats Fed Diet with Soy Protein Concentrate Derived from Conventional Soybean (Control) or Transgenic Soybean Line A2704-12 (Test) ($M \pm m$, $n = 6–8$)

| Parameter | 30 Days | | 180 Days | |
|---|---|---|---|---|
| | Control | Test | Control | Test |
| Superoxide dismutase (SOD), Unit/min/g Hb | 1544.5 ± 0.4 | 1633.6 ± 43.9 | 1558.8 ± 458 | 1579.5 ± 39.4 |
| Glutathione reductase, μmol/min/g Hb | 39.33 ± 2.50 | 37.69 ± 3.30 | 39.55 ± 1.90 | 40.19 ± 3.50 |
| Glutathione peroxidase, μmol/min/g Hb | 47.90 ± 0.80 | 52.86 ± 1.80* | 56.63 ± 1.20 | 60.88 ± 2.00 |
| Catalase, mmol/min/g Hb | 461.0 ± 13.0 | 446.0 ± 10.6 | 410.7 ± 14.8 | 428.4 ± 19.6 |

*$p < 0.05$ in comparison with control.

**Table 5.50** Hematological Parameters of Peripheral Blood Drawn from Rats Fed Diet with Protein Concentrate Derived from Conventional Soybean or Transgenic Soybean Line A 2704-12 ($M \pm m$, $n = 6–8$)

| Parameter | 30 Days | | 180 Days | |
|---|---|---|---|---|
| | Control | Test | Control | Test |
| Hemoglobin concentration, g/L | 138.4 ± 4.0 | 134.3 ± 3.4 | 159.2 ± 3.5 | 154.9 ± 3.4 |
| Total erythrocyte count, $\times 10^{12}$/L | 6.38 ± 0.11 | 6.40 ± 0.10 | 6.63 ± 0.24 | 6.58 ± 0.18 |
| Hematocrit, vol.% | 50.8 ± 0.3 | 51.0 ± 0.3 | 50.5 ± 0.5 | 50.1 ± 0.5 |
| MCH, pg | 21.7 ± 0.80 | 21.1 ± 0.87 | 23.8 ± 0.40 | 23.4 ± 0.4 |
| MCHC, % | 27.2 ± 0.8 | 26.4 ± 0.7 | 31.1 ± 0.4 | 30.8 ± 0.5 |
| MCV, μm$^3$ | 79.7 ± 1.0 | 78.7 ± 1.1 | 76.6 ± 2.1 | 76.3 ± 1.2 |
| Total leukocyte count, $\times 10^9$/L | 11.1 ± 0.6 | 11.6 ± 0.9 | 13.4 ± 1.0 | 12.8 ± 1.1 |

Note: Here and in Table 5.51 the differences are not significant ($p > 0.05$).

### Hematological Assessments

The addition of the protein concentrate derived from transgenic soybean line A2704-12 to the rat diet did not affect the concentration of hemoglobin, total erythrocyte count, hematocrit, MCH, MCHC, MCV, and the leukogram parameters values in comparison with the control values of the rats fed soy protein concentrate derived from conventional soybean (Tables 5.50 and 5.51).

### Morphological Assessment

Macroscopic examination of internal organs of the rats fed diet with soy protein concentrate derived from conventional soybean and transgenic soybean line A2704-12 revealed no pathological alteration in 30 and 180 days from the

**Table 5.51** Leukogram Parameters of Rats Fed Diet with Protein Concentrate Derived from Conventional Soybean or Transgenic Soybean Line A2704-12 ($M \pm m$, $n = 6$–$8$)

| Parameter | 30 Days | | 180 Days | |
|---|---|---|---|---|
| | Control | Test | Control | Test |
| *Segmentonuclear neutrophils* | | | | |
| rel., % | 27.0 ± 0.4 | 27.8 ± 0.2 | 23.6 ± 1.4 | 23.8 ± 1.4 |
| abs., ×10$^9$/L | 3.0 ± 0.5 | 3.2 ± 0.2 | 3.3 ± 0.4 | 3.1 ± 0.5 |
| *Eosinophils* | | | | |
| rel., % | 1.0 ± 0.3 | 1.2 ± 0.5 | 1.2 ± 0.3 | 1.5 ± 0.4 |
| abs., ×10$^9$/L | 0.16 ± 0.13 | 0.12 ± 0.04 | 0.14 ± 0.06 | 0.19 ± 0.06 |
| *Lymphocytes* | | | | |
| rel., % | 71.8 ± 0.7 | 71.0 ± 0.7 | 74.5 ± 1.4 | 74.3 ± 1.4 |
| abs., ×10$^9$/L | 7.9 ± 0.4 | 8.2 ± 0.6 | 9.9 ± 0.5 | 9.4 ± 0.7 |

**Table 5.52** Microscopic Examination of Internal Organs in Rats Fed Diet with Protein Concentrate Derived from Conventional Soybean (Control) or Transgenic Soybean Line A2704-12 (Test)

| Organ | Control | Test |
|---|---|---|
| Liver | Clear trabecular structure; no alterations in hepatocytes and in the portal ducts | No differences from control |
| Kidneys | Usual appearance of cortical and medullar substance; no alterations in glomeruli or pelvis epithelium | No differences from control |
| Lung | Alveolar space is air-filled; no alterations in bronchi and blood vessels | No differences from control |
| Spleen | Large or middle-size folliculi with wide clear marginal zones and reactive centers | No differences from control |
| Small intestine | Preserved villous epithelium; usual infiltration in villus stroma | No differences from control |
| Testicle | Usual size and appearance of seminiferous tubules; clearly definable spermiogenesis | No differences from control |

*Note: Here are the summary data obtained on experimental days 30 and 180.*

onset of the experiment. Similarly, histological examination detected no significant differences in the internal organs of the rats in both groups (Table 5.52).

### Assessment of Biological Value and Digestibility of Protein Derived from Transgenic Soybean Line A2704-12

Experiments were carried out on growing male Wistar rats ($n = 40$) with an initial body weight of 53–56 g. For 28 days, all rats were fed an isocaloric diet

(420 kcal/100 g dry feed) containing soybean protein (9% w/w), sunflower-seed oil (11.5% w/w), maize starch (70% w/w), salt mix (4% w/w) and a mix of fat- and water-soluble vitamins. The test rats were fed a diet with soy protein concentrate derived from transgenic soybean line A2704-12, while the control group was fed a diet with an equivalent amount of soy protein concentrate derived from conventional soybean. During the entire experiment, rats were kept in individual cages with feed and water *ad libitum*. The rats were weighed every other day, and the amount of feed consumed by each animal was recorded. During the last 5 days of the experiments, referred to as "the metabolic assessment period", the amount of consumed feed and the weight of excrement were recorded. The content of total nitrogen in the feed and excrement was determined with the Kjeldahl semi-micromethod [8]. Statistical analysis of the data was carried out with standard software.

Biological value and digestibility of the proteins were calculated with standard methods with due account for endogenous nitrogen loss. To obtain the correct values, these parameters were calculated with the use of protein efficiency ratio (PER) and Net Protein Efficiency Ratio (NPER). The resulting absolute values were compared with the values for control protein and expressed as a percentage (relative biological value) [8].

Daily feed intake, daily protein intake, daily body weight gain, and the parameters of biological value (PER and NPER) in rats fed diet with protein concentrate derived from GM soybean line A2704-12 did not significantly differ from the corresponding values for the rats fed diet with protein concentrate prepared from the conventional soybean (Table 5.53).

The relative biological values of the proteins derived from transgenic soybean line A2704-12 and the conventional soybean were 60–67% relative to the baseline control group fed diet with casein. This difference is explained by a greater content of sulfur amino acids (their biological requirement is enhanced in the rats because of the need to maintain their coat) in casein in

**Table 5.53** Comparative Biological Value and Digestibility of the Feed with Protein Concentrate Derived from Conventional Soybean or Transgenic Soybean Line A2704-12 Fed to Rats for 28 Days ($M \pm m$, $n = 12$)

| Parameter | Control | Test |
|---|---|---|
| DFI, g/day | 9.35 ± 0.31 | 9.56 ± 0.64 |
| DPI, g/day | 0.84 ± 0.028 | 0.87 ± 0.058 |
| DBWG, g/day | 1.33 ± 0.05 | 1.60 ± 0.148 |
| PER | 1.57 ± 0.04 | 1.74 ± 0.108 |

*DBWG, daily body weight gain; DFI, daily feed intake (g/day); DPI, daily protein intake (g/day); PER, protein efficiency ratio.*

**Table 5.54** Comparative Biological Value and Digestibility of the Feed with Protein Concentrate Derived from Conventional Soybean or Transgenic Soybean Line A2704-12 Fed to Rats during 5-Day Metabolism Assessment Period ($M \pm m$, $n = 12$)

| Parameter | Control | Test |
|---|---|---|
| DFI, g/day | 12.93 ± 0.26 | 13.40 ± 0.87 |
| DPI, g/day | 1.18 ± 0.024 | 1.21 ± 0.08 |
| DBWG, g/day | 2.33 ± 0.116 | 2.51 ± 0.26 |
| PER | 1.96 ± 0.085 | 2.02 ± 0.097 |
| NPER | 2.68 ± 0.112 | 2.70 ± 0.056 |

*Note: Here and in Tables 5.55 and 5.56 the differences were not significant ($p > 0.05$).*

**Table 5.55** True Digestibility of Protein (Nitrogen) in Rats Fed Diet with Protein Concentrate Derived from Conventional Soybean or Transgenic Soybean Line A2704-12 ($M \pm m$, $n = 12$)

| Group | True Digestibility |
|---|---|
| Control | 84.60 ± 0.20 |
| Test | 86.77 ± 1.58 |

comparison with soybean proteins and smaller daily feed intake by the rats fed diet with soy proteins. There were no significant differences between the biological values of the proteins derived from conventional and transgenic soybean.

During the metabolic assessment period the differences in the values of daily feed intake, daily protein intake, daily body weight gain, and biological value of the proteins in the rats fed soy protein concentrate derived from both soybean varieties, were insignificant (Table 5.54). Comparison of the parameters obtained during the metabolic assessment period and during the entire length of the experiment showed that, in the metabolic assessment period, all the rats consumed more feed and protein (about 1.4-fold). This was reflected in the relative biological values, which were smaller (76–90%) than similar values for the rats fed diet with casein.

The values of true digestibility of the proteins derived from conventional and transgenic soybean did not significantly differ (Table 5.55).

The experiments were terminated after 18 h of fasting. The rats were decapitated under ether anesthesia. Visual inspection of internal organs of the control and test rats fed diet with soybean-derived proteins did not reveal any structural alterations in comparison with the rats fed diet with casein

**Table 5.56** Absolute Weight of Internal Organs of Control and Test Rats on Experimental Day 29 ($M \pm m$, $n = 12$)

| Weight (g) | Control | Test |
| --- | --- | --- |
| Body weight | 89.90 ± 2.59 | 96.70 ± 6.28 |
| Liver | 4.07 ± 0.19 | 4.14 ± 0.25 |
| Kidneys | 0.83 ± 0.032 | 0.84 ± 0.053 |
| Heart | 0.49 ± 0.022 | 0.53 ± 0.033 |
| Spleen | 0.60 ± 0.068 | 0.67 ± 0.08 |
| Testicles | 0.86 ± 0.13 | 0.88 ± 0.17 |

(baseline control). The weight of the internal organs of the control and test rats did not significantly differ (Table 5.56). Comparative analysis of biological value of the proteins derived from conventional soybean and transgenic soybean line A2704-12 attested to equivalence of both soybean varieties by this criterion.

## Assessment of Potential Impact of Transgenic Soybean Line A2704-12 on Immune System in Studies on Mice
### Potential Effect on Humoral Component of Immune System

The potential immunomodulating effect of GM soybean on the humoral component of the immune system was examined on two oppositely reacting mice lines, C57Bl/6 and CBA, by determining the level of hemagglutination to sheep erythrocytes (SE). The experimental conditions are described in section 5.1.1.

In CBA mice fed diet with soy protein concentrate derived from transgenic soybean line A2704-12 or from conventional soybean, the antibody titer was 1:20 starting from post-immunization Day 14. Under the same conditions, the antibodies appeared in C57Bl/6 mice on post-immunization Day 14 (1:16–1:64) and they could be detected on Day 21 after immunization (1:32). Thus, the synthesis rate of the antibodies raised against SE in C57Bl/6 and CBA mice lines fed diet with soy protein concentrate derived from transgenic soybean and in mice of the same lines fed on conventional soy protein concentrate was similar.

### Potential Effect on Cellular Component of Immune System

The potential immunomodulating effect of transgenic soybean was assessed by delayed hypersensitivity reaction to SE. The experimental conditions are described in section 5.1.1. The studies demonstrated the absence of any significant differences in the parameters of local inflammatory reaction in mice of both examined lines fed diet with soy protein concentrate derived either from transgenic soybean line A2704-12 or from conventional soybean. Thus, delayed hypersensitivity reaction to SE showed that the transgenic soybean line A2704-12 produced no effect on the cellular component of the immune system.

### Assessment of Potential Sensitizing Effect of Transgenic Soybean Line A2704-12

Examination of possible sensitizing action of transgenic soybean on the immune response to endogenous metabolic products was carried out in the analysis of mouse sensitivity to histamine. For 21 days, the control and test mice were fed diets with protein concentrate derived from conventional or transgenic soybean. Mice of both groups were injected intraperitoneally with 2.5 mg histamine hydrochloride dissolved in 0.5 mL physiological solution. The reaction was examined at 1 h and 24 h post-injection by the percent mortality. The transgenic soybean line A2704-12 was used as a sensitizing agent to the metabolic products in this test. It did not affect behavior or mortality of the animals, which attests to the absence of sensitizing agent in the examined product.

### Potential Effect of Soybean on Susceptibility of Mice to *Salmonella typhimurium*

The effect of transgenic soybean on susceptibility of mice to infection by salmonella of murine typhus was examined in the experiments on mice injected intraperitoneally with various doses of Strain 415 *Salmonella typhimurium*. Three weeks prior to infection, the diet of control and test mice was supplemented with protein concentrate derived from conventional and transgenic soybean, respectively. The injected doses ranged from $10^2$ to $10^5$ microbial cells per mouse and varied on a 10-fold basis. The post-injection observation period was 21 days.

In both groups, first mortality was observed in post-injection Week 1, and all infected mice died by post-injection Day 18. The lifetimes of the test and control mice were 15.4 and 16.1 days, correspondingly. $LD_{50}$ values of control and test mice were 154 and 76 microbial cells per mouse, respectively.

These data showed that *Salmonella typhimurium* produced a typical infection in all mice. The infectious disease in test mice was identical to that in the control group. According to the values of $LD_{50}$ and lifetime, the infectious process developed similarly in both groups. Thus, feeding mice with transgenic soybean line A2704-12 did not affect their susceptibility to salmonella of murine typhus.

Introduction of protein concentrate derived from transgenic soybean line A2704-12 to mouse diet produced no effect on the humoral and cellular components of the immune system, did not sensitize mouse organism, and did not affect the natural resistance against typical infection such as murine typhus.

### Assessment of Potential Impact of Transgenic Soybean Line A2704-12 on Immune System in Studies on Rats

The study was carried out on male Wistar rats ($n = 46$) weighing initially $140 \pm 10$ g. After a 7-day adaptation period to standard vivarium diet, the rats were fed diet supplemented with protein concentrate derived from conventional soybean (control group) or from transgenic soybean line A2704-12 (test group) for the next 28 days. The soy protein concentrates from both varieties were dissolved in boiled water to the consistency of dense curd and supplemented with sunflower-seed oil to improve intake. The test feed was used instead of equally caloric amount of oatmeal (composition of the test diet is described in Table 5.11).

The model of generalized anaphylaxis was developed according to [8] as described in section 5.1.1.

During the entire experiment the rats of both groups fed diet with protein concentrate derived from conventional and transgenic soybean line A2704-12 grew normally, which indicates nutritional adequacy of the diets used. On experimental Day 29, the body weights of test and control rats were $223 \pm 7$ g and $220 \pm 5$ g, respectively ($p > 0.05$).

Intensity of humoral immune response was assessed according to concentration of circulating specific immunoglobulin antibodies (the sum of $IgG_1$ and $IgG_4$ fractions) by the method of indirect solid-phase enzyme-linked immunosorbent assay (standard ELISA) on polystyrene. The details of assay and mathematical processing of the data are described in section 5.1.1. The levels of severity of active anaphylactic shock did not significantly differ between control and test groups (Table 5.57).

Table 5.58 shows the mean parameters of $D_{492}$, the concentration of antibodies, and common logarithm of this concentration in the control and test groups. All values did not significantly differ in both groups ($p > 0.05$), therefore intensity of immune response was equal in control and test rats.

**Table 5.57** Severity of Active Anaphylactic Shock in Rats Fed Diet with Protein Concentrate Derived from Conventional Soybean or Transgenic Soybean Line A2704-12

| Group of Rats | AI | Severe Reactions, % | Mortality, % |
|---|---|---|---|
| Control ($n = 23$) | 2.61 | 52.2 | 47.8 |
| Test ($n = 24$) | 3.00 | 59.1 | 59.1 |

Note: Here and in Table 5.58 the differences are not significant ($p > 0.05$).

**Table 5.58** Values of Humoral Immune Response in Rats Fed Diet with Protein Concentrate Derived from Conventional Soybean or Transgenic Soybean Line A2704-12

| Group of Rats | $D_{492}$ | Concentration of Antibodies, mg/mL | Logarithm of Antibody Concentration |
|---|---|---|---|
| Control ($n = 23$) | $1.087 \pm 0.035$ | $4.2 \pm 0.9$ | $0.459 \pm 0.076$ |
| Test ($n = 23$) | $1.032 \pm 0.052$ | $4.1 \pm 0.8$ | $0.385 \pm 0.102$ |

Taken together, these data support the conclusion that the degree of sensitization by ovalbumin in test rats fed diet with soy protein concentrate derived from transgenic soybean line A2704-12 did not increase in comparison with the control rats fed diet with protein concentrate derived from conventional soybean. The studies showed that the protein concentrate prepared from glufosinate-tolerant transgenic soybean line A2704-12 did not significantly change the allergenic reactivity and sensitization by a model allergen in test rats in comparison with the control rats fed conventional soybean.

### Assessment of Potential Genotoxicity of Transgenic Soybean Line A2704-12

The genotoxic studies were carried out on male C57Bl/6 mice and hybrid female CBA mice. The animals were fed the standard diet supplemented with a mix of soft feed with soy protein concentrate derived from transgenic (test group) or conventional (control group) soybean.

The cytogenetic analysis was carried out with metaphasic method [8]. The details of experiments are described in section 5.1.1.

Genetic alterations in the sex cells were examined by revealing the dominant lethal mutations in C57Bl/6 male mice. The control and test mice were fed diet with soy protein concentrate during 45 days. Then the test ($n = 15$) and control ($n = 12$) male mice were caged with virgin CBA female mice in 1:2 ratio. The details of experiments are described in section 5.1.1. The test mice had no overt chromosomal abnormalities. Only single segments and the gaps were detected, and their number did not surpass 2% (the level of spontaneous mutation characteristic of this species). These chromosomal aberrations are not preserved in mitosis and are eliminated during the following divisions of cell nucleus (Table 5.59).

To examine the dominant lethal mutations in gametes, the test female mice ($n = 60$) were dissected to count and analyze 332 embryos and 363 corpus lutea. The pre-implantation mortality was approximately equal in the control and test groups. At the stages of early and late spermatids or mature

**Table 5.59** Cytogenetic Parameters of Bone Marrow in Mice Fed a 45-Day Diet with Protein Concentrate Derived from Transgenic Soybean Line A2704-12 (Test) and Conventional Soybean (Control)

| Parameter | Control ($n = 5$) | Test ($n = 5$) |
|---|---|---|
| Number of analyzed metaphases | 312 | 353 |
| Number of cells, % | | |
| with chromosomal aberrations | 0.64 ± 0.45 | 0.56 ± 0.4 |
| with gaps | 0.84 ± 0.45 | 0.85 ± 0.48 |
| with polyploid chromosome set | 1.28 ± 0.55 | 1.18 ± 0.56 |

Note: The differences are not significant ($p > 0.05$).

**Table 5.60** Dominant Lethal Mutations in Sex Cells of Mice Fed Diet with Protein Concentrate Derived from Transgenic Soybean Line A 2704-12 (Test) and Conventional Soybean (Control)

| Parameter, % | Week 1, Mature Spermatozoa | | Week 2, Late Spermatids | | Week 3, Early Spermatids | |
|---|---|---|---|---|---|---|
| | Control | Test | Control | Test | Control | Test |
| Pre-implantation mortality, % | 9.90 | 8.51 | 9.75 | 10.10 | 5.80 | 6.40 |
| Post-implantation mortality, % | 6.42 | 3.87 | 6.30 | 5.20 | 6.20 | 5.10 |
| Survival rate, % | 84.29 | 87.94 | 84.50 | 85.10 | 88.30 | 88.80 |
| Induced mortality, % | 0 | 0 | 0 | 0 | 0 | 0 |

spermatozoa, the post-implantation embryonic mortality (the most reliable index of mutagenic activity of examined substance) was smaller in the test group than in the control.

There was no induced mortality at these stages, attesting to the absence of dominant lethal mutations in sex cells and any negative effects on spermiogenesis in the mice fed diet with protein concentrate derived from transgenic soybean line A2704-12 (Table 5.60). The above-mentioned data support the conclusion that glufosinate-tolerant soybean line A2704-12 produced no mutagenic effect in the described experiments.

### *Assessment of Technological Parameters*
The assessment of technological parameters was carried out in Moscow State University of Applied Biotechnology (Ministry of Science and Education of Russian Federation).

The following parameters were determined to characterize the seed samples of the transgenic soybean line A2704-12 and its conventional counterpart: the yield of protein from the defatted flour during preparative isolation; amino acid and fractional composition of the obtained protein preparations; denaturation temperature of individual globulin fractions in the resulting protein preparations; the yield of oil and its fatty acid composition.

The study resulted in the following conclusions:

1. The composition of the seeds of transgenic soybean line A2704-12 did not differ from that of the conventional soybean.
2. The yield of protein, its amino acid and fractional composition, and the thermodynamic parameters of individual fractions were virtually identical in the seed samples of transgenic soybean line A2704-12 and conventional soybean.
3. The lipids extracted from the seed samples of transgenic soybean line A2704-12 and conventional soybean demonstrated similar fatty acid composition characteristic of this crop.

Thus, there were no significant differences in the properties of the seeds of transgenic soybean line A2704-12 and its conventional counterpart.

### *Conclusions*
By all examined parameters, the data of complex safety assessment of glufosinate-tolerant transgenic soybean line A2704-12 attest to the absence of any toxic, genotoxic, immune system modulating, or allergenic effects in this soybean variety. Analysis of the biochemical composition of transgenic soybean line A2704-12 and the protein concentrate derived from it demonstrated its identity to the composition of conventional soybean.

Based on the results of the studies, the State Sanitation Service of the Russian Federation (Department of State Sanitation and Epidemiological Inspectorate) granted the Registration Certificate which allows the transgenic soybean line A2704-12 to be used in the food industry and to be placed on the market without restrictions.

## 5.1.3 GLUFOSINATE-TOLERANT SOYBEAN LINE A5547-127

### Molecular Characterization of Soybean Line A5547-127
*Recipient Organism*
Family legumes (*Leguminosae*), the genus *Glycine*, species *max*, has a very long history of cultivation and use as human food (about 4000 years). Soybean *Glycine max* A5547 is a late-ripening soybean variety, which combines a high crop capacity, resistance to lodging, and good germinating capability.

### 5.1.3 Glufosinate-Tolerant Soybean Line A5547-127

*Donor Organism*

The donor of the *pat* gene responsible for tolerance to ammonium glufosinate is *Streptomyces viridochromogenes* strain Tu 494 [12].

*Method of Genetic Transformation*

The plasmid DNA was inserted into the plant genome by biolistic transformation. The *pat* gene was reliably integrated into the genome of soybean line A5547 in a vector pB2/35SAcK, which has the following basic genetic elements: synthetic *pat* gene providing tolerance to ammonium glufosinate, 35S promoter from cauliflower mosaic virus (CaMV) to control expression of *pat* gene in the plant tissue, CaMV 35S terminator to control termination of the transcription, and *bla* gene isolated from *E. coli* encoding synthesis of β-lactamase to provide tolerance to some antibiotics used as a selective marker gene and not expressed in the A5547-127 soybean plant [19].

## Global Registration Status of Soybean Line A5547-127

Table 5.61 shows the registration status of transgenic soybean line A5547-127 at the time of registration in Russia [19].

## Safety Assessment of Soybean Line A5547-127 Conducted in the Russian Federation

The studies were conducted in accordance with the requirements of the Ministry of Health of the Russian Federation authorized for risk and safety assessment of food derived from GM sources [8]. PCR analysis of the soybean test and control samples was performed to confirm the identify, the transformation event and its absence in the conventional control line.

*Biochemical Composition of Grain from Soybean Line A5547-127*

## Biochemical Composition of Transgenic Soybean and Soybean Protein Concentrate

The content of proteins and amino acid composition in the beans of transgenic soybean line A5547-127 and soy protein concentrate derived from

**Table 5.61** Registration Status of the Transgenic Soybean Line A5547-127 in Various Countries

| Country | Date of approval | Scope |
| --- | --- | --- |
| Canada | 2000 | Food, feed, environmental release |
| USA | 1998 | Food, feed, environmental release |

Note: The current registration status of the transgenic crop is on http://www.biotradestatus.com/

**Table 5.62** Protein Content (%) and Amino Acid Composition (g/100 g protein) in Soybean and Soybean Protein Concentrate

| Ingredient | Conventional Soybean | Transgenic Soybean A5547-127 | Protein Concentrate (Conventional Soybean) | Protein Concentrate (Transgenic Soybean A5547-127) |
|---|---|---|---|---|
| Protein | 35.09 | 36.23 | 62.22 | 66.50 |
| *Amino acids* | | | | |
| Lysine | 6.50 | 6.48 | 6.51 | 6.39 |
| Histidine | 3.06 | 2.91 | 3.02 | 3.00 |
| Arginine | 6.86 | 7.08 | 7.46 | 7.61 |
| Aspartic acid | 10.47 | 10.18 | 10.35 | 10.20 |
| Threonine | 3.85 | 3.40 | 3.29 | 3.20 |
| Serine | 4.91 | 4.66 | 4.29 | 4.18 |
| Glutamic acid | 16.25 | 16.95 | 17.75 | 18.33 |
| Proline | 4.83 | 4.72 | 4.95 | 5.01 |
| Glycine | 4.65 | 4.46 | 4.38 | 4.31 |
| Alanine | 3.73 | 3.94 | 3.55 | 3.48 |
| Cysteine | 1.25 | 1.22 | 1.21 | 1.23 |
| Valine | 4.71 | 4.78 | 5.11 | 5.08 |
| Methionine | 1.29 | 1.33 | 1.27 | 1.32 |
| Isoleucine | 4.23 | 4.19 | 4.54 | 4.70 |
| Leucine | 6.79 | 6.78 | 7.08 | 7.05 |
| Tyrosine | 3.66 | 3.08 | 3.08 | 3.34 |
| Phenylalanine | 4.71 | 5.09 | 5.09 | 5.07 |

transgenic soybean line A5547-127 did not differ from the corresponding values for the conventional soybean (Table 5.62).

The content of carbohydrates in the beans of transgenic soybean line A5547-127 and soy protein concentrate derived from transgenic soybean line A5547-127 did not significantly differ from that in the conventional soybean (Table 5.63). The revealed changes in the content of raffinose and stachyose were within the mean statistical variations characteristic of these parameters, which are 0–3.0 (raffinose) and 0–5.0 (stachyose) according to the data of the State Research Institute of Nutrition, RAMS.

The content of lipids and fatty acid composition in the beans of both soybean varieties and soy protein concentrates also did not significantly differ (Table 5.64).

### 5.1.3 Glufosinate-Tolerant Soybean Line A5547-127

**Table 5.63** Carbohydrates Content (g/100 g product) in Soybean and Soybean Protein Concentrate

| Carbohydrate | Conventional Soybean | Transgenic Soybean A5547-127 | Protein Concentrate (Conventional Soybean) | Protein Concentrate (Transgenic Soybean A5547-127) |
|---|---|---|---|---|
| Fructose | 0.22 | 0.20 | 0.09 | 0.08 |
| Sucrose | 2.60 | 2.73 | 0.05 | 0.22 |
| Starch | 2.70 | 3.10 | 0.70 | 0.71 |
| Cellulose | 4.67 | 5.15 | 4.51 | 4.11 |
| Raffinose | 0.69 | 0.70 | 0.01 | 0.01 |
| Stachyose | 2.63 | 2.20 | 0.07 | 0.06 |

**Table 5.64** Lipids and Fatty Acids Content in Soybean and Soybean Protein Concentrate

| Ingredient | Conventional Soybean | Transgenic Soybean A5547-127 | Protein Concentrate (Conventional Soybean) | Protein Concentrate (Transgenic Soybean A5547-127) |
|---|---|---|---|---|
| Lipids, % | 17.9 | 17.5 | 0.5 | 0.35 |
| *Fatty acids, rel. %:* | | | | |
| Lauric 12:0 | 0.01 | 0.01 | – | – |
| Myristic 14:0 | 0.08 | 0.12 | 0.25 | 0.30 |
| Pentadecanoic 15:0 | 0.02 | 0.02 | 0.09 | 0.10 |
| Palmitic 16:0 | 0.02 | 0.02 | 27.40 | 26.99 |
| Palmitoleic 16:1 | 0.09 | 0.09 | 1.23 | 1.30 |
| Margaric (heptadecanoic) 17:0 | 0.09 | 0.09 | 0.19 | 0.16 |
| Heptadecenoic 17:1 | 0.04 | 0.04 | – | – |
| Stearic 18:0 | 3.80 | 3.40 | 5.72 | 5.34 |
| cis-9-Oleic 18:1 | 20.81 | 19.10 | 13.13 | 13.31 |
| trans-11-Vaccenic 18:1 | 1.12 | 1.27 | 1.80 | 1.70 |
| Linoleic 18:2 | 55.45 | 53.97 | 43.34 | 43.99 |
| γ-Linolenic 18:3 | 0.02 | 0.02 | 1.88 | 1.75 |
| α-Linolenic 18:3 | 7.88 | 8.41 | 4.28 | 4.37 |
| Arachidic 20:0 | 0.27 | 0.29 | 0.22 | 0.15 |
| Gondoic 20:1 | 0.17 | 0.18 | 0.27 | 0.20 |
| Behenic 22:0 | 0.03 | 0.14 | – | – |
| Erucic 22:1 | 0.28 | 0.33 | 0.16 | 0.30 |

**Table 5.65** Vitamins Content (mg/100 g product) in Soybean and Soybean Protein Concentrate

| Ingredient | Conventional Soybean | Transgenic Soybean A5547-127 | Protein Concentrate (Conventional Soybean) | Protein Concentrate (Transgenic Soybean A5547-127) |
|---|---|---|---|---|
| Vitamin $B_1$ | 0.76 | 0.41 | – | – |
| Vitamin $B_2$ | 0.19 | 0.14 | – | – |
| Vitamin $B_6$ | 0.60 | 0.63 | 0.51 | 0.56 |
| Vitamin A | Not detected | Not detected | – | – |
| Vitamin E ($\alpha$-tocopherol) | 4.7 | 4.4 | – | – |
| $\gamma$-Tocopherol | – | – | 0.61 | 0.50 |
| Carotenoids (lutein) | 0.21 | 0.21 | Traces | Traces |

**Table 5.66** Mineral Content in Soybean and Soybean Protein Concentrate

| Ingredient | Conventional Soybean | Transgenic Soybean A5547-127 | Protein Concentrate (Conventional Soybean) | Protein Concentrate (Transgenic Soybean A5547-127) |
|---|---|---|---|---|
| Sodium, mg/kg | 40.6 | 82.6 | 7905 | 7365 |
| Calcium, mg/kg | 1728 | 1865 | 2091 | 1987 |
| Magnesium, mg/kg | 3476 | 3547 | 1699 | 1409 |
| Iron, mg/kg | 68.7 | 71.1 | 107 | 149 |
| Potassium, mg/kg | 19 124 | 18 682 | 6454 | 3755 |
| Zinc, mg/kg | 60.8 | 60.7 | 28.5 | 38.4 |
| Copper, mg/kg | 4.84 | 4.83 | 4.94 | 4.55 |
| Selenium, µg/kg | 175 | 181 | 175 | 180 |

The vitamin composition of conventional and transgenic soybean line A5547-127 and soy protein concentrate derived from conventional and transgenic soybean line A5547-127 was approximately equal in both groups (Table 5.65).

The mineral content in conventional and transgenic soybean line A5547-127 and soy protein concentrate derived from conventional and transgenic soybean line A5547-127 did not significantly differ in both groups. The changes in the content of sodium were within the mean statistical variations characteristic of soybean (25–500 mg/kg) according to the data of the State Research Institute of Nutrition, RAMS. The content of heavy metals (cadmium, lead, copper, and zinc) in transgenic soybean line A5547-127 did not surpass the acceptable levels valid in Russia (Tables 5.66 and 5.67) [5].

### 5.1.3 Glufosinate-Tolerant Soybean Line A5547-127

**Table 5.67** Analysis of Toxic Elements in Soybean and Soybean Protein Concentrate

| Ingredient | Conventional Soybean | Transgenic Soybean A5547-127 | Protein Concentrate (Conventional Soybean) | Protein Concentrate (Transgenic Soybean A5547-127) |
|---|---|---|---|---|
| Aflatoxin B1, mg/kg | Not detected | Not detected | Not detected | Not detected |
| Cadmium, mg/kg | 0.103 | 0.108 | 0.091 | 0.120 |
| Lead, mg/kg | <0.001 | <0.001 | 0.067 | 0.091 |

Thus, the above-mentioned data showed that biochemical composition of glufosinate-tolerant transgenic soybean line A5547-127 and the corresponding protein concentrate did not significantly differ from conventional counterparts. The revealed variations of the examined parameters remained within the range characteristic of soybean [4,17,18,22,35,37,42]. The safety parameters of the beans of conventional and transgenic soybean line A5547-127 as well as the corresponding soy protein concentrates, met the requirements of the regulations valid in the Russian Federation [5]. The content of antinutrient ingredients in the protein concentrate derived from transgenic soybean line A5547-127 or conventional soybean did not surpass the acceptable levels established in the Russian Federation for children's diet and medical dietary nutrition [5].

#### *Toxicological Assessment of Soybean Line A5547-127*

The chronic experiment (180 days) was carried out on male Wistar rats with an initial body weight of 70–80 g. After admission to the vivarium of the State Research Institute of Nutrition, the rats were placed in quarantine for 10 days. At the onset of feeding the experimental diet, the body weight of rats was 85–95 g. In the test group, the rats were maintained on the diet with soy protein concentrate derived from transgenic soybean line A5547-127. In the control group, the standard semi-synthetic diet was supplemented with an equivalent amount of protein concentrate derived from conventional soybean line A5547 (composition of the diets is described in Table 5.68).

The biochemical, hematological, and morphological studies were conducted in accordance with the requirements of the Ministry of Health of the Russian Federation authorized for risk and safety assessment of food derived from GM sources [8]. Samples were collected on days 30 and 150 of the experiment. During the entire length of the experiment, no mortality was observed and the general condition of the rats was similar in the control and test groups.

#### **Assessment of Proximate Parameters**

The comparative dynamics of body weight of rats fed diet with soy protein concentrate derived from conventional and transgenic soybean are shown

**Table 5.68** Standard Semi-Synthetic Rat Diet with Soy Protein Concentrate

| Ingredient | Mass, g/100 g feed |
|---|---|
| Soybean protein concentrate | 22.5 |
| Starch | 57.4 |
| Vegetable oil | 5.0 |
| Lard | 5.0 |
| Salt mix[a] | 4.0 |
| Liposoluble vitamins[b] | 1.0 |
| Vitamin mix[c] | 0.1 |
| Bran | 5.0 |
| Total | 100.0 |

[a]See Table 5.39.
[b]see Table 5.40.
[c]see Table 5.41.

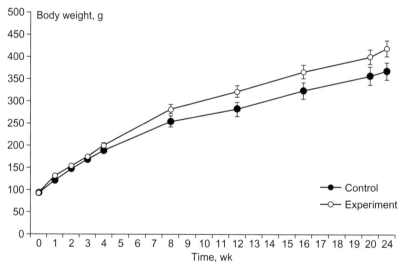

**FIGURE 5.4**
Comparative dynamics of body weight of rats fed diet containing protein concentrate derived from transgenic soybean line A5547-127 (test) or its conventional counterpart (control).

in Figure 5.4. The body weight of rats fed protein concentrate derived from transgenic soybean line A5547-127 did not significantly differ from that of the control rats fed diet with an equivalent amount of protein concentrate derived from conventional soybean line A5547. The differences detected in the end of experiment (16–24 wk) remained within the physiological

**Table 5.69** Body Weight of Rats (g) Fed Diet with Protein Concentrate Derived from Conventional Soybean or Transgenic Soybean Line A5547-127 ($M \pm m$; $n = 6–8$)

| Duration of Experiment, Weeks | Control | Test |
|---|---|---|
| 0 | 93.4 ± 4.0 | 93.4 ± 3.5 |
| 1 | 121.4 ± 4.4 | 131.8 ± 2.6 |
| 2 | 147.0 ± 3.9 | 153.4 ± 3.6 |
| 3 | 166.2 ± 3.3 | 174.4 ± 2.4 |
| 4 | 188.7 ± 5.4 | 200.7 ± 4.8 |
| 8 | 254.0 ± 12.9 | 282.5 ± 10.4 |
| 12 | 282.5 ± 15.8 | 322.0 ± 12.9 |
| 16 | 323.0 ± 17.5 | 366.5 ± 14.7 |
| 20 | 357.5 ± 21.1 | 400.0 ± 15.6 |
| 24 | 368.5 ± 18.7 | 419.5 ± 17.8 |

Note: The differences are not significant ($p > 0.05$).

variations characteristic of the rats of corresponding age and sex, i.e., 320–430 g (data of the State Research Institute of Nutrition, RAMS; Table 5.69).

The absolute and relative weights of internal organs of test rats fed diet with soy protein concentrate derived from GM soybean line A5547-127 did not significantly differ from the corresponding values for the control rats fed similar concentrate made of conventional soybean line A5547 (Table 5.70).

## Assessment of Biochemical Parameters

During the entire experiment, the content of total protein, glucose, and activity of aspartate aminotransferase in blood serum of test group rats did not significantly differ from the corresponding parameters of control group rats (Table 5.71). At the initial stage of the experiment, activity of alkaline phosphatase in test rats was significantly lower than that in the control rats; however, this decrease remained within the physiological boundaries of 6.00–15.00 μcat/L when assayed with this method.

The content and daily urinary excretion of creatinine, pH and the relative density of urine did not significantly differ between test and control groups (Table 5.72).

## Assessment of Sensitive Biomarkers

Analyses of activity of the enzymes involved in degradation of xenobiotics in the liver for both groups of rats fed experimental diets for 180 days revealed no significant differences between groups except for a significant elevation

**Table 5.70** Absolute and Relative Weight of Internal Organs of Rats Fed Diet with Protein Concentrate Derived from GM Soybean Line A5547-127 or Conventional Soybean ($M \pm m$, $n = 6–8$)

| Organ | | 30 Days | | 180 Days | |
|---|---|---|---|---|---|
| | | Control | Test | Control | Test |
| Kidneys | Abs.[a], g | 1.62 ± 0.12 | 1.61 ± 0.10 | 2.23 ± 0.09 | 2.24 ± 0.07 |
| | Rel.[b], g/100g | 0.59 ± 0.02 | 0.55 ± 0.02 | 0.51 ± 0.02 | 0.53 ± 0.02 |
| Liver | Abs., g | 10.30 ± 0.57 | 9.67 ± 0.65 | 12.10 ± 0.50 | 12.04 ± 0.58 |
| | Rel., g/100g | 3.79 ± 0.10 | 3.27 ± 0.07 | 2.78 ± 0.33 | 2.81 ± 0.09 |
| Spleen | Abs., g | 1.05 ± 0.03 | 1.16 ± 0.06 | 1.41 ± 0.09 | 1.33 ± 0.11 |
| | Rel., g/100g | 0.40 ± 0.02 | 0.39 ± 0.02 | 0.33 ± 0.02 | 0.31 ± 0.02 |
| Heart | Abs., g | 0.91 ± 0.04 | 0.98 ± 0.04 | 1.17 ± 0.08 | 1.17 ± 0.08 |
| | Rel., g/100g | 0.33 ± 0.003 | 0.33 ± 0.01 | 0.27 ± 0.02 | 0.28 ± 0.02 |
| Testicles | Abs., g | 2.59 ± 0.07 | 2.70 ± 0.11 | 2.84 ± 0.15 | 3.05 ± 0.15 |
| | Rel., g/100g | 0.96 ± 0.02 | 0.92 ± 0.03 | 0.66 ± 0.03 | 0.72 ± 0.04 |
| Hypophysis | Abs., mg | 7.67 ± 0.78 | 7.33 ± 0.56 | 7.60 ± 1.02 | 9.30 ± 1.40 |
| | Rel., mg/100g | 2.40 ± 0.19 | 2.49 ± 0.13 | 1.70 ± 0.24 | 2.15 ± 0.29 |
| Adrenal glands | Abs., mg | 21.00 ± 1.77 | 24.7 ± 2.96 | 25.5 ± 1.90 | 21.83 ± 2.01 |
| | Rel., mg/100g | 7.85 ± 0.82 | 8.43 ± 1.06 | 5.94 ± 0.57 | 5.10 ± 0.48 |
| Seminal vesicles | Abs., mg | 357.3 ± 71.3 | 419.0 ± 63.8 | 591.7 ± 82.0 | 696.6 ± 30.6 |
| | Rel., mg/100g | 127.9 ± 22.73 | 143.08 ± 22.47 | 134.8 ± 16.2 | 163.7 ± 9.15 |
| Prostate | Abs., mg | 201.3 ± 44.49 | 163.2 ± 27.2 | 331.6 ± 26.5 | 413.3 ± 34.4 |
| | Rel., mg/100g | 71.56 ± 14.30 | 54.46 ± 7.23 | 90.0 ± 8.40 | 108.1 ± 7.43 |

[a] Absolute weight of internal organs, g.
[b] Relative weight of internal organs (g/100g body weight).

**Table 5.71** Biochemical Parameters of Blood Serum of Rats Fed Diet with Protein Concentrate Derived from Conventional Soybean (Control) or Transgenic Soybean Line A5547-127 (Test) ($M \pm m$, $n = 6–8$)

| Parameter | 30 Days | | 180 Days | |
|---|---|---|---|---|
| | Control | Test | Control | Test |
| Total protein, g/L | 90.30 ± 6.72 | 94.50 ± 6.88 | 79.20 ± 6.34 | 81.00 ± 6.11 |
| Glucose, mmol/L | 7.51 ± 0.85 | 6.37 ± 0.88 | 8.00 ± 0.49 | 6.86 ± 0.38 |
| Alanine aminotransferase, μcat/L | 0.60 ± 0.05 | 0.50 ± 0.03 | 1.02 ± 0.05 | 1.17 ± 0.07 |
| Aspartate aminotransferase, μcat/L | 0.82 ± 0.03 | 0.75 ± 0.02 | 1.23 ± 0.06 | 1.01 ± 0.05 |
| Alkaline phosphatase, μcat/L | 13.40 ± 0.18 | 10.50 ± 0.24* | 9.36 ± 0.43 | 9.61 ± 0.35 |

*$p < 0.05$ in comparison with control.

### 5.1.3 Glufosinate-Tolerant Soybean Line A5547-127

**Table 5.72** Biochemical Parameters of Urine of Rats Fed Diet with Protein Concentrate Derived from Conventional Soybean or Transgenic Soybean Line A5547-127 ($M \pm m$, $n = 6$–$8$)

| Parameter | 30 Days | | 180 Days | |
| --- | --- | --- | --- | --- |
| | Control | Test | Control | Test |
| pH | 7.0 | 7.0 | 7.0 | 7.0 |
| Daily diuresis, mL | 4.1 ± 0.9 | 3.9 ± 0.8 | 3.7 ± 0.2 | 5.0 ± 0.6 |
| Relative density, g/mL | 1.15 ± 0.03 | 1.18 ± 0.04 | 1.12 ± 0.01 | 1.10 ± 0.01 |
| Creatinine, mg/mL | 2.24 ± 0.31 | 2.37 ± 0.21 | 2.72 ± 0.16 | 2.05 ± 0.32 |
| Creatinine, mg/day | 8.02 ± 1.08 | 8.62 ± 1.51 | 10.11 ± 0.69 | 10.27 ± 2.04 |

*Note: The differences are not significant ($p > 0.05$).*

**Table 5.73** Activity of Enzymes Involved in Metabolism of Xenobiotics and Protein Content in Liver Microsomes in Rats Fed Diet with Protein Concentrate Derived from Conventional or Transgenic Soybean for 180 Days ($M \pm m$, $n = 6$–$8$)

| Parameter | Control | Test |
| --- | --- | --- |
| Cytochrome P450, nM/mg protein | 0.75 ± 0.03 | 0.95 ± 0.04* |
| Cytochrome $b_5$, nM/mg protein | 0.71 ± 0.02 | 0.78 ± 0.03 |
| Aminopyrine N-demethylation, nM/min/mg protein | 9.23 ± 0.32 | 10.22 ± 0.35 |
| Benzpyrene hydroxylation, Fl-units/min/mg protein | 8.07 ± 0.27 | 8.87 ± 0.42 |
| Epoxide hydrolase, nM/min/mg protein | 4.42 ± 0.20 | 4.27 ± 0.24 |
| UDP-glucuronosil transferase, nM/min/mg protein | 26.5 ± 1.1 | 27.1 ± 2.4 |
| CDNB-glutathione transferase, µM/min/mg protein | 1.19 ± 0.03 | 1.16 ± 0.04 |
| Microsomal protein, mg/g | 15.3 ± 0.6 | 16.8 ± 0.7 |
| Cytosolic protein, mg/g | 92.9 ± 1.8 | 89.2 ± 1.6 |

*$p < 0.05$ in comparison with control.*

of Cytochrome P450 activity in the test rats relative to the control value (Table 5.73). Whereas detected changes occurred under conditions in the absence of complex stress manifestations of enzymes activity, it was suggested that prolonged intake of GM soybean has no effect on the xenobiotics degradation system.

The total and non-sedimentable activity of hepatic lysosomal enzymes of the test rats did not significantly differ from the corresponding values for the control rats (Table 5.74).

**Table 5.74** Total Activity of Hepatic Lysosomal Enzymes of Rats Fed Diet with Protein Concentrate Derived from Conventional or Transgenic Soybean for 180 days ($M \pm m$, $n = 6-8$)

| Parameters | Group | |
|---|---|---|
| | Control | Test |
| *Total activity, µM/min/g tissue* | | |
| Arylsulfatase A, B | 2.04 ± 0.04 | 2.02 ± 0.03 |
| β-Galactosidase | 2.27 ± 0.08 | 2.32 ± 0.10 |
| β-Glucuronidase | 2.16 ± 0.03 | 2.18 ± 0.04 |
| *Non-sedimentable activity, % total activity* | | |
| Arylsulfatase A, B | 3.97 ± 0.05 | 3.86 ± 0.09 |
| β-Galactosidase | 5.64 ± 0.39 | 5.50 ± 0.34 |
| β-Glucuronidase | 4.81 ± 0.15 | 4.62 ± 0.16 |

Note: Here and in Tables 5.75 – 5.78 the differences are not significant ($p > 0.05$).

**Table 5.75** Content of LPO Products in Blood and Liver of Rats Fed Diet with Soy Protein Concentrate Derived from Conventional Soybean (Control) or Transgenic Soybean Line A5547-127 (Test) ($M \pm m$, $n = 6-8$)

| Parameter | 30 Days | | 180 Days | |
|---|---|---|---|---|
| | Control | Test | Control | Test |
| *Erythrocytes* | | | | |
| DC, nM/mL | 3.51 ± 0.4 | 3.26 ± 0.3 | 4.86 ± 0.4 | 4.55 ± 0.3 |
| MDA, nM/mL | 2.51 ± 0.5 | 1.84 ± 0.1 | 2.29 ± 0.2 | 1.89 ± 0.1 |
| *Blood serum* | | | | |
| DC, nM/mL | 2.43 ± 0.2 | 3.14 ± 0.3 | 3.28 ± 0.3 | 3.42 ± 0.3 |
| MDA, nM/mL | 4.60 ± 0.3 | 4.49 ± 0.2 | 5.81 ± 0.1 | 5.96 ± 0.2 |
| *Liver* | | | | |
| DC, Unit | 1.07 ± 0.01 | 1.04 ± 0.01 | 1.04 ± 0.01 | 1.00 ± 0.02 |
| MDA, nM/g | 326.2 ± 24.1 | 299.2 ± 11.1 | 429.3 ± 5.7 | 427.9 ± 13.1 |

The studies showed that a long-term addition of protein concentrate derived from transgenic soybean line A5547-127 to the diet of rats produced no significant changes in activity of enzymes involved in phase I and phase II xenobiotic degradation, and did not affect activity of lysosomal marker enzymes in comparison with the corresponding parameters of the rats fed diet with conventional soy protein concentrate.

The parameters characterizing LPO level and activity of the enzymes of the antioxidant protection system in test group rats did not significantly differ from the corresponding values for the control group rats (Table 5.75).

**Table 5.76** Activity of Enzymes of Erythrocytic Antioxidant Protection System in Rats Fed Diet with Soy Protein Concentrate Derived from Conventional Soybean (Control) or Transgenic Soybean Line A5547-127 (Test) ($M \pm m$, $n = 6–8$)

| Parameter | 30 Days | | 180 Days | |
|---|---|---|---|---|
| | Control | Test | Control | Test |
| Superoxide dismutase (SOD), U/min/g Hb | 1414.9 ± 37.0 | 1506.5 ± 68.4 | 1702.5 ± 63.0 | 1716.8 ± 40.8 |
| Glutathione reductase, µmol/min/g Hb | 38.20 ± 1.6 | 34.86 ± 3.2 | 32.27 ± 1.2 | 36.31 ± 3.0 |
| Glutathione peroxidase, µmol/min/g Hb | 45.08 ± 2.3 | 50.23 ± 3.3 | 52.13 ± 1.9 | 54.64 ± 1.5 |
| Catalase, mmol/min/g Hb | 396.4 ± 13.0 | 442.6 ± 25.1 | 420.1 ± 15.6 | 453.1 ± 13.3 |

**Table 5.77** Hematological Parameters of Peripheral Blood Drawn from Control and Test Rats ($M \pm m$, $n = 6–8$)

| Parameter | 30 Days | | 180 Days | |
|---|---|---|---|---|
| | Control | Test | Control | Test |
| Hemoglobin concentration, g/L | 151.0 ± 6.2 | 150.0 ± 6.9 | 158.1 ± 2.6 | 154.8 ± 3.3 |
| Total erythrocyte count, ×10$^{12}$/L | 6.36 ± 0.16 | 6.70 ± 0.20 | 6.30 ± 0.10 | 6.51 ± 0.20 |
| Hematocrit, vol.% | 50.5 ± 0.5 | 50.8 ± 0.3 | 50.7 ± 0.3 | 50.7 ± 0.4 |
| MCH, pg | 23.8 ± 1.17 | 24.3 ± 1.30 | 23.6 ± 0.3 | 23.7 ± 0.9 |
| MCHC, % | 28.1 ± 1.2 | 30.1 ± 1.5 | 31.2 ± 0.3 | 30.8 ± 0.9 |
| MCV, µm$^3$ | 79.5 ± 1.3 | 80.7 ± 1.6 | 75.8 ± 2.0 | 77.1 ± 2.4 |
| Total leukocyte count, ×10$^9$/L | 11.9 ± 0.83 | 11.3 ± 0.6 | 12.6 ± 0.6 | 12.1 ± 0.9 |

On experimental Days 30 and 180, activity of the antioxidant protection enzymes did not significantly differ in the control and test groups. These data support the conclusion that the transgenic soybean line A5547-127 produces no effect on the antioxidant status of the rats and does not contain ingredients with pro-oxidant action (Table 5.76).

### Hematological Assessments
Addition of the protein concentrate derived from transgenic soybean line A5547-127 to the rat diet did not induce significant changes in concentration of hemoglobin, total erythrocyte count, hematocrit, MCH, MCHC, MCV, and the leukogram parameters in comparison with the control values of the rats fed soy protein concentrate derived from conventional soybean (Tables 5.77 and 5.78).

### Morphological Assessments
Macroscopic examination of internal organs of the rats fed diet with soy protein concentrate derived from conventional soybean and transgenic soybean

**Table 5.78** Leukogram Parameters of Rats Fed Diet with Protein Concentrate Derived from Conventional Soybean or Transgenic Soybean Line A5547-127 ($M \pm m$, $n = 6–8$)

| Parameter | 30 Days | | 180 Days | |
| --- | --- | --- | --- | --- |
| | Control | Test | Control | Test |
| *Segmentonuclear neutrophils* | | | | |
| rel., % | 27.2 ± 0.4 | 28.3 ± 0.7 | 23.8 ± 1.8 | 23.5 ± 1.8 |
| abs., ×10$^9$/L | 3.2 ± 0.2 | 3.2 ± 0.3 | 3.0 ± 0.4 | 2.9 ± 0.4 |
| *Eosinophils* | | | | |
| rel., % | 1.2 ± 0.5 | 1.5 ± 0.5 | 1.0 ± 0.3 | 0.9 ± 0.4 |
| abs., ×10$^9$/L | 0.13 ± 0.07 | 0.16 ± 0.05 | 0.11 ± 0.04 | 0.09 ± 0.04 |
| *Lymphocytes* | | | | |
| rel., % | 71.7 ± 0.8 | 70.1 ± 0.9 | 75.2 ± 1.4 | 75.7 ± 1.4 |
| abs., ×10$^9$/L | 8.5 ± 0.6 | 7.9 ± 0.5 | 9.4 ± 0.3 | 9.1 ± 0.5 |

**Table 5.79** Microscopic Examination of Internal Organs in Rats Fed Diet with Protein Concentrate Derived from Conventional Soybean or Transgenic Soybean Line A5547-127

| Organ | Control | Test |
| --- | --- | --- |
| Liver | Clear trabecular structure; no alterations in hepatocytes and in the portal ducts | No differences from the control |
| Kidneys | Usual appearance of cortical and medullar substance; no alterations in glomeruli or pelvis epithelium | No differences from the control |
| Lung | Alveolar space is air-filled; no alterations in bronchi and blood vessels | No differences from the control |
| Spleen | Large or middle-size folliculi with wide clear marginal zones and reactive centers | No differences from the control |
| Small intestine | Preserved villous epithelium; usual infiltration in villous stroma | No differences from the control |
| Testicle | Usual size and appearance of seminiferous tubules; clearly definable spermiogenesis; no focal alterations | No differences from the control |

Note: Shown are the summary data obtained on experimental days 30 and 180.

line A5547-127 revealed no pathological alterations in 30 and 180 days from the onset of the experiment. Similarly, histological examination detected no significant differences in the internal organs of the rats in both groups (Table 5.79).

## Assessment of Potential Impact of Transgenic Soybean Line A5547-127 on Immune System in Studies on Mice

### Potential Effect on Humoral Component of Immune System

The immunomodulating effect of transgenic soybean on the humoral component of the immune system was examined on two oppositely reacting mice lines, C57Bl/6 and CBA, by determining the level of hemagglutination to sheep erythrocytes (SE). The experimental conditions are described in section 5.1.1. In CBA mice fed diet with soy protein concentrate derived from transgenic soybean line A5547-127 or from conventional soybean, the antibody appeared on post-immunization Day 14 at the titer of 1:16.

### Potential Effect on Cellular Component of Immune System

The immunomodulating effect of transgenic soybean was assessed by delayed hypersensitivity reaction to SE.

In this experiment, both lines of mice were examined. The experimental conditions are described in section 5.1.1. The studies revealed no significant difference in the index of local inflammatory reaction in CBA mice fed diet with soy protein concentrate derived either from transgenic soybean line A5547-127 or from conventional soybean. However, test C57Bl/6 mice fed diet with soy protein concentrate derived from transgenic soybean line A5547-127 demonstrated a small but statistically significant ($p < 0.05$) elevation of this parameter in comparison with the control group fed diet with conventional soy protein concentrate, but no significant differences were observed when compared with the control mice maintained on the standard vivarium diet.

### Assessment of Potential Sensibilization Effect of Transgenic Soybean Line A5547-127

Assessment of possible sensitizing action of transgenic soybean on the immune response to endogenous metabolic products was carried out in the test of mouse sensitivity to histamine. The experimental conditions are described in section 5.1.1. There were no differences in behavior or mortality of test and control mice, which attests to the absence of sensitizing agent in the examined transgenic product.

### Potential Effect of Soybean on Susceptibility of Mice to *Salmonella typhimurium*

The effect of transgenic soybean on susceptibility of mice to infection by salmonella of murine typhus was examined in experiments on mice injected intraperitoneally with various doses of Strain 415 *Salmonella typhimurium*. Three weeks prior to infection, the diet of control and test mice was supplemented with protein concentrate derived from conventional soybean and transgenic soybean line A5547-127, respectively. The experimental conditions are described in section 5.1.1.

In both groups, mortality of mice was first observed in post-injection Week 1, and all infected mice died prior to post-injection Day 18. The lifetimes of the test and control mice were 11.3 and 14.7 d, correspondingly. $LD_{50}$ values of control and test mice were 72 and 69 microbial cells per mouse, respectively. These data showed that *Salmonella typhimurium* produced a typical infection in both groups of mice. Manifestations of infection were identical in test mice fed diet with soy protein concentrate derived from transgenic soybean line A5547-127 and in the control mice fed conventional soy protein concentrate. The course of disease, $LD_{50}$, and the lifetime of infected mice indicate that the infectious process developed similarly in both groups. Thus, feeding mice with transgenic soybean line A5547-127 did not affect their resistance against salmonella of murine typhus.

Therefore, introduction of protein concentrate derived from transgenic soybean line A5547-127 to mouse diet produced no effect on the humoral and cellular components of the immune system, did not sensitize the mouse organism, and did not affect susceptibility to typical infection such as murine typhus.

### Assessment of Potential Impact of Transgenic Soybean Line A5547-127 on Immune System in Studies on Rats

The study was carried out on male Wistar rats ($n = 47$) with body weight $140 \pm 10\,g$ at the start of the experiment. After a 7-day adaptation period to standard vivarium diet, the rats were fed for the next 28 days diet supplemented with protein concentrate derived from conventional soybean (control group) or from transgenic soybean line A5547-127 (test group). The soy protein concentrates from both varieties were dissolved in warm previously boiled water to the consistency of a dense curd and supplemented with sunflower-seed oil to improve intake. The experimental feed was used instead of an equally caloric amount of oatmeal (composition of experimental diet is described in Table 5.11).

The model of generalized anaphylaxis was developed according to [8] as described in section 5.1.1.

During the entire experiment the rats of both groups fed diet with protein concentrate derived from conventional and transgenic soybean line A5547–127 grew normally, which indicates nutritional adequacy of the diets used. On experimental Day 29, the body weights of test and control rats were $253 \pm 7\,g$ and $244 \pm 7\,g$, respectively ($p > 0.1$).

### 5.1.3 Glufosinate-Tolerant Soybean Line A5547-127

**Table 5.80** Severity of Active Anaphylactic Shock in Rats Fed Diet with Protein Concentrate Derived from Conventional Soybean (Control) or Transgenic Soybean Line A5547-127 (Test)

| Group of Rats | AI | Severe Reactions, % | Mortality, % |
|---|---|---|---|
| Control ($n = 23$) | 2.48 | 52.2 | 52.2 |
| Test ($n = 24$) | 2.96 | 62.5 | 62.5 |

**Table 5.81** Parameters of Humoral Immune Response in Rats Fed Diet with Protein Concentrate Derived from Conventional Soybean (Control) or Transgenic Soybean Line A5547-127 (Test)

| Group of Rats | $D_{492}$ | Concentration of Antibodies, mg/mL | Logarithm of Antibody Concentration |
|---|---|---|---|
| Control ($n = 23$) | $1.087 \pm 0.037$ | $8.3 \pm 1.1$ | $0.830 \pm 0.061$ |
| Test ($n = 24$) | $0.963 \pm 0.041$ | $5.2 \pm 0.7$ | $0.631 \pm 0.06$ |

In both groups, severity of active anaphylactic shock was assessed (Table 5.80). No significant differences were observed between the two groups of rats fed soybean-based diet.

Table 5.81 shows the mean values of $D_{492}$, the concentration of antibodies, and the common logarithm of this concentration in the control and test groups. All parameters for test rats indicated significantly smaller intensity of the humoral immune response in comparison with the control rats. Thus, transgenic soy protein concentrate moderated the degree of sensitization to a soybean product.

Taken together, these data support the conclusion that the degree of sensitization by ovalbumin in test rats fed diet with soy protein concentrate derived from transgenic soybean line A5547-127 did not increase in comparison with the control rats fed diet with protein concentrate derived from conventional soybean.

The studies showed that the protein concentrate derived from glufosinate-tolerant transgenic soybean line A5547-127 did not significantly change or increase allergic reactivity. In contrast, there was a certain moderation of sensitization to a model allergen in test rats in comparison with the control rats fed conventional soy protein concentrate. Therefore, feeding rats with transgenic glufosinate-tolerant soybean line A5547-127 did not enhance sensitization and allergic reactivity in comparison with the rats fed diet with conventional soybean line A5547.

**Table 5.82** Bone Marrow Cytogenetic Parameters in Mice of Parental Generation Fed a 45-Day Diet with Protein Concentrate Derived from Transgenic Soybean Line A5547-127 (Test) and Conventional Soybean (Control)

| Parameter | Control ($n = 5$) | Test ($n = 5$) |
|---|---|---|
| Number of analyzed metaphases | 356 | 350 |
| Share of cells, % | | |
| with chromosomal aberrations | $0.84 \pm 0.48$ | $0.57 \pm 0.4$ |
| with gaps | $0.84 \pm 0.48$ | $1.14 \pm 0.56$ |
| with polyploid chromosome set | $1.12 \pm 0.55$ | $0.85 \pm 0.49$ |

Note: The differences are not significant ($p > 0.05$).

### Assessment of Potential Genotoxicity of Transgenic Soybean Line A5547-127

Genotoxic studies were carried out on male C57Bl/6 mice and hybrid female CBA mice. The mice, weighing 20–25 g, were maintained on standard vivarium diet (Table 5.11). The cytogenetic analysis was carried out with the metaphasic method [8]. The details of experiments are described in section 5.1.1.

Genetic alterations in the sex cells were examined by revealing the dominant lethal mutations in C57Bl/6 male mice. The control and test mice were fed diet with soy protein concentrate for 45 days. For mating, the test ($n = 15$) and control ($n = 12$) male mice were placed in a cage with virgin CBA female mice in 1:2 ratio. The details of experiments are described in section 5.1.1.

Taking into consideration a particular sensitivity of the developing young organism to adverse (mutagenic included) factors, potential for mutagenic effect of transgenic soybean on the generation of mice whose parents were fed a long-term diet with GM soybean (males for 1.5 months, females during the entire lactation period) was assessed. The weanling mice were continuously fed diet with soybean for 30–35 days; thereafter the bone marrow was isolated from both femoral bones for cytogenetic examination with the above method. The results of examination of the control and test mice are shown in Table 5.82.

The number of cells with chromosomal abnormalities in parental F(0) mice and in the first generation F(1) mice fed diet with soy protein concentrate derived from the transgenic soybean line A5547-127 did not differ from that of control mice. The revealed chromosomal abnormalities (single segments and gaps) spontaneously appear in mice in similar amount, so they are not related to the effect of the examined transgenic soybean (Table 5.83).

To examine the dominant lethal mutations in gametes, the test female mice ($n = 64$) were dissected and 393 embryos and 425 corpus lutea were counted

**Table 5.83** Bone Marrow Cytogenetic Parameters in F(1) Mice Fed Diet with Protein Concentrate Derived from Transgenic Soybean Line A5547-127 (Test) and Conventional Soybean (Control)

| Parameter | Control (n = 5) | Test (n = 5) |
|---|---|---|
| Number of analyzed metaphases | 312 | 353 |
| Share of cells, % | | |
| with chromosomal aberrations | 0.64 ± 0.45 | 0.56 ± 0.40 |
| with gaps | 0.64 ± 0.48 | 0.85 ± 0.48 |
| with polyploid chromosome set | 1.28 ± 0.53 | 1.18 ± 0.56 |

Note: The differences are not significant ($p > 0.05$).

**Table 5.84** Dominant Lethal Mutations in Sex Cells of Mice Fed a 45-Day Diet with Protein Concentrate Derived from Conventional Soybean (Control) or Transgenic Soybean Line A5547-127 (Test)

| Parameter, % | Week 1, Mature Spermatozoa | | Week 2, Late Spermatids | | Week 3, Early Spermatids | |
|---|---|---|---|---|---|---|
| | Control | Test | Control | Test | Control | Test |
| Pre-implantation mortality | 9.40 | 7.85 | 5.48 | 8.20 | 6.10 | 6.92 |
| Post-implantation mortality | 3.77 | 4.65 | 6.62 | 6.61 | 4.87 | 4.65 |
| Survival rate | 84.29 | 87.94 | 84.50 | 85.10 | 88.30 | 88.80 |
| Induced mortality | 0 | 0 | 1 | 1 | 0 | 0 |

and analyzed. In control female mice ($n = 56$), a similar analysis was performed with 350 embryos and 376 corpus lutea. The pre-implantation mortality was similar in the control and test groups. At the stages of early and late spermatids or mature spermatozoa, the post-implantation embryonic mortality (the most reliable index of mutagenic activity of an examined substance) varied within equal range in the test and control groups.

The induced mortality at these stages was in the range 0–1%, attesting to the absence of mutagenic effect of transgenic soybean line A5547-127 on spermiogenesis in mice (Table 5.84).

The above data support the conclusion that glufosinate-tolerant soybean line A5547-127 produced no mutagenic effect in the described experiments.

## Assessment of Technological Parameters

The study of technological parameters was carried out in Moscow State University of Applied Biotechnology (Ministry of Science and Education of Russian Federation).

The following parameters were determined to characterize the seed samples of transgenic soybean line A5547-127 and its conventional counterpart: the yield of protein from the defatted flour during preparative isolation, amino acid and fractional composition of the obtained protein preparations, denaturation temperature of individual globulin fractions in the resulting protein preparations, and the yield of oil and its fatty acid composition.

The study resulted in the following conclusions:

1. The composition of the seeds of transgenic soybean line A5547-127 did not differ from that if the conventional soybean.
2. The yield of protein, its amino acid and fractional composition, and the thermodynamic parameters of individual fractions were virtually identical in the seed samples of transgenic soybean line A5547-127 and conventional soybean.
3. The lipids extracted from the seed samples of transgenic soybean line A5547-127 and conventional soybean had similar fatty acid composition characteristic of this crop.

Thus, there were no significant differences in the properties of the seeds of transgenic soybean line A5547-127 and its conventional counterpart.

### Conclusions

By all examined parameters, the data of complex safety assessment of glufosinate-tolerant transgenic soybean line A5547-127 attest to the absence of any toxic, genotoxic, immune system modulating, or allergenic effects of this soybean variety. Analysis of the biochemical composition of transgenic soybean line A5547-127 and the protein concentrate derived from it showed its identity to the composition of their conventional counterparts.

Based on the results of the studies, the State Sanitation Service of the Russian Federation (Department of State Sanitation and Epidemiological Inspectorate) granted the Registration Certificate which allows the transgenic soybean line A5547-127 to be used in the food industry and placed on the market without restrictions.

## 5.1.4 GLYPHOSATE-TOLERANT SOYBEAN LINE MON 89788

### Molecular Characteristics of GM Soybean Line MON 89788

#### Recipient Organism

Soybean *Glycine max (L.) Merill* (family *Leguminosae*, genus *Glycine Willd*, species *Glycine max*) has a long-term history of safe use as human food and animal feed.

### 5.1.4 Glyphosate-Tolerant Soybean Line MON 89788

#### *Donor Organism and Method of Genetic Transformation*

GM-soybean line MON 89788 was developed by agrobacterium transformation of the genome of a high-yielding soybean variety with the binary vector PV-GMGOX20, containing expression cassette of the gene 5-*Enolpyruvylshikimate-3-phosphate-synthase* (*cp4 epsps*).

The expression cassette of gene CS-*cp4 epsps* includes the following genetic elements: chimeric constitutive promoter P-FMV/Tsf1, containing *enhancer* sequence of *35S* promoter of the figwort mosaic virus and the promoter of the gene *Tsf1*, encoding the elongation factor EF-1 alpha in Arabidopsis plants (*Arabidopsis thaliana*); untranslated leader sequence (L-*Tsf1*) and the intron (I-*Tsf1*) of gene *Tsf1*; the sequence (TS-*CTP2*) of Arabidopsis genome (*Arabidopsis thaliana*), encoding synthesis of chloroplast transit peptide of *rybuloso-1.5*-bisphosphate-carboxylase-oxygenase (CTP) used to transport CP4 EPSPS protein into chloroplasts; gene CS-*cp4 epsps*, with coding region containing optimized coding triplets, isolated from *Agrobacterium* sp., line CP-4; T-E9—terminator coding sequence (3′-untranslated gene region of *rybuloso-1.5*-bisphosphate-carboxylase), isolated from subunit *RbcS2* of the gene *E9* of pea (*Pisum sativum*).

The presence of the CS-*cp4 epsps* gene, encoding CP4 EPSPS protein synthesis (Mr ~ 47.6 kD, 455 amino-acid residues), confers plant tolerance to glyphosate, making plants insensitive to the effect of the herbicide.

#### *Registration Status of GM Soybean Line MON 89788*

Table 5.85 shows the registration status of soybean line MON 89788 in various countries at the time of registration in Russia [19].

**Table 5.85** Registration Status of GM Soybean Line MON 89788 in Various Countries

| Country | Registration Date | Application |
| --- | --- | --- |
| Australia | 2008 | Food, feed |
| European Union | 2008 | Food, feed |
| Canada | 2007 | Food, feed, environmental release |
| China | 2008 | Food, feed |
| Korea | 2009 | Food, feed |
| Mexico | 2008 | Food, feed |
| USA | 2007 | Food, feed, environmental release |
| Taiwan | 2007 | Food |
| Philippines | 2007 | Food, feed |
| Japan | 2007 | Food |

Note: The current registration status of the transgenic crop is on http://www.biotradestatus.com/

**Table 5.86** Quality Assessment of Defatted Flour Samples ($M \pm m$ from $n = 3$)

| Parameter, % | Defatted Flour Produced from Traditional Soybean | Defatted Flour Produced From GM Soybean Line MON 89788 |
|---|---|---|
| Crude protein | 50.68 ± 0.29 | 50.08 ± 0.19 |
| Lipids | 1.997 ± 0.007 | 2.410 ± 0.020 |
| Carbohydrates | 24.69 ± 0.87 | 24.78 ± 0.62 |
| Ash | 8.353 ± 0.272 | 8.475 ± 0.227 |
| Moisture | 9.208 ± 0.090 | 9.183 ± 0.056 |
| Dietary fibers, $\Sigma$ | 5.003 ± 0.149 | 5.103 ± 0.228 |
| Dietary fibers, insoluble fraction | 3.813 ± 0.084 | 3.377 ± 0.222 |
| Dietary fibers, soluble fraction | 1.190 ± 0.099 | 1.727 ± 0.020 |

### Safety Assessment of GM Soybean Line MON 89788 Conducted in the Russian Federation

Research was conducted in accordance with the requirements outlined in methodological instructions for regulations MY 2.3.2.2306-07 "Medico-biological safety assessment of genetically-engineered and modified organisms of plant origin" [9]. PCR analysis of the soybean test and control samples was performed to confirm the identity of the transformation event and its absence in the conventional control line.

### Quality and Safety Assessment of Defatted Flour, Produced from GM Soybean Line MON 89788 and its Conventional Counterpart

Compliance of soybean flour samples with the safety and quality requirements adopted in the Russian Federation was assessed by analyses of quality factors and safety determination [5,9]. Results of the analyses of defatted flour samples, produced from GM soybean line MON 89788 and conventional soybean variety, are shown in Tables 5.86 and 5.87.

The data presented in Tables 5.86 and 5.87 show equivalent content of nutrient materials in the experimental defatted flour samples produced from GM soybean line MON 89788 and the conventional soybean variety. The safety parameters of defatted flour samples did not exceed the acceptable levels, established by SanRN 2.3.2.1078-01 (s.it. 1.9.1., 1.9.1.1.) [5].

Thus, the results of the sanitary-hygienic analyses demonstrated conformity of defatted flour samples, produced from GM soybean line MON 89788, to the safety requirements and nutrition value adopted in the Russian Federation.

### Assessment of Potential Toxicity of GM Soybean Line MON 89788 in Chronic Experiment in Rats

A 182-day study was conducted on male Wistar rats with an initial weight of 110–140 g. Rats were randomly divided into two groups (50 animals in each

**Table 5.87** Safety Assessment of Defatted Soybean Flour Samples ($M \pm m$ from $n = 3$)

| Parameter | | Defatted Flour Produced from Traditional Soybean | Defatted Flour Produced from GM Soybean Line MON 89788 | Acceptable levels (SanRN 2.3.2.1078-01 (s.it. 1.9.1, 1.9.1.1) |
|---|---|---|---|---|
| *Toxic elements, mg/kg* | | | | |
| Lead | | 0.002 ± 0.001 | 0.004 ± 0.003 | 1.0 |
| Arsenic | | ≤0.10 ± 0.00 | ≤0.10 ± 0.00 | 1.0 |
| Cadmium | | ≤0.001 ± 0.000 | ≤0.001 ± 0.000 | 0.2 |
| Mercury | | n/d | n/d | 0.03 |
| *Pesticides, mg/kg* | | | | |
| Hexachlorocyclohexane | | n/d | n/d | 0.2 |
| DDT and its metabolites | | n/d | n/d | 0.05 |
| Aldrin | | n/d | n/d | – |
| Heptachlor | | n/d | n/d | – |
| Kelthane | | n/d | n/d | – |
| *Mycotoxins, mg/kg* | | | | |
| Aflatoxin $B_1$ | | n/d | n/d | 0.005 |
| *Content of anti-alimentary substances, g/100 g* | | | | |
| Oligo-sugar | Stachyose | 0.822 ± 0.021 | 0.863 ± 0.027 | 2.0 |
| | Raffinose | 0.383 ± 0.005 | 0.401 ± 0.012 | |
| Trypsin inhibitor | | 0.018 ± 0.008 | 0.026 ± 0.006 | 0.5 |
| *Microbiological factors* | | | | |
| QMAFAnM, CFU/g | | 340.0 ± 30.6 | 166.7 ± 44.1 | $<5 \times 10^4$ |
| CGB (coliforms) in 0.1 g | | not detected | not detected | not acceptable |
| S. aureus in 0.1 g | | not detected | not detected | not acceptable |
| B. cereus, CFU/g | | <20 | <20 | – |
| Pathogenic, including salmonella in 25 g | | not detected | not detected | not acceptable |
| Sulfite-reducing clostridia in 0.1 g | | not detected | not detected | not acceptable |
| Yeast, CFU/g | | <5 | <5 | <100 |
| Mold, CFU/g | | <10 | <15 | |

group): animals of the test group received GM soybean line MON 89788 with their feed; animals of the control group received a counterpart of conventional soybean variety [9]. Soybean in the form of defatted flour was included in the feed at the rate of ~8 g/rat/24 h, replacing diet ingredients with account for proteins, fats, and carbohydrate content in the introduced product, observing isocaloric principle. Rats had free access to feed and water and were kept in plastic cages with wooden underlay, in a ventilated location, 3–4 animals per cage. The composition of the diet is described in Table 5.88.

**Table 5.88** Composition of the rat diet

| Ingredients, g | Amount, g/100 g Diet |
|---|---|
| Soybean flour | 38.0 |
| Maize starch | 44.7 |
| Sunflower oil | 4.2 |
| Lard | 5.0 |
| Salt max[a] | 4.0 |
| Water-soluble vitamins mixture[a] | 1.0 |
| Fat-soluble vitamins mixture[a] | 0.1 |
| Bran | 3.0 |
| Total | 100 |

[a]in accordance with [9].

Analyses of the organs were carried out after 30 and 182 days. During the experiment, palatability of the feed, body mass, and overall conditions of the animals were monitored.

During the experiment, no rat mortality was observed in test or control group, and overall condition of the animals was satisfactory. Feed palatability constituted 20–22 g/rat/24 h. The general appearance of the animals, including coat condition, behavior, growth rate, and body weight, were similar between the test and control groups (Figure 5.5, Table 5.89.). Weekly weight gain of both groups of rats corresponded to the gain level characteristic of animals of this breed and age [20,23,38].

Results of morphological, hematological, and biochemical studies are described in Tables 5.90 to 5.100. The obtained data indicate absence of any toxic effect of GM soybean line MON 89788. Compared with similar parameters of the control group rats, no differences in integral, morphological, or hematological parameters, nor of biochemical parameters of blood serum and urine, were detected. Comparative evaluation of the system biomarkers, characterizing organism potential adaptation condition, did not detect diagnostically significant differences between the groups. Values of all parameters fell within the limits of physiological fluctuations characteristic for rats.

Evaluation for potential toxicity of GMO includes studies of a great number of parameters; moreover, in certain cases statistically non-uniform distribution of values of some parameters in the groups, influencing the size of the average ($M \pm m$), takes place. As shown in Table 5.90, at the 30th day of the experiment, the mass of hypophysis in rats of the test group was slightly higher than in rats of the control group: absolute by 20%, relative by 19% ($p < 0.05$). The range of fluctuation of hypophysis mass in rats of

### 5.1.4 Glyphosate-Tolerant Soybean Line MON 89788

**Table 5.89** Body Weight of Rats (g) Fed Diet with Conventional Soybean (Control) or Transgenic Soybean Line MON 89788 (Test) ($M \pm m$; $n = 30$)

| Duration of Experiment, Days | Weight, g | |
|---|---|---|
| | Control | Test |
| 0   | 129.0 ± 1.6 | 131.5 ± 1.4 |
| 7   | 182.2 ± 2.2 | 188.8 ± 2.6 |
| 14  | 218.2 ± 2.8 | 225.3 ± 2.6 |
| 21  | 259.3 ± 3.1 | 263.7 ± 3.2 |
| 28  | 289.1 ± 3.2 | 285.1 ± 3.8 |
| 35  | 309.9 ± 5.4 | 308.4 ± 5.7 |
| 42  | 329.0 ± 5.3 | 328.0 ± 5.7 |
| 49  | 345.8 ± 4.9 | 347.7 ± 4.7 |
| 56  | 360.3 ± 6.5 | 364.1 ± 5.3 |
| 63  | 380.0 ± 6.5 | 370.7 ± 6.4 |
| 70  | 387.8 ± 7.7 | 382.2 ± 6.9 |
| 77  | 407.0 ± 6.7 | 407.4 ± 5.0 |
| 84  | 414.4 ± 7.0 | 415.4 ± 5.5 |
| 91  | 431.8 ± 7.1 | 429.9 ± 5.4 |
| 98  | 446.6 ± 7.4 | 438.5 ± 6.3 |
| 105 | 447.4 ± 7.9 | 442.9 ± 6.1 |
| 112 | 456.3 ± 7.9 | 451.5 ± 6.2 |
| 119 | 462.6 ± 8.4 | 459.6 ± 6.2 |
| 126 | 467.7 ± 7.9 | 464.7 ± 6.0 |
| 133 | 473.6 ± 8.3 | 469.6 ± 6.8 |
| 140 | 479.5 ± 8.3 | 472.8 ± 6.8 |
| 147 | 480.6 ± 8.6 | 473.9 ± 6.8 |
| 154 | 489.5 ± 8.7 | 483.3 ± 7.0 |
| 161 | 491.6 ± 8.8 | 482.2 ± 7.0 |
| 168 | 490.4 ± 8.9 | 486.3 ± 8.4 |
| 175 | 491.5 ± 9.9 | 491.2 ± 8.2 |
| 182 | 498.1 ± 9.9 | 495.8 ± 7.4 |

the control group was 0.0081–0.0151 g and 0.0028–0.0046 g/100 g; in rats of the test group it was 0.0076–0.0164 g and 0.0027–0.0053 g/100 g body mass. According to data of the State Institute of Nutrition of RAMS (sample size more than 300 animals), physiological fluctuation of hypophysis mass in growing male rats (age ~60 days) is 0.004–0.015 g and 0.001–0.005 g/100 g body mass. By Day 182 of the experiment, the difference in hypophysis mass was no longer evident, but a difference in thymus mass was detected: in rats of the test group this factor was lower than in rats of the control group, with absolute mass being lower by 25% and relative mass by 23% ($p < 0.05$).

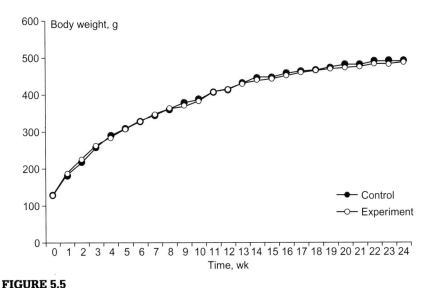

**FIGURE 5.5**
Comparative dynamics of body weight of rats fed diet containing transgenic soybean line MON 89788 (test) or its conventional counterpart (control).

The range of physiological fluctuations of thymus mass in adult Wistar rats is 0.245–0.844 g and 0.040–0.137 g/100 g body mass [28], therefore thymus mass values were within the norm. Macro- and microscopic examination of internal organs of the test group rats and control group rats revealed no pathological alterations in 30 and 182 days from the onset of the experiment. Similarly, morphometric examination detected no significant differences in the internal organs of the rats in both groups (Tables 5.91 to 5.93).

According to the data provided in Table 5.94, blood parameters conformed to general norms. Calculated values, characterizing erythrocyte condition, were an exception: after 30 days of the experiment, mean corpuscular volume (MCV) of erythrocytes in rats of the test group was lower, and mean corpuscular hemoglobin concentration (MCHC) was higher, than in control animals ($p < 0.05$). The content of segmentonuclear neutrophils in blood of test rats was 33% lower than in controls ($p < 0.05$). After 182 days of the experiment there was no difference between the groups. Based on the results of hematologic studies, conducted with account of reference values for rats (mean volume of erythrocytes and mean hemoglobin concentration vary within 85% and 49%, respectively, and segmentonuclear neutrophils content varies within >1000 % [28,45]), as well as taking into account that differences between the groups were not maintained, the following conclusion was made: the use of GM soybean line MON 89788 in feed does not influence rat peripheral blood structure.

### 5.1.4 Glyphosate-Tolerant Soybean Line MON 89788

**Table 5.90** Mass of the Internal Organs in Rats Fed Diet with Conventional Soybean (Control) or Transgenic Soybean Line MON 89788 (Test) ($M \pm m$ from $n = 15$)

| Parameter | | 30 Days Control | 30 Days Test | 182 Days Control | 182 Days Test |
|---|---|---|---|---|---|
| Liver, g | abs.[a] | 10.96 ± 0.23 | 10.46 ± 0.34 | 11.26 ± 0.23 | 11.45 ± 0.30 |
| | rel.[b] | 3.688 ± 0.081 | 3.522 ± 0.067 | 2.433 ± 0.034 | 2.535 ± 0.044 |
| Kidneys, g | abs. | 2.753 ± 0.060 | 2.842 ± 0.056 | 3.432 ± 0.092 | 3.566 ± 0.073 |
| | rel. | 0.926 ± 0.018 | 0.963 ± 0.023 | 0.742 ± 0.019 | 0.792 ± 0.018 |
| Spleen, g | abs. | 1.914 ± 0.115 | 1.771 ± 0.182 | 1.401 ± 0.081 | 1.453 ± 0.048 |
| | rel. | 0.645 ± 0.040 | 0.599 ± 0.063 | 0.302 ± 0.017 | 0.323 ± 0.011 |
| Heart, g | abs. | 1.153 ± 0.034 | 1.123 ± 0.023 | 1.576 ± 0.032 | 1.537 ± 0.041 |
| | rel. | 0.387 ± 0.009 | 0.379 ± 0.006 | 0.341 ± 0.007 | 0.341 ± 0.008 |
| Lungs, g | abs. | 2.066 ± 0.080 | 1.890 ± 0.062 | 2.583 ± 0.126 | 2.637 ± 0.173 |
| | rel. | 0.697 ± 0.028 | 0.640 ± 0.023 | 0.559 ± 0.028 | 0.585 ± 0.038 |
| Thymus, g | abs. | 0.582 ± 0.025 | 0.613 ± 0.033 | 0.371 ± 0.021 | 0.280 ± 0.026* |
| | rel. | 0.196 ± 0.008 | 0.206 ± 0.009 | 0.080 ± 0.004 | 0.062 ± 0.005* |
| Brain, g | abs. | 1.891 ± 0.026 | 1.950 ± 0.021 | 2.084 ± 0.024 | 2.054 ± 0.033 |
| | rel. | 0.637 ± 0.011 | 0.660 ± 0.011 | 0.452 ± 0.010 | 0.457 ± 0.011 |
| Testis, g | abs. | 2.877 ± 0.105 | 2.836 ± 0.150 | 3.494 ± 0.198 | 3.622 ± 0.174 |
| | rel. | 0.968 ± 0.034 | 0.966 ± 0.059 | 0.754 ± 0.041 | 0.802 ± 0.038 |
| Hypophysis, g | abs. | 0.0108 ± 0.0005 | 0.0130 ± 0.0007* | 0.0166 ± 0.0010 | 0.0141 ± 0.0007 |
| | rel. | 0.0036 ± 0.0001 | 0.0043 ± 0.0002* | 0.0036 ± 0.0004 | 0.0031 ± 0.0003 |
| Supramental capsules, g | abs. | 0.066 ± 0.003 | 0.073 ± 0.002 | 0.104 ± 0.008 | 0.112 ± 0.005 |
| | rel. | 0.022 ± 0.001 | 0.025 ± 0.001 | 0.023 ± 0.002 | 0.025 ± 0.001 |
| Prostate, g | abs. | 1.391 ± 0.075 | 1.487 ± 0.069 | 0.667 ± 0.049 | 0.683 ± 0.032 |
| | rel. | 0.467 ± 0.024 | 0.501 ± 0.020 | 0.144 ± 0.010 | 0.151 ± 0.007 |

[a]Absolute mass of internal organs, g.
[b]Relative mass of internal organs (per 100 g body mass).
*Here and in Tables 5.91 to 5.100 an asterisk indicates difference from control significant at $p < 0.05$.

**Table 5.91** Histo-topographical Characteristics of the Small Intestine Walls in Rats Fed Diet with Conventional Soybean (Control) or Transgenic Soybean Line MON 89788 (Test) ($M \pm m$ from $n = 10$)

| Structural Component, μm | | 30 Days Control | 30 Days Test | 182 Days Control | 182 Days Test |
|---|---|---|---|---|---|
| Whole intestine wall | | 814.3 ± 125.19 | 735.04 ± 123.3 | 651.09 ± 61.81 | 633.98 ± 57.38 |
| Muscular layer | | 41.94 ± 7.34 | 43.28 ± 6.21 | 41.21 ± 8.76 | 36.33 ± 9.31 |
| Submucous layer | | 29.82 ± 9.01 | 32.07 ± 8.04 | 23.19 ± 4.93 | 22.83 ± 5.20 |
| Mucous layer: | | 692.33 ± 81.17 | 668.27 ± 76.91 | 582.51 ± 59.71 | 575.47 ± 55.51 |
| including | intestinal crypt | 181.75 ± 25.14 | 180.53 ± 23.74 | 161.83 ± 25.14 | 153.98 ± 20.67 |
| | villi | 477.28 ± 89.37 | 491.45 ± 91.61 | 501.82 ± 42.01 | 492.91 ± 38.67 |

**Table 5.92** Absolute Quantity of Various Types of Epithelial Cells as a Part of Intestinal Villi and Small Intestine Crypt in Rats Fed Diet with Conventional Soybean (Control) or Transgenic Soybean Line MON 89788 (Test) ($M \pm m$ from $n = 10$)

| Structural Component, abs. U/100 μm | 30 Days | | 182 Days | |
|---|---|---|---|---|
| | Control | Test | Control | Test |
| *Quantity of villus epithelial cells* | | | | |
| Limbic cells | 15.89 ± 1.46 | 15.37 ± 1.64 | 21.96 ± 2.76 | 19.08 ± 2.04 |
| Goblet cells | 4.84 ± 0.78 | 4.16 ± 0.91 | 6.07 ± 1.01 | 5.11 ± 0.97 |
| Indifferent cells | 5.35 ± 0.67 | 5.51 ± 0.61 | 5.73 ± 0.99 | 5.01 ± 0.74 |
| *Quantity of crypt epithelial cells* | | | | |
| Limbic cells | 13.91 ± 1.51 | 14.46 ± 1.22 | 16.45 ± 2.54 | 17.34 ± 3.01 |
| Goblet cells | 4.92 ± 0.42 | 5.36 ± 0.74 | 5.94 ± 1.12 | 5.61 ± 1.75 |
| Indifferent cells | 2.81 ± 1.01 | 3.00 ± 0.99 | 4.08 ± 0.99 | 3.76 ± 1.09 |

**Table 5.93** Histo-topographical Characteristics of Large Intestine Walls in Rats Fed Diet with Conventional Soybean (Control) or Transgenic Soybean Line MON 89788 (Test) ($M \pm m$ from $n = 10$)

| Structural Component, μm | 30 Days | | 182 Days | |
|---|---|---|---|---|
| | Control | Test | Control | Test |
| Whole intestine walls | 587.54 ± 95.19 | 596.65 ± 85.30 | 500.22 ± 86.16 | 518.75 + 91.50 |
| Muscular layer | 123.76 ± 17.43 | 119.05 ± 13.64 | 125.62 ± 17.40 | 109.86 + 13.85 |
| Submucous layer | 62.97 ± 20.16 | 66.22 ± 17.68 | 43.51 ± 19.80 | 64.74 + 20.84 |
| Mucous layer | 292.74 ± 71.75 | 285.99 ± 75.83 | 306.60 ± 61.65 | 325.20 + 53.79 |

Biochemical analyses of blood serum and urine detected statistically significant difference of some factors, varying within the limits of physiological fluctuations typical for rats [28,45,48]. Thus, after 30 days of the experiment (Table 5.95), content of total protein, albumin, globulin, triglycerides, sodium, and phosphorus in blood serum of rats of the test group was higher than in rats of the control group ($p < 0.05$); after 182 days, content of albumin, total bilirubin, urea, and sodium in blood serum of rats of the test group was lower than in control animals ($p < 0.05$). It should be noted that most of the differences registered on the 30th day of the experiment disappeared on the 182nd day (total protein, globulin, triglycerides, phosphorus) or reversed in the opposite value ratio (albumin, sodium). Taking into account the absence of a tendency to maintain differences between groups,

**Table 5.94** Hematological Blood Values in Rats Fed Diet with Conventional Soybean (Control) or Transgenic Soybean Line MON 89788 (Test) ($M \pm m$ from $n = 15$)

| Parameter | 30 Days | | 182 Days | | Normal Values According to [28,45,48] |
|---|---|---|---|---|---|
| | Control | Test | Control | Test | |
| Total erythrocyte count, ×10$^{12}$/L | 7.213 ± 0.154 | 7.330 ± 0.151 | 8.295 ± 0.097 | 8.524 ± 0.132 | 4.4–8.9 |
| Hemoglobin concentration, g/L | 138.6 ± 2.6 | 139.9 ± 3.1 | 153.6 ± 1.6 | 154.5 ± 2.1 | 86–173 |
| Hematocrit, vol.% | 38.33 ± 0.62 | 37.53 ± 0.77 | 39.94 ± 0.74 | 41.23 ± 0.54 | 31.4–51.9 |
| MCV, fL | 53.32 ± 0.56 | 50.66 ± 0.98* | 48.94 ± 0.40 | 48.49 ± 0.47 | 50.6–93.8 |
| MCH, pg | 18.79 ± 0.42 | 19.04 ± 0.25 | 18.48 ± 0.17 | 18.10 ± 0.15 | 13.4–26.1 |
| MCHC, g/dL | 36.10 ± 0.23 | 37.21 ± 0.16* | 37.86 ± 0.19 | 37.43 ± 0.16 | 24.7–36.8 |
| ESR, mm/h | 1.467 ± 0.133 | 1.733 ± 0.228 | 2.375 ± 0.287 | 2.625 ± 0.515 | 0–5 |
| Leukocytes, ×10$^3$/μL | 11.69 ± 0.77 | 12.45 ± 1.31 | 10.94 ± 0.61 | 11.51 ± 0.88 | 1.4–34.3 |
| Basophilic cells, % | 0 | 0 | 0 | 0 | 0 |
| Eosinophils, % | 1.333 ± 0.465 | 2.067 ± 0.859 | 2.875 ± 0.625 | 2.500 ± 0.342 | 0.0–5.5 |
| Band neutrophils, % | 2.867 ± 0.456 | 2.533 ± 0.551 | 0.813 ± 0.262 | 1.250 ± 0.371 | 0.0–3.8 |
| Segmented neutrophils, % | 26.73 ± 2.95 | 18.00 ± 1.67* | 25.38 ± 1.06 | 26.13 ± 1.97 | 0.4–50.8 |
| Metamyelocytes, % | 0 | 0 | 0 | 0 | 0 |
| Lymphocytes, % | 67.67 ± 3.25 | 74.20 ± 2.39 | 69.88 ± 1.47 | 69.31 ± 2.43 | 42.3–98.0 |
| Monocytes, % | 1.533 ± 0.576 | 3.200 ± 0.705 | 1.188 ± 0.306 | 0.750 ± 0.171 | 0.0–7.9 |
| Thrombocytes, ×10$^3$/μL | 450.8 ± 22.8 | 462.5 ± 20.8 | 500.1 ± 20.2 | 507.7 ± 29.1 | 409–937 |

changes within normal range can be ascribed to individual fluctuations of the animals' biochemical status.

At urine examination, after 30 days of the experiment, there was a significant decrease in glucose levels in rats of the test group in comparison with the control group (Table 5.96); however, glucose content in rat urine in the control group did not exceed upper normal level [28,45,48]. After 182 days of experiment no difference between groups was detected.

Factors characterizing antioxidant status in general fell within the limits of physiological variations (Table 5.97). Decrease of MDA content in the liver of rats of the test group detected on the 30th day, and an increase of MDA content in blood serum of animals in the test group on the 182nd day of the experiment fell within the limits of physiological fluctuations characteristic of animals of the given age [11]. The nature of the detected changes, and the absence of complex antioxidant protection system strain effects in rats of the test group, makes it possible to draw the following conclusion: long-term use of GM soybean with feed does not influence antioxidant status.

**Table 5.95** Biochemical Parameters of Blood Serum in Rats Fed Diet with Conventional Soybean (Control) or Transgenic Soybean Line MON 89788 (Test) ($M \pm m$ from $n = 15$)

| Parameter | 30 Days Control | 30 Days Test | 182 Days Control | 182 Days Test | Normal Values According to [28,45,48] |
|---|---|---|---|---|---|
| Total protein, g/L | 60.63 ± 0.93 | 71.57 ± 2.82* | 76.59 ± 1.49 | 74.28 ± 1.37 | 56–82 |
| Albumin, g/L | 37.54 ± 0.48 | 40.40 ± 0.82* | 39.70 ± 0.34 | 38.20 ± 0.60* | 25–48 |
| Globulin, g/L | 23.09 ± 0.62 | 31.17 ± 2.10* | 36.89 ± 1.33 | 35.46 ± 1.22 | 12–57 |
| Triglycerides, mmol/L | 0.490 ± 0.022 | 0.607 ± 0.040* | 0.746 ± 0.042 | 0.842 ± 0.077 | 0.3–1.6 |
| Total bilirubin, µmol/L | 2.800 ± 0.355 | 2.733 ± 0.419 | 4.077 ± 0.309 | 2.688 ± 0.198* | 1–4 |
| Direct bilirubin, µmol/L | 0 | 0 | 0 | 0 | 0 |
| Urea, mmol/L | 8.067 ± 0.320 | 8.593 ± 0.283 | 6.563 ± 0.207 | 5.313 ± 0.197* | 4–10 |
| Glucose, mmol/L | 6.224 ± 0.129 | 6.397 ± 0.254 | 5.493 ± 0.146 | 5.498 ± 0.137 | 4.5–10.0 |
| Cholesterine, mmol/L | 1.575 ± 0.074 | 1.602 ± 0.057 | 1.843 ± 0.068 | 1.831 ± 0.057 | 0.6–4.3 |
| Gamma GTP, U/L | 1.473 ± 0.353 | 2.240 ± 0.282 | 1.688 ± 0.334 | 1.356 ± 0.244 | 0–3 |
| Alpha-amylase, U/L | 1329 ± 62 | 1442 ± 62 | 939.6 ± 33.8 | 901.3 ± 51.2 | 766–1850 |
| Alkaline phosphatase, U/L | 710.8 ± 55.3 | 762.4 ± 33.8 | 332.6 ± 20.8 | 331.1 ± 23.5 | 112–814 |
| ALT U/L | 72.20 ± 4.42 | 78.87 ± 4.84 | 77.88 ± 3.53 | 77.44 ± 3.22 | 33–120 |
| AST U/L | 186.7 ± 7.1 | 195.6 ± 7,4 | 177.8 ± 10.5 | 186.0 ± 9.0 | 60–236 |
| Lipase, E/L | 47.53 ± 5.02 | 48.07 ± 3.33 | 26.81 ± 4.31 | 27.94 ± 5.39 | up to 30 |
| Calcium, mmol/L | 3.533 ± 0.149 | 3.807 ± 0.204 | 2.738 ± 0.090 | 3.138 ± 0.201 | 1.1–6.6 |
| Sodium, mmol/L | 159.5 ± 3.8 | 176.4 ± 4.5* | 169.7 ± 4.4 | 134.7 ± 3.5* | 137–159 |
| Phosphorus, mmol/L | 2.439 ± 0.060 | 2.709 ± 0.059* | 2.114 ± 0.047 | 2.073 ± 0.043 | 1.3–2.7 |

**Table 5.96** Biochemical Parameters of Urine in Rats Fed Diet with Conventional Soybean (Control) or Transgenic Soybean Line MON 89788 (Test) ($M \pm m$ from $n = 14$)

| Parameter | 30 Days Control | 30 Days Test | 182 Days Control | 182 Days Test | Normal Values According to [28,45,48] |
|---|---|---|---|---|---|
| Daily urine, mL | 4.46 ± 0.33 | 4.33 ± 0.40 | 4.40 ± 0.36 | 5.11 ± 0.36 | 3–34 |
| Color | Yellow | Yellow | Yellow | Yellow | Yellow |
| Transparency | Cloudy | Cloudy | Cloudy | Cloudy | – |
| pH | 8.13 ± 0.15 | 8.24 ± 0.22 | 8.22 ± 0.29 | 7.82 ± 0.14 | 6.5–8.5 |
| Relative density, g/mL | 1.09 ± 0.01 | 1.09 ± 0.01 | 1.05 ± 0.03 | 1.04 ± 0.03 | 1.0–1.1 |
| Creatinine, mmol/L | 12.7 ± 0.7 | 12.7 ± 0.9 | 23.7 ± 1.3 | 23.0 ± 1.7 | 1.7–21.0 |
| Protein, g/L | 0.66 ± 0.04 | 0.64 ± 0.04 | 0.68 ± 0.03 | 0.59 ± 0.04 | 0.333–0.600 |
| Glucose, mmol/L | 0.52 ± 0.07 | 0.32 ± 0.06* | 0.49 ± 0.06 | 0.43 ± 0.05 | up to 0.82 |

### 5.1.4 Glyphosate-Tolerant Soybean Line MON 89788

**Table 5.97** Antioxidant Status in Rats Fed Diet with Conventional Soybean (Control) or Transgenic Soybean Line MON 89788 (Test) ($M \pm m$ from $n = 15$)

| Parameter | 30 Days | | 182 Days | |
|---|---|---|---|---|
| | Control | Test | Control | Test |
| *Enzymatic activity of antioxidant protection system* | | | | |
| Glutathione reductase, µM/min·g Hb | 46.42 ± 1.49 | 46.51 ± 1.75 | 33.32 ± 1.34 | 34.91 ± 1.73 |
| Glutathione peroxidase, µM/min·g Hb | 68.86 ± 1.92 | 64.83 ± 1.79 | 60.58 ± 0.94 | 60.11 ± 1.88 |
| Catalase, mmol/min·g Hb | 449.1 ± 16.5 | 405.7 ± 14.5 | 499.2 ± 14.9 | 466.4 ± 12.0 |
| Superoxide dismutase, AU/min·g Hb | 1993 ± 34 | 1916 ± 59 | 1902 ± 26 | 1690 ± 48* |
| *Content of lipid peroxidation products* | | | | |
| Erythrocyte MDA, nmol/mL | 3.531 ± 0.132 | 3.512 ± 0.144 | 4.868 ± 0.356 | 4.653 ± 0.214 |
| Serum MDA, nmol/mL | 5.987 ± 0.195 | 6.051 ± 0.282 | 6.142 ± 0.321 | 7.500 ± 0.299* |
| Liver MDA, nmol/g | 478.1 ± 16.7 | 399.7 ± 9.6* | 392.8 ± 8.3 | 372.0 ± 5.9 |

**Table 5.98** Activity of Enzymes of Xenobiotic Metabolism and Protein Content in Liver of Rats Fed Diet with Conventional Soybean (Control) or Transgenic Soybean Line MON 89788 (Test) ($M \pm m$ from $n = 9$)

| Parameter | 30 Days | | 182 Days | |
|---|---|---|---|---|
| | Control | Test | Control | Test |
| Cytochrome P450, nmol/mg protein | 0.62 ± 0.05 | 0.70 ± 0.03 | 0.80 ± 0.03 | 0.92 ± 0.08 |
| Cytochrome $b_5$, nmol/mg protein | 0.66 ± 0.02 | 0.71 ± 0.03 | 0.90 ± 0.02 | 0.94 ± 0.02 |
| EROD, pmol/min·mg protein | 56.3 ± 3.3 | 59.7 ± 5.7 | 71.9 ± 2.5 | 78.9 ± 3.9 |
| 7-Pentoxyresorufin-O-deetilase, pmol/min·mg protein | 12.3 ± 1.0 | 11.8 ± 1.3 | 15.1 ± 0.5 | 14.2 ± 0.7 |
| UDP-Glucuronosyl transferase, nmol/min·mg protein | 32.1 ± 2.2 | 32.8 ± 1.9 | 26.1 ± 1.1 | 26.4 ± 1.1 |
| HDNB-glutathione S-transferase, µM/min·mg protein | 1.51 ± 0.04 | 1.39 ± 0.05 | 1.74 ± 0.07 | 1.68 ± 0.06 |
| Microsomal protein, mg/g | 14.1 ± 0.3 | 14.4 ± 0.4 | 14.6 ± 0.5 | 15.7 ± 0.4 |
| Cytosolic protein, mg/g | 80.5 ± 1.8 | 82.3 ± 1.2 | 84.4 ± 2.5 | 86.3 ± 1.8 |

New parameters characterizing activity of apoptosis processes in liver were added to the tested system biomarkers. The method of morphological determination of apoptosis bodies in liver cells allows detection of apoptosis regardless of the inducer characteristics [7], therefore in this study histological verification of apoptosis death was used. As shown in Table 5.100, level of apoptosis in liver cells of rats of control and test groups had no statistically significant difference throughout the experiment.

**Table 5.99** Activity of Enzymes of Liver Lysosomes in Rats Fed Diet with Conventional Soybean (Control) or Transgenic Soybean Line MON 89788 (Test) ($M \pm m$ from $n = 9$)

|  | 30 Days | | 182 Days | |
|---|---|---|---|---|
| Parameter | Control | Test | Control | Test |
| *Total activity, μM/min/g tissue* | | | | |
| Arylsulfatases A and B | 2.16 ± 0.04 | 2.19 ± 0.05 | 2.14 ± 0.04 | 2.17 ± 0.05 |
| β-Galactosidase | 2.25 ± 0.07 | 2.32 ± 0.05 | 2.20 ± 0.06 | 2.26 ± 0.05 |
| β-Glucuronidase | 2.13 ± 0.05 | 2.20 ± 0.07 | 2.24 ± 0.06 | 2.32 ± 0.06 |
| *Non-sedimentating activity, % of total* | | | | |
| Arylsulfatases A and B | 3.41 ± 0.17 | 3.32 ± 0.16 | 2.81 ± 0.05 | 2.81 ± 0.12 |
| β-Galactosidase | 6.36 ± 0.19 | 6.54 ± 0.28 | 6.65 ± 0.45 | 6.32 ± 0.41 |
| β-Glucuronidase | 6.39 ± 0.37 | 6.76 ± 0.31 | 5.62 ± 0.27 | 5.53 ± 0.33 |

**Table 5.100** Level of Hepatocytes Apoptosis in Rats Fed Diet with Conventional Soybean (Control) or Transgenic Soybean Line MON 89788 (Test) ($M \pm m$ from $n = 7000$)

|  | 30 Days | | 182 Days | | Normal Values According to [7] |
|---|---|---|---|---|---|
| Parameter | Control | Test | Control | Test | |
| Number of apoptosis bodies, % | 0.328 ± 0.022 | 0.332 ± 0.024 | 0.325 ± 0.022 | 0.327 ± 0.022 | 0.1–0.5 |

Thus, the results of the 182-day toxicological experiment in rats that received maximum possible quantities of GM soybean line MON 89788 with their feed (33% caloric value) demonstrate the absence of any toxic effect. Values of all studied parameters fell within the limits of physiological fluctuations characteristic of rats.

### Genotoxicity Studies of GM Soybean Line MON 89788 in Experiment in Mice

A 30-day study was conducted in male mice of C57Bl/6 line, sensitive to genotoxic influence, with initial body mass of 17–19 g. Mice received semisynthetic casein diet (diet composition described in Table 5.88). Soybean (in the form of defatted flour) was included in the feed at the rate of 3–3.2 g/mouse/24 h. Evaluation of potential genotoxicity included identification of DNA damage by the method of alkaline gel-electrophoresis of isolated cells of bone marrow, kidneys, liver, rectum (DNA-comet assay), and mutagenic activity detection by the method of chromosomal aberrations accounting in metaphase cells of mice bone marrow [2].

After 30 days from the beginning of the experiment, mice body mass was 19–23 g. According to the results of cytogenetic analysis of mice bone marrow, average parameters of chromosomal aberrations in mice bone marrow cells of control and test groups showed no significant difference (Table 5.101) and did not exceed spontaneous mutagenesis level characteristic of mice of C57Bl/6 line [6]. DNA structure damage level in bone marrow, kidneys, liver, and rectum in mice of the test group did not differ from similar parameters in mice of the control group (Table 5.102).

Thus, the results of DNA integrity and level of chromosomal aberrations in the experiment in mice demonstrate absence of genotoxic effect of GM soybean line MON 89788 compared with its conventional soybean.

### Assessment of Potential Impact on Immune System of GM Soybean Line MON 89788 in Experiment in Mice

An experiment of 45 days duration was conducted on mice lines CBA and C57Bl/6 with initial body mass of 16–18 g. Mice received a semisynthetic casein diet (Table 5.88). Soybean (in the form of defatted flour) was included in the feed at the rate of 3–3.2 g/mouse/24 h. Evaluation of immunomodulating and sensitizing properties was carried out in four tests: (1) effect on the humoral component of the immune system was detected with hemagglutinin levels in response to sheep erythrocytes; (2) effect on the cellular component of the immune system, through delayed hypersensitivity reaction of the response to sheep erythrocytes; (3) sensibilization effect, through a histamine sensitivity test; (4) response to infection by *Salmonella typhimurium* (salmonella of mouse typhus).

The assessment of the state of the humoral component of the immune system demonstrated that development of antibodies in response to injection of sheep erythrocytes in mice of the control group was similar to that in the experimental group (both in mice of CBA and C57Bl/6 lines), which demonstrates absence of immunomodulating effect of GM soybean line MON 89788 compared with its traditional counterpart (Table 5.103).

In the assessment of the condition of the cellular component of the immune system in terms of delayed hypersensitivity test, no immunomodulating effect of GM soybean line MON 89788 was detected (Table 5.104).

In the assessment of sensibilization effect and effect on mice response to *Salmonella typhimurium*, no negative influence of GM soybean was detected. When infected with *Salmonella typhimurium*, mice of control and experimental groups had a typical infection; mice of the CBA line (insensitive to *Salmonella typhimurium*) were more resistant to the infection than mice of the C57Bl/6 line (sensitive to *Salmonella typhimurium*).

Thus, results of the assessment of potential impact of GM soybean line MON 89788 on the immune system of mice of oppositely reacting lines

**Table 5.101** Analyses of Potential Chromosomal Damage in Bone Marrow Cells in Mice Fed Diet with Conventional Soybean (Control) or Transgenic Soybean Line MON 89788 (Test) ($M \pm m$)

| Group | Cell Number | DNA Damaged, % | Genes | Per 100 cells | | | | Total of Damaged Metaphases, % |
|---|---|---|---|---|---|---|---|---|
| | | | | Individual Fragments | Paired Fragments | Exchanges | Cells with MI[a] | |
| Control | 500 | 0.3 | | 1.5 | – | – | – | 1.5 ± 0.6 |
| Test | 500 | 0.4 | | 0.8 | – | 0.2 | 0.2 | 1.6 ± 0.6 |

*Note: Differences not significant ($p > 0.05$).*
[a] More than five chromosomal aberrations in a cell (MI, multiple injuries).

**Table 5.102** Analyses of a Potential DNA Structure Damage in Organs in Mice Fed Diet with Conventional Soybean (Control) or Transgenic Soybean Line MON 89788 (Test) ($M \pm m$)

| | Bone marrow | | Liver | | Kidneys | | Rectum | |
|---|---|---|---|---|---|---|---|---|
| Group | Cell Number | DNA Damaged, % | Cell Number | DNA Damaged, % | Cell Number | DNA Damaged, % | Cell Number | DNA Damaged, % |
| Control | 500 | 6.99 ± 0.10 | 500 | 5.60 ± 0.18 | 500 | 6.24 ± 0.15 | 500 | 7.40 ± 0.20 |
| Test | 500 | 7.10 ± 0.19 | 500 | 5.63 ± 0.13 | 500 | 6.41 ± 0.18 | 500 | 7.51 ± 0.18 |

*Note: Differences not significant ($p > 0.05$).*

**Table 5.103** Level of Antibodies to Sheep Erythrocytes Content in Mice Fed Diet with Conventional Soybean (Control) or Transgenic Soybean Line MON 89788 (Test)

| Number of Days Post-Injection | CBA Line | | C57Bl/6 Line | |
|---|---|---|---|---|
| | Control | Test | Control | Test |
| 7 | 160.0 ± 21.9 | 166.4 ± 19.6 | 100.8 ± 14.3 | 108.8 ± 9.8 |
| 14 | 110.4 ± 20.6 | 128.0 ± 16.5 | 64.0 ± 11.7 | 62.4 ± 12.3 |
| 21 | 96.0 ± 13.5 | 99.2 ± 12.1 | 33.6 ± 5.6 | 30.4 ± 4.4 |

*Average data shown, $M \pm m$ from $n = 10$; $p > 0.05$.*

**Table 5.104** Delayed Hypersensitivity Reaction in Mice Fed Diet with Conventional Soybean (Control) or Transgenic Soybean Line MON 89788 (Test)

| | CBA Line | | C57Bl/6 Line | |
|---|---|---|---|---|
| | Control | Test | Control | Test |
| Delayed hypersensitivity reaction index | 8.45 ± 1.13 | 10.10 ± 0.95 | 4.75 ± 0.76 | 6.73 ± 0.59 |

*Average data shown, $M \pm m$ from $n = 10$; $p > 0.05$.*

demonstrate absence of any immunomodulating and sensibilization effect of the GM soybean line compared with its traditional counterpart.

## Assessment of Potential Allergenicity of GM Soybean Line MON 89788 in Experiments in Rats

An experiment of duration 29 days was conducted in male Wistar rats with an initial body mass of 150 ± 10 g. Throughout the experiment rats received standard vivarium diet. Soybean (in the form of defatted flour) was added to the feed at the rate of 3.3 g/rat/24 h, excluding equivalent in caloric value and nutrient materials content quantity of oatmeal and grain mixture. The base composition of the diet is shown in Table 5.11.

The model of generalized anaphylaxis was developed according to [8] as described in section 5.1.1.

The body mass of rats of control and experimental groups at the 29th day of experiment was 263 ± 5 and 255 ± 4 g, respectively ($p > 0.1$). Severity of anaphylactic shock reaction in rats of the test group had no statistically significant differences from the severity of reaction in rats of the control group: the obtained results were within the range of typical values (30–60% lethality),

**Table 5.105** Severity of Reaction to Active Anaphylactic Shock in Rats Fed Diet with Conventional Soybean (Control) or Transgenic Soybean Line MON 89788 (Test)

| Group[a] | Anaphylactic Index | Severe Reactions, % | Mortality, % |
|---|---|---|---|
| Control | 2.21 | 33.3 | 33.3 |
| Test | 2.68 | 52.0 | 52.0 |
| P | >0.1[b] | >0.1[c] | >0.1[c] |
|  | $P = 0.285$ | $P = 0.184$ | $P = 0.184$ |

[a]24 rats in the control group, 25 rats in the test group.
[b]Mann-Whitney nonparametric rank test.
[c]Bi-directional test U – Fisher angular transformation.

**Table 5.106** Intensity of Humoral Immune Response in Rats Fed Diet with Conventional Soybean (Control) or Transgenic Soybean Line MON 89788 (Test)

| Group[a] | $D_{492}$ | AT Level, mg/mL | Lg of AT Level |
|---|---|---|---|
| Control | $0.938 \pm 0.030$ | $3.42 \pm 0.27$ | $0.486 \pm 0.048$ |
| Test | $0.912 \pm 0.024$ | $3.14 \pm 0.36$ | $0.441 \pm 0.043$ |
| *Statistical analysis* | | | |
| t-Student test | >0.1 | >0.1 | >0.1 |
| Test for homogeneity of distribution, ANOVA, P | >0.1 | >0.1 | >0.1 |
| Mann-Whitney nonparametric rank test | >0.1 | | |

Average data shown, $M \pm m$.
[a]24 rats in the control group, 25 rats in the experimental group.

which are usually observed at administration of anaphylaxis-inducing dose of ovalbumin to sensitized rats (Table 5.105). Comparison of the factors characterizing intensity of humoral immune response (Table 5.106) did not detect significant differences between the groups. Analysis of factors distribution in the groups, carried out with the use of ANOVA criterion, shows their homogeneity ($p > 0.1$).

Thus, results of allergenicity studies on GM soybean line MON 89788 in experiment in rats demonstrate absence of an allergenic effect of the given GM soybean line compared with its traditional counterpart.

### Assessment of Technological Parameters of Soybean Line MON 89788

Assessment of technological properties of GM soybean line MON 89788 was conducted according to the requirements outlined in methodological regulations MU 2.3.2.2306-07 "Medico-biologic safety assessment of genetically-engineered and modified organisms of plant origin".

### Results
The seed of GM soybean line MON 89788 does not differ in composition from its conventional counterpart. Protein yield, its amino-acid and fractional composition, as well as thermodynamic parameters of separate fractions practically coincide for all studied soy seed samples. Lipids, extracted from all studied soy seed samples, have similar fatty acid composition, characteristic of the given kind of raw materials, and differ within the limits of measurement error.

Thus, the results of comparative study of technological properties of GM soybean line MON 89788 and its traditional counterpart demonstrate absence of significant difference between the samples.

### Conclusion
Expert assessment of the data provided by the applicant, and results of complex biomedical research of GM soybean line MON 89788, glyphosate tolerant, demonstrate the absence of any toxic, genotoxic, sensitizing, immunomodulating, or allergenic effect in this soybean line, as well as its composition equivalence to its traditional counterpart.

Based on the results of the studies, the State Sanitation Service of the Russian Federation (Department of State Sanitation and Epidemiological Inspectorate) granted the Registration Certificate which allows the transgenic soybean line MON 89788 to be used in the food industry and placed on the market without restrictions.

## Subchapter 5.2

# Maize

## 5.2.1 GLYPHOSATE-TOLERANT MAIZE LINE GA 21
### Molecular Characterization of Maize Line GA 21
*Recipient Organism*
Maize (*Zea mays L.*) is characterized by a long history of safe use as human food and animal feed.

### Donor Organism
Wild-type maize was used to isolate the gene encoding synthesis of 5-enolpyruvilshikimate-3-phosphate synthase (EPSPS), the key enzyme in synthesis of aromatic amino acids in plants and microorganisms. This native gene was subjected to directional mutagenesis *in vitro* to obtain the synthetic gene, which encodes synthesis of EPSPS tolerant to glyphosate. The resulting enzyme is a polypeptide composed of 445 amino acids, which is identical to the prototype enzyme by 99.3% [27].

### Method of Genetic Transformation
The genetic modification of maize embryonic cells was performed by biolistic transformation using plasmid pDPG434, which incorporates the modified gene. The vector contained promoter r-act (5′ region of actin gene of rice) [30] and NOS3′ terminator. This sequence was isolated from the Ti-plasmid of Agrobacterium. The gene *bla* coding tolerance to β-lactamase played the role of selective bacterial marker.

## Global Registration Status of Maize Line GA 21
Table 5.107 shows the registration status for the use of transgenic maize line GA 21 at the time of registration in Russia [19].

## Safety Assessment of Transgenic Maize Line GA 21 Conducted in the Russian Federation
The studies were conducted in accordance with the requirements of the Ministry of Health of the Russian Federation authorized for risk and safety assessment of food derived from GM sources [8]. PCR analysis of the maize

**Table 5.107** Registration Status of Transgenic Maize Line GA 21 in Various Countries

| Country | Date of Approval | Application |
|---|---|---|
| Australia | 2000 | Food |
| Argentina | 1998 | Environmental release |
| Canada | 1997 | Feed, environmental release |
| | 1998 | Food |
| USA | 1997 | Environmental release |
| | 1996 | Food, feed |
| Japan | 1998 | Environmental release |
| | 1999 | Food, feed |

*Note: The current registration status of the transgenic crop is on http://www.biotradestatus.com/*

**Table 5.108** Protein Content and Amino Acid Composition (g/100 g protein) in Maize Grain

| Ingredient | Conventional Maize | Transgenic Maize Line GA 21 |
|---|---|---|
| Protein, % | 9.06 | 8.18 |
| *Amino acids* | | |
| Lysine | 2.61 | 2.78 |
| Histidine | 2.31 | 3.09 |
| Arginine | 8.92 | 5.40 |
| Aspartic acid | 9.69 | 7.41 |
| Threonine | 3.07 | 3.70 |
| Serine | 4.46 | 4.47 |
| Glutamic acid | 21.50 | 21.30 |
| Proline | 11.23 | 8.33 |
| Glycine | 2.46 | 2.93 |
| Alanine | 5.69 | 5.86 |
| Cysteine | 2.61 | 3.70 |
| Valine | 2.92 | 4.01 |
| Methionine | 0.92 | 2.62 |
| Isoleucine | 2.00 | 2.93 |
| Leucine | 9.85 | 10.8 |
| Tyrosine | 4.77 | 5.56 |
| Phenylalanine | 4.92 | 5.09 |

**Table 5.109** Carbohydrates Content (g/100 g product) in Maize Grain

| Carbohydrate | Conventional Maize | Transgenic Maize Line GA 21 |
|---|---|---|
| Sucrose | 1.33 | 1.31 |
| Starch | 54.3 | 52.6 |
| Cellulose | 1.96 | 1.83 |

test and control samples was performed to confirm the identity of the transformation event and its absence in the conventional control line.

### Biochemical Composition of Maize Grain

The content of proteins and amino acid composition in the grain of transgenic maize line GA 21 did not significantly differ from the corresponding values for the conventional maize (Table 5.108).

Similarly, the content of carbohydrates and lipids in the grain of transgenic maize line GA 21 did not significantly differ from the corresponding values for the conventional maize (Tables 5.109 and 5.110).

**Table 5.110** Content of Lipids (%) in Maize Grain Crop

| Ingredient | Conventional Maize | Transgenic Maize Line GA 21 |
|---|---|---|
| Lipids | 4.7 | 4.9 |

**Table 5.111** Vitamins Content (mg/100 g product) in Maize Grain

| Ingredient | Conventional Maize | Transgenic Maize Line GA 21 |
|---|---|---|
| Vitamin $B_1$ | 0.298 | 0.204 |
| Vitamin $B_2$ | 0.182 | 0.151 |
| Vitamin $B_6$ | 0.18 | 0.21 |
| Vitamin E ($\alpha$-tocopherol) | 8.0 | 2.2 |
| $\beta$-Carotene | 0.03 | 0.04 |
| Total carotenoids | 0.55 | 1.29 |

**Table 5.112** Mineral Content (mg/kg) in Maize Grain

| Ingredient | Conventional Maize | Transgenic Maize Line GA 21 |
|---|---|---|
| Copper | 1.42 | 1.48 |
| Zinc | 28.1 | 24.5 |
| Iron | 37.8 | 27.5 |
| Sodium | 30.9 | 125 |
| Potassium | 3815 | 3434 |
| Calcium | 11.8 | 3.00 |
| Magnesium | 1322 | 1568 |

The vitamin composition in the grain of transgenic maize line GA 21 did not differ from the corresponding values for the conventional maize (Table 5.111). However, there were significant differences for the content of vitamin E and overall carotenoids, but these values did not surpass the limits of physiological variations characteristic for maize according to the data of the State Research Institute of Nutrition, RAMS: 1.0–10.0 mg/100 g (vitamin E) and 0.2–1.5 mg/100 g (total carotenoids).

The content of minerals in conventional and transgenic maize was almost identical (Table 5.112). The revealed changes in the content of sodium, potassium, and calcium were within the physiological variations for maize assessed by the State Research Institute of Nutrition, RAMS: 20–200 mg/kg (sodium), 3000–4800 mg/kg (potassium), and 1–25 mg/kg (calcium).

The content of heavy metals and mycotoxins in the grains of conventional and transgenic maize line GA 21 did not surpass the acceptable limits according to the regulations valid in the Russian Federation (Table 5.113) [5].

**Table 5.113** Sanitary and Chemical Safety Parameters (mg/kg) of Maize Grain

| Ingredient | Conventional Maize | Transgenic Maize Line GA 21 |
|---|---|---|
| Aflatoxin $B_1$ | Not detected | Not detected |
| Deoxynivalenol | Not detected | Not detected |
| Zearalenone | Not detected | Not detected |
| T2-toxin | Not detected | Not detected |
| Fumonisin $B_1$ | Not detected | Not detected |
| Lead | 0.048 | <0.001 |
| Cadmium | 0.008 | 0.002 |

**Table 5.114** Composition of Rat Diet

| Ingredient | Weight, g |
|---|---|
| Maize cereals | 9.00 |
| Bread, second grade | 9.20 |
| Grain mix | 29.0 |
| Curd | 4.61 |
| Fish flour | 1.15 |
| Meat of second grade | 9.23 |
| Carrot | 18.5 |
| Greens | 18.5 |
| Cod-liver oil | 0.23 |
| Yeast | 0.23 |
| NaCl | 0.35 |
| Total | 100 |

Thus, the above-mentioned data showed that, by biochemical composition, the grain of transgenic maize line GA 21 and conventional maize did not significantly differ. The revealed differences in the content of metals (calcium, sodium, and potassium) and vitamins (vitamin E and carotenoids) fell within the range characteristic of maize [15,43,49,50,51]. The safety parameters of the grain of conventional and transgenic maize line GA 21 met the requirements of the regulations valid in the Russian Federation [5].

### *Toxicological Assessment of Transgenic Maize Line GA 21*

The chronic (180-day) experiment was carried out on male Wistar rats with an initial body weight of 65–80 g. In the control group, the rats were fed a daily diet including 3 g of grain derived from conventional maize. The test rats were fed a diet with an equivalent amount of grain from transgenic maize line GA 21 (Table 5.114).

# CHAPTER 5: Human and Animal Health Safety Assessment

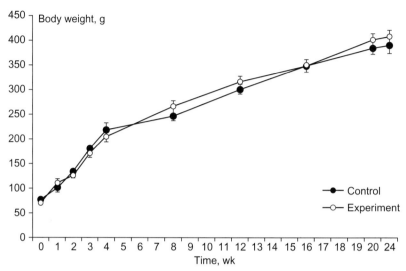

**FIGURE 5.6**
Comparative dynamics of body weight of rats fed diet containing transgenic maize line GA 21 (test) or its conventional counterpart (control).

**Table 5.115** Dynamics of Body Weight (g) of Rats Fed Diet with Conventional Maize (Control) or Transgenic Maize Line GA 21 (Test) ($M \pm m$; $n = 6$–$8$)

| Duration of Experiment, Weeks | Control | Test |
|---|---|---|
| 0 | 76.0 ± 2.2 | 70.6 ± 1.7 |
| 1 | 101.8 ± 10.2 | 110.3 ± 9.4 |
| 2 | 133.3 ± 6.2 | 126.0 ± 4.8 |
| 3 | 178.9 ± 7.2 | 170.0 ± 7.8 |
| 4 | 218.3 ± 14.0 | 204.2 ± 9.6 |
| 8 | 247.6 ± 9.9 | 266.3 ± 11.2 |
| 12 | 301.8 ± 9.3 | 315.2 ± 11.7 |
| 16 | 348.6 ± 14.3 | 348.8 ± 11.4 |
| 20 | 385.1 ± 13.5 | 400.3 ± 14.1 |
| 24 | 390.8 ± 16.5 | 407.5 ± 12.9 |

*Note: The differences are not significant ($p > 0.05$).*

During the entire duration of the experiment, there was no mortality in either group of animals. General condition of the rats was satisfactory. The appearance of the rats in control and test groups was similar. Figure 5.6 and Table 5.115 show the comparative dynamics of body weight of control and test rats.

Table 5.116 Absolute and Relative Weight of Internal Organs of Rats Fed Diet with Conventional Maize (Control) or Transgenic Maize Line GA 21 (Test) ($M \pm m$, $n = 6–8$)

| Organ | | 30 Days | | 180 Days | |
|---|---|---|---|---|---|
| | | Control | Test | Control | Test |
| Kidneys | Abs.[a], g | 1.61 ± 0.07 | 1.480 ± 0.08 | 2.30 ± 0.20 | 2.70 ± 0.10 |
| | Rel.[b], g/100g | 0.74 ± 0.03 | 0.73 ± 0.03 | 0.59 ± 0.05 | 0.68 ± 0.03 |
| Liver | Abs., g | 12.00 ± 0.48 | 9.70 ± 0.51* | 12.70 ± 0.60 | 13.30 ± 0.75 |
| | Rel., g/100g | 5.60 ± 0.34 | 4.80 ± 0.32 | 3.30 ± 0.10 | 3.30 ± 0.04 |
| Spleen | Abs., g | 1.30 ± 0.07 | 0.95 ± 0.11* | 1.63 ± 0.13 | 1.60 ± 0.12 |
| | Rel., g/100g | 0.41 ± 0.10 | 0.46 ± 0.08 | 0.42 ± 0.03 | 0.41 ± 0.10 |
| Heart | Abs., g | 0.95 ± 0.04 | 0.86 ± 0.03 | 1.20 ± 0.09 | 1.20 ± 0.05 |
| | Rel., g/100g | 0.44 ± 0.01 | 0.43 ± 0.01 | 0.30 ± 0.03 | 0.29 ± 0.01 |
| Testicles | Abs., g | 2.50 ± 0.25 | 2.40 ± 0.22 | 3.20 ± 0.16 | 3.30 ± 0.03 |
| | Rel., g/100g | 1.19 ± 0.23 | 1.33 ± 0.28 | 0.83 ± 0.06 | 0.83 ± 0.05 |
| Hypophysis | Abs., mg | 7.60 ± 0.47 | 5.80 ± 0.72 | 8.40 ± 0.20 | 10.6 ± 1.5 |
| | Rel., mg/100g | 3.50 ± 0.19 | 2.80 ± 0.28 | 2.10 ± 0.25 | 2.70 ± 0.47 |
| Adrenal glands | Abs., mg | 23.10 ± 2.8 | 25.10 ± 2.2 | 27.6 ± 3.4 | 24.8 ± 3.1 |
| | Rel., mg/100g | 10.80 ± 1.90 | 12.50 ± 1.50 | 6.90 ± 1.00 | 6.30 ± 0.99 |
| Seminal vesicles | Abs., mg | 326.1 ± 17.4 | 328.5 ± 42.6 | 474.5 ± 35.6 | 641.2 ± 67.6* |
| | Rel., mg/100g | 164.8 ± 12.4 | 163.6 ± 24.1 | 121.5 ± 8.6 | 162.3 ± 14.5* |
| Prostate | Abs., mg | 170.60 ± 9.4 | 172.6 ± 24.0 | 218.3 ± 21.6 | 340.0 ± 33.4* |
| | Rel., mg/100g | 67.3 ± 8.9 | 56.4 ± 7.6 | 55.8 ± 5.4 | 85.9 ± 5.5* |

[a]Absolute weight of internal organs.
[b]Relative weight of internal organs (per 100 g body weight).
Note: Here and in Tables 5.117 to 5.123 *$p < 0.05$ in comparison with control.

The body weight of rats fed grain from transgenic maize line GA 21 did not significantly differ from that of the control rats fed diet with an equivalent amount of grain from conventional maize.

The absolute weights of internal organs of test rats did not significantly differ from the corresponding values for the control rats (Table 5.116). Some decrease in absolute weights of liver and spleen observed in the test rats on Day 30 remained within the physiological variations characteristic of the rats of the corresponding age, i.e., 8.00–15.00 g (liver) and 0.75–1.50 g (spleen). The relative weights of internal organs of test rats did not significantly differ from the corresponding values for the control rats fed conventional maize.

The increments in absolute and relative weight of seminal vesicles and prostate observed in test rats on Day 180 also remained within the physiological variations characteristic of rats of corresponding age: i.e., 300–700 mg and 60–160 g/100 g (seminal vesicles); 200–500 mg and 40–120 g/100 g (prostate).

**Table 5.117** Biochemical Parameters of Blood Serum of Rats Fed Diet with Conventional Maize (Control) or Transgenic Maize Line GA 21 (Test) ($M \pm m$, $n = 6$–$8$)

| Parameter | 30 Days | | 180 Days | |
|---|---|---|---|---|
| | Control | Test | Control | Test |
| Total protein, g/L | 60.4 ± 3.9 | 62.4 ± 2.1 | 61.1 ± 6.0 | 57.6 ± 1.9 |
| Glucose, mM/L | 8.4 ± 0.6 | 8.6 ± 0.5 | 10.0 ± 0.49 | 10.4 ± 0.42 |
| Alanine aminotransferase, μcat/L | 0.74 ± 0.05 | 0.66 ± 0.03 | 0.78 ± 0.04 | 0.65 ± 0.03* |
| Aspartate aminotransferase, μcat/L | 0.31 ± 0.03 | 0.38 ± 0.06 | 0.46 ± 0.05 | 0.48 ± 0.06 |
| Alkaline phosphatase, μcat/L | 7.4 ± 1.9 | 10.4 ± 1.8 | 5.66 ± 0.64 | 6.79 ± 0.71 |

**Table 5.118** Urinary Biochemical Parameters of Control and Test Groups ($M \pm m$, $n = 6$–$8$)

| Parameter | 30 Days | | 180 Days | |
|---|---|---|---|---|
| | Control | Test | Control | Test |
| pH | 7.3 | 7.5 | 7.0 | 7.0 |
| Daily diuresis, mL | 6.2 ± 0.7 | 6.7 ± 1.3 | 9.1 ± 1.2 | 12.6 ± 1.5 |
| Relative density, g/mL | 1.01 ± 0.06 | 1.02 ± 0.01 | 1.07 ± 0.01 | 1.05 ± 0.01 |
| Creatinine, mg/mL | 0.72 ± 0.11 | 0.80 ± 0.16 | 1.60 ± 0.23 | 0.95 ± 0.14* |
| Creatinine, mg/day | 4.44 ± 0.75 | 4.90 ± 0.81 | 13.48 ± 0.83 | 11.15 ± 0.73 |

### Assessment of Biochemical Parameters

During the entire experiment, the content of total protein and glucose, activity of alanine aminotransferase, aspartate aminotransferase, and alkaline phosphatase in the blood serum of rats fed diet with transgenic maize line GA 21 (test group) did not significantly differ from the corresponding parameters of the control rats fed diet with conventional maize. The decrease in activity of aspartate aminotransferase observed in blood serum of test rats remained within the physiological variations characteristic of rats of the corresponding age, ranging within 0.5–0.9 μcat/L when measured by this method (Table 5.117).

Daily diuresis, pH, relative density of urine, and daily urinary excretion of creatinine did not significantly differ in rats fed diet with transgenic maize line GA 21 from the corresponding parameters in control rats fed diet with conventional maize. The decrease in urinary creatinine observed in test rats on Day 180 in comparison with control rats remained within the physiological variations characteristic of rats of the corresponding age, ranging 0.7–2.0 mg/mL (Table 5.118).

On experimental Day 30, lipid peroxidation (LPO) intensity in blood of test group rats did not significantly differ from the corresponding parameters in

**Table 5.119** Content of LPO Products in Blood and Liver of Rats Fed Diet with Conventional Maize (Control) or Transgenic Maize Line GA 21 (Test) ($M \pm m$, $n = 6-8$)

| Parameter | 30 Days | | 180 Days | |
|---|---|---|---|---|
| | Control | Test | Control | Test |
| *Erythrocytes* | | | | |
| DC, nM/mL | 6.118 ± 0.8 | 4.542 ± 0.2 | 4.365 ± 0.2 | 3.049 ± 0.2* |
| MDA, nM/mL | 5.272 ± 0.5 | 4.446 ± 0.1 | 6.256 ± 0.4 | 3.990 ± 0.3* |
| *Blood serum* | | | | |
| DC, nM/mL | 5.204 ± 0.4 | 5.208 ± 0.8 | 4.541 ± 0.4 | 4.423 ± 0.5 |
| MDA, nM/mL | 4.957 ± 1.5 | 4.602 ± 0.4 | 4.397 ± 0.5 | 5.721 ± 0.5 |
| *Liver* | | | | |
| DC, Unit | 1.070 ± 0.05 | 1.158 ± 0.04* | 0.958 ± 0.02 | 1.051 ± 0.02* |
| MDA, nM/g | 331.0 ± 9.9 | 361.8 ± 16.3 | 356.9 ± 12.2 | 411.8 ± 11.2* |

**Table 5.120** Activity of Enzymes of Antioxidant Protection System in Rats Fed Diet with Maize (Control) or Transgenic Maize Line GA 21 (Test) ($M \pm m$, $n = 6-8$)

| Parameter | 30 Days | | 180 Days | |
|---|---|---|---|---|
| | Control | Test | Control | Test |
| Glutathione reductase, μmol/min/g Hb | 28.63 ± 1.6 | 30.96 ± 1.4 | 28.36 ± 1.1 | 31.35 ± 1.1 |
| Glutathione peroxidase, μmol/min/g Hb | 57.37 ± 2.1 | 44.35 ± 1.8* | 55.07 ± 2.4 | 48.96 ± 2.6 |
| Catalase, mmol/min/g Hb | 353.5 ± 20.9 | 360.3 ± 10.9 | 452.0 ± 10.7 | 395.3 ± 15.8* |
| Superoxide dismutase (SOD), U/min/g Hb | 1949.2 ± 43.8 | 1951.4 ± 44.4 | 2094.0 ± 63.3 | 2135.3 ± 69.0 |

control group rats (Table 5.119). There was only a small (8%) increase in the content of DC in rat liver in the test group in comparison with the control ($p < 0.05$).

On experimental Day 180, the content of DC and MDA in the erythrocytes of test rats was lower than in control by 30% and 36% ($p < 0.05$), respectively. At the same time, there was significant elevation in hepatic concentrations of DC and MDA in test rats in comparison with control by 10% and 15% ($p < 0.05$), respectively.

On experimental Day 30, activity of glutathione peroxidase in test rats was lower by 23% than in the control. On 180 day, activity of catalase in control rats was higher than that in the test rats by 13% (Table 5.120).

Thus, analysis of parameters describing activity of LPO processes and that of antioxidant enzymatic protection system showed that there was no pro-oxidant load in rats fed diet with grain from transgenic maize line GA 21. The

Table 5.121 Activity of Enzymes Involved in Metabolism of Xenobiotics and Protein Content in Hepatic Microsomes of Rats Fed Diet with Maize (Control) or Transgenic Maize Line GA 21 (Test) ($M \pm m$, $n = 6–8$)

| Parameter | 30 Days | | 180 Days | |
|---|---|---|---|---|
| | Control | Test | Control | Test |
| Cytochrome P450, nM/mg protein | 0.75 ± 0.04 | 0.84 ± 0.04 | 0.79 ± 0.03 | 0.80 ± 0.03 |
| Cytochrome $b_5$, nM/mg protein | 0.61 ± 0.03 | 0.65 ± 0.03 | 0.78 ± 0.03 | 0.81 ± 0.04 |
| Aminopyrine N-demethylation, nM/min/mg protein | 10.2 ± 0.6 | 9.8 ± 0.5 | 7.62 ± 0.50 | 7.37 ± 0.53 |
| Benzpyrene hydroxylation, Fl-units/min/mg protein | 8.48 ± 0.54 | 8.32 ± 0.14 | 8.37 ± 0.61 | 8.63 ± 0.41 |
| Acetylesterase, μM/min/mg protein | 8.26 ± 0.35 | 8.55 ± 0.32 | 7.42 ± 0.32 | 7.85 ± 0.40 |
| Epoxide hydrolase, nM/min/mg protein | 8.83 ± 0.94 | 10.04 ± 0.97 | 7.98 ± 0.73 | 7.34 ± 0.38 |
| UDP-glucuronosil transferase, nM/min/mg protein | 32.4 ± 1.6 | 30.6 ± 4.1 | 26.9 ± 1.2 | 26.9 ± 0.8 |
| CDNB-glutathione transferase, μM/min/mg protein | 48.3 ± 1.8 | 50.5 ± 1.4 | 61.7 ± 2.1 | 63.3 ± 3.0 |
| Protein, mg/g | 12.3 ± 0.9 | 14.6 ± 0.6 | 12.3 ± 0.6 | 13.2 ± 0.4 |

Note: Here and in Tables 5.124 to 5.126 the differences are not significant ($p > 0.05$).

antioxidant status of the rats in both groups was in dynamic equilibrium, as indicated by the content of LPO products in erythrocytes and the liver. On experimental Day 180, the content of DC and MDA in the erythrocytes of test rats was lower than that in the control group by 30% and 36% ($p < 0.05$), respectively. At the same time, there was significant elevation in hepatic concentrations of DC and MDA in test rats in comparison with control by 10% and 15%, respectively. In this case, the opposite changes in the intensity of LPO processes in erythrocytes and liver attest to compensation of the integral antioxidant status, otherwise the alterations would be unidirectional and persistent during the entire length of the experiment. Some variations of certain parameters could reflect the individual peculiarities in the intensity of LPO and *de novo* enzyme synthesis. On experimental Days 30 and 180, there were no differences in activity of the enzymes involved in degradation of xenobiotics (Table 5.121). Similarly, the total and non-sedimentable activity of hepatic lysosomal enzymes did not significantly differ between the groups over the entire duration of the experiment (Table 5.122).

Thus, addition of transgenic maize line GA 21 to the rat diet produced no effect on the activity of lysosomal enzymes or the enzymes involved in xenobiotic degradation.

### Hematological Assessments

The hematological parameters of the test rats fed diet with transgenic maize line GA 21 did not significantly differ from those in control rats fed conventional maize. The leukogram parameters of the control and test rats did not

**YBP Library Services**

GENETICALLY MODIFIED FOOD SOURCES: SAFETY
ASSESSMENT AND CONTROL; ED. BY V.A. TUTELYAN.
                                                 Cloth    338 P.
SAN DIEGO: ELSEVIER ACADEMIC PRESS, 2013

ED: RUSSIAN ACADEMY OF MEDICAL SCIENCES.
COLLECTION OF NEW PAPERS. TRANSLATED FROM RUSSIAN.

   ISBN 0124058787     Library PO#  GENERAL APPROVAL
                                                     List     124.95  USD
   5461 UNIV OF TEXAS/SAN ANTONIO     Disc     17.0%
   App. Date  9/18/13  BIO.APR     6108-11  Net     103.71  USD

SUBJ: 1. GENETICALLY MODIFIED FOODS.
2. FOOD--BIOTECHNOLOGY.

CLASS TP248.65       DEWEY# 664.          LEVEL ADV-AC

---

**YBP Library Services**

GENETICALLY MODIFIED FOOD SOURCES: SAFETY
ASSESSMENT AND CONTROL; ED. BY V.A. TUTELYAN.
                                                 Cloth    338 P.
SAN DIEGO: ELSEVIER ACADEMIC PRESS, 2013

ED: RUSSIAN ACADEMY OF MEDICAL SCIENCES.
COLLECTION OF NEW PAPERS. TRANSLATED FROM RUSSIAN.

   ISBN 0124058787     Library PO#  GENERAL APPROVAL
                                                     List     124.95  USD
   5461 UNIV OF TEXAS/SAN ANTONIO     Disc     17.0%
   App. Date  9/18/13  BIO.APR     6108-11  Net     103.71  USD

SUBJ: 1. GENETICALLY MODIFIED FOODS.
2. FOOD--BIOTECHNOLOGY.

CLASS TP248.65       DEWEY# 664.          LEVEL ADV-AC

**Table 5.122** Total and Non-sedimentable Activity of Hepatic Lysosomal Enzymes of Rats Fed Diet with Conventional Maize (Control) or Transgenic Maize Line GA 21 (Test) ($M \pm m$, $n = 6$–$8$)

| Parameters | 30 Days | | 180 Days | |
|---|---|---|---|---|
| | Control | Test | Control | Test |
| *Total activity, µM/min/g tissue* | | | | |
| Arylsulfatase A, B | 2.47 ± 0.05 | 2.44 ± 0.06 | 2.22 ± 0.04 | 2.17 ± 0.07 |
| β-Galactosidase | 2.60 ± 0.08 | 2.39 ± 0.07 | 2.39 ± 0.08 | 2.48 ± 0.08 |
| β-Glucuronidase | 2.17 ± 0.06 | 2.12 ± 0.03 | 2.16 ± 0.04 | 2.15 ± 0.04 |
| *Non-sedimentable activity, % total activity* | | | | |
| Arylsulfatase A, B | 3.51 ± 0.08 | 3.58 ± 0.07 | 3.35 ± 0.14 | 3.10 ± 0.07 |
| β-Galactosidase | 5.65 ± 0.15 | 5.27 ± 0.25 | 5.00 ± 0.25 | 4.74 ± 0.15 |
| β-Glucuronidase | 4.68 ± 0.25 | 4.81 ± 0.17 | 4.50 ± 0.24 | 4.28 ± 0.33 |

**Table 5.123** Hematological Parameters of Peripheral Blood Drawn from Rats Fed Diet with Conventional Maize (Control) or Transgenic Maize Line GA 21 (Test) ($M \pm m$, $n = 6$–$8$)

| Parameter | 30 Days | | 180 Days | |
|---|---|---|---|---|
| | Control | Test | Control | Test |
| Hemoglobin concentration, g/L | 160.1 ± 2.75 | 161.9 ± 3.9 | 163.6 ± 3.83 | 163.3 ± 7.2 |
| Total erythrocyte count, ×$10^{12}$/L | 6.20 ± 0.18 | 6.20 ± 0.17 | 5.82 ± 0.15 | 6.10 ± 0.32 |
| Hematocrit, vol.% | 50.00 ± 0.57 | 49.60 ± 0.57 | 49.60 ± 0.57 | 50.20 ± 0.57 |
| MCH, pg | 25.82 ± 0.88 | 26.09 ± 0.53 | 28.23 ± 1.29 | 26.79 ± 0.50 |
| MCHC, % | 32.00 ± 0.57 | 32.64 ± 0.28 | 33.01 ± 1.01 | 32.48 ± 1.08 |
| MCV, µm$^3$ | 80.56 ± 1.44 | 80.0 ± 1.28 | 85.4 ± 1.86 | 82.98 ± 3.45 |
| Total leukocyte count, ×$10^9$/L | 14.37 ± 1.49 | 12.72 ± 0.77 | 14.00 ± 1.25 | 14.15 ± 0.96 |

significantly differ during the entire duration of the experiment (Tables 5.123 and 5.124).

## Morphological Assessments

Macroscopic examination of internal organs of the control and test rats revealed no pathological alterations in 30 and 180 days from the onset of the experiment. The data of macro- and microscopic morphological examinations of the internal organs revealed no differences between the control and test groups during the entire period of the experiment (Table 5.125).

Thus, the biochemical, hematological, and morphological studies performed during the chronic 180-day toxicological experiment on animals revealed no adverse effects of transgenic maize.

**Table 5.124** Leukogram Parameters of Rats Fed Diet with Conventional Maize (Control) or Transgenic Maize Line GA 21 (Test) ($M \pm m$, $n = 6$–$8$)

| Parameter | 30 Days | | 180 Days | |
|---|---|---|---|---|
| | Control | Test | Control | Test |
| **Neutrophils** | | | | |
| rel., % | 11.6 ± 1.5 | 10.6 ± 1.2 | 14.2 ± 1.3 | 15.2 ± 2.7 |
| abs., ×10$^9$/L | 1.72 ± 0.34 | 1.43 ± 0.20 | 2.01 ± 0.35 | 2.17 ± 1.06 |
| **Eosinophils** | | | | |
| rel., % | 1.20 ± 0.38 | 1.20 ± 0.38 | 1.40 ± 0.38 | 1.40 ± 0.38 |
| abs., ×10$^9$/L | 0.15 ± 0.06 | 0.15 ± 0.05 | 0.20 ± 0.07 | 0.20 ± 0.07 |
| **Lymphocytes** | | | | |
| rel., % | 87.2 ± 1.1 | 88.2 ± 1.2 | 84.4 ± 1.3 | 83.8 ± 2.7 |
| abs., ×10$^9$/L | 12.49 ± 1.17 | 11.20 ± 0.68 | 11.79 ± 1.08 | 11.80 ± 0.72 |

**Table 5.125** Microscopic Examination of Internal Organs in Rats Fed Diet with Conventional Maize (Control) or Transgenic Maize Line GA 21 (Test) (Combined Data Obtained on Experimental Days 30 and 180)

| Organ | Control | Test |
|---|---|---|
| Liver | Clear trabecular structure; no alterations in hepatocytes and in the portal ducts | No differences from control |
| Kidneys | Usual appearance of cortical and medullar substance; no alterations in glomeruli or pelvis epithelium | No differences from control |
| Lung | Alveolar space is air-filled; no alterations in bronchi and blood vessels | No differences from control |
| Spleen | Large folliculi with wide clear marginal zones and reactive centers; splenic pulp is moderately plethoric | No differences from control |
| Small intestine | Preserved villous epithelium; usual infiltration in villus stroma | No differences from control |
| Testicle | Clearly definable spermatogenesis and age-related spermiogenesis | No differences from control |

### Assessment of Potential Impact of Transgenic Maize GA 21 on Immune System in Studies on Mice
#### Potential Effect on Humoral Component of Immune System

The immunomodulating effect of transgenic maize line GA 21 on the humoral component of the immune system was examined on two oppositely reacting mice lines, C57Bl/6 and CBA, by determining the level of hemagglutination to sheep erythrocytes (SE). The experimental conditions and the diet are described in section 5.1.1 and Table 5.114, correspondingly.

In test and control groups of CBA mice (highly sensitive animals), the antibody appeared at the titer of 1:2–1:8 in a week after immunization. In test and control groups of C57Bl/6 mice (animals with low sensitivity), the antibody appeared on post-immunization Day 7 at the titers of 1:30 and 1:10, correspondingly. In the following days, the antibody titer decreased in all groups to the level of 1:4–1:10. Thus, the antibodies against SE appeared on post-immunization Day 7 in all examined groups.

**Potential Effect on Cellular Component of Immune System**
The immunomodulating effect of transgenic maize line GA 21 on the cellular component of the immune system was assessed by delayed hypersensitivity reaction to SE. The experimental conditions are described in section 5.1.1. These studies revealed no elevation in RI (reaction index) in CBA mice, which was $19 \pm 2$ and $18 \pm 2$ in test and control mice, respectively. Similarly, RI did not change in C57Bl/6 mice: $23 \pm 3$ in test and $28 \pm 3$ in the control group. These data support the conclusion that transgenic maize line GA 21 produced no effect on the cellular component of the immune system.

**Assessment of Sensibilization Effect of Transgenic Maize**
Examination of possible sensitizing action of transgenic maize on the immune response to endogenous metabolic products was carried out in the test of mouse sensitivity to histamine. For 21 days, the control and test mice were fed diets with conventional or transgenic maize, correspondingly. After 21 days the mice of both groups were injected intraperitoneally with 2.5 mg histamine hydrochloride dissolved in 0.5 mL physiological solution. The reaction was assessed by mortality (in percentage) counted at 1 h and 24 h after injection. There was no death or differences in behavior in test and control mice, which attests to the absence of a sensitizing agent in transgenic maize.

**Potential Effect of Transgenic Maize on Susceptibility of Mice to *Salmonella typhimurium***
The effect of transgenic maize on susceptibility of mice to infection by salmonella of murine typhus was examined in experiments on mice injected intraperitoneally with various doses of Strain 415 *Salmonella typhimurium*. Four weeks prior to infection, the diet of control and test mice was supplemented with conventional or transgenic maize line GA 21, respectively. The injected doses ranged from $10^2$ to $10^5$ microbial cells per mouse and varied on a 10-fold basis. The post-injection observation period was 21 days. The following data were obtained:

- In both groups, the mortality of mice started on post-injection week 1, and all mice died by post-injection Day 18;
- The mean lifetime of mice was approximately equal (14.2 ± 8.0 d);
- $LD_{50}$ values in test and control groups were 301 and 84 microbial cells, respectively.

These data showed that *Salmonella typhimurium* produced a typical infection both in control and test mice, and the course of disease was similar in both groups. Thus, transgenic maize line GA 21 did not modify resistance of mice against salmonella of murine typhus.

Therefore, transgenic maize line GA 21 had no sensitizing potency and produced no stimulating effect on humoral and cellular components of the immune system in the oppositely reacting mice lines.

### Assessment of Potential Impact of Transgenic Maize Line GA 21 on Immune System in Studies on Rats

The study was carried out on male Wistar rats ($n = 52$) weighing initially 160 ± 10 g. After a 7-day adaptation period to the standard vivarium diet, during the next 28 days the rats were fed diet supplemented with conventional maize (control group) or transgenic maize line GA 21 (test group). Processed maize grain samples were dissolved in boiled water to the consistency of dense curd and supplemented with sunflower-seed oil to improve intake. The feed was used instead of the equally caloric amount of oatmeal (see Table 5.114).

The model of generalized anaphylaxis was developed according to [8] as described in section 5.1.1.

During the entire experiment, the rats of both groups grew normally, which indicates nutritional adequacy of the diets used. On experimental Day 29, the body weights of test and control rats were 248 ± 6 and 250 ± 5 g, respectively ($p > 0.1$). In both groups, severity of active anaphylactic shock was assessed (Table 5.126). There were no significant differences ($p > 0.1$) between groups of rats fed transgenic or conventional maize.

**Table 5.126** Severity of Active Anaphylactic Shock in Rats Fed Diet with Conventional Maize (Control) or Transgenic Maize Line GA 21 (Test)

| Group of Rats | AI | Severe Reactions, % | Mortality, % |
|---|---|---|---|
| Control ($n = 26$) | 3.27 | 77 | 73 |
| Test ($n = 26$) | 3.23 | 77 | 77 |

**Table 5.127** Values of Humoral Immune Response (Level of Specific IgG Antibodies Raised Against Ovalbumin) in Rats Fed Diet with Conventional Maize (Control) or Transgenic Maize Line GA 21 (Test)

| Group of Rats | $D_{492}$ | Concentration of Antibodies, mg/mL | Logarithm of Antibody Concentration |
|---|---|---|---|
| Control ($n = 26$) | $0.802 \pm 0.032$ | $8.8 \pm 1.5$ | $0.783 \pm 0.074$ |
| Test ($n = 25$) | $0.803 \pm 0.032$ | $8.7 \pm 1.6$ | $0.787 \pm 0.067$ |

Table 5.127 shows the mean values of $D_{492}$, the concentration of antibodies and the common logarithm of this concentration in the control and test groups. According to the examined parameters, the difference between test rats fed diet with transgenic maize line GA 21 and the control rats maintained on conventional maize was insignificant ($p > 0.1$). Thus, the intensity of humoral immune response was practically identical in both groups of animals.

On the whole, these data conclude that the degree of sensitization by ovalbumin in test group rats did not increase in comparison with the control group rats.

Feeding rats with transgenic maize did not enhance sensitization and allergic reactivity in comparison with the rats fed diet with conventional maize.

### Assessment of Potential Genotoxicity of Transgenic Maize GA 21

Genotoxic studies were carried out on male C57Bl/6 mice and hybrid female CBA mice fed diet shown in Table 5.114. During the experiment, the animals were fed diet composed of a soft feed with milled maize of test or control variety.

The cytogenetic analysis was carried out by the metaphasic method. The details of experiments are described in section 5.1.1. Genetic alterations in the sex cells were examined by revealing the dominant lethal mutations in C57Bl/6 male mice [8]. The details of experiments are described in section 5.1.1.

In the test mice, the number of cells with structural alterations was insignificant. Both control and test mice exhibited cells with chromosomal abnormalities (single segments and gaps), known to be transient disturbances that are eliminated in the subsequent nuclear divisions. The number of cells with polyploid chromosome set did not differ between control and test groups (Table 5.128).

**Table 5.128** Cytogenetic Parameters of Bone Marrow of Mice Fed a 45-Day Diet with Conventional Maize (Control) or Transgenic Maize Line GA 21 (Test) ($M \pm m$)

| Number of Cells, % | Control | Test |
|---|---|---|
| with chromosomal aberrations | 0.57 ± 0.40 | 0.55 ± 0.38 |
| with gaps | 1.14 ± 0.56 | 0.82 ± 0.47 |
| with polyploid chromosome set | 1.42 ± 0.63 | 1.09 ± 0.54 |

Note: Here and in Tables 5.129 and 5.130 the differences are not significant ($p > 0.05$). The numbers of analyzed metaphases were: 364 (test) and 350 (control).

**Table 5.129** Dominant Lethal Mutations in Sex Cells of Mice Fed Diet with Conventional Maize (Control) or Transgenic Maize Line GA 21 (Test) ($M \pm m$)

| Parameter, % | Week 1, Mature Spermatozoa | | Week 2, Late Spermatids | | Week 3, Early Spermatids | |
|---|---|---|---|---|---|---|
| | Control | Test | Control | Test | Control | Test |
| Pre-implantation mortality | 8.30 | 8.66 | 7.0 | 7.92 | 7.90 | 6.93 |
| Post-implantation mortality | 3.30 | 2.91 | 4.79 | 3.97 | 6.93 | 4.08 |
| Survival rate, % | 88.70 | 88.66 | 88.53 | 88.40 | 87.70 | 89.20 |
| Induced mortality, % | 0 | 0 | 0 | 0 | 0 | 0 |

Note: The number of females in the test group was 90 (435 embryos and 472 corpus lutea were analyzed). In the control group (78 females), 450 embryos and 488 corpus lutea were analyzed.

At the stages of early and late spermatids or mature spermatozoa, the post-implantation embryonic mortality in the test group did not surpass the control level during the entire period of the experiment. There was no induced mortality, which attests to the absence of mutagenic effect of long-term feeding of mice with transgenic maize on spermiogenesis (Table 5.129).

Taking into consideration enhanced sensitivity of the developing (young) organism to adverse (mutagenic included) factors, the possible mutagenic effect of transgenic maize line GA 21 was examined by feeding it to gravid mice (prenatal development of the fetus) and then to newborn females during lactation period and 30 days after its termination. Then the bone marrow of the first generation was isolated from both femoral bones for cytogenetic examination (the details are given in section 5.1.1). The results of examination of the control and test mice are presented in Table 5.130.

In the test mice, the number of cells with chromosomal aberrations was negligible and did not significantly differ from that of the control. These chromosomal aberrations (single segments and gaps) could appear spontaneously; they

**Table 5.130** Cytogenetic Parameters of Bone Marrow of C57Bl/6 Line Mice of First Generation Fed Diet with Conventional Maize (Control) or Transgenic Maize Line GA 21 (Test) ($M \pm m$)

| Parameter | Control ($n = 5$) | Test ($n = 5$) |
|---|---|---|
| Number of analyzed metaphases | 348 | 323 |
| Share of cells, % | | |
| with chromosomal aberrations | $0.86 \pm 0.49$ | $0.62 \pm 0.43$ |
| with gaps | $0.57 \pm 0.40$ | $0.62 \pm 0.43$ |
| with polyploid chromosome set | $1.14 \pm 0.56$ | $0.92 \pm 0.53$ |

were unstable and disappeared in the subsequent nuclear divisions. The data support the conclusion that feeding transgenic maize to gravid and lactating mice and then to pups during 30 days produced no mutagenic effect on the developing organism of the young animals.

Therefore, the above study showed that glyphosate-tolerant transgenic maize line GA 21 exhibited no mutagenic effects under the experimental conditions described.

### Assessment of Technological Parameters

The study of technological parameters was carried out in Moscow State University of Applied Biotechnology (Ministry of Science and Education of Russian Federation).

To compare the samples of glyphosate-tolerant transgenic maize line GA 21 and conventional maize, the moisture and ash contents were determined. Maize starch was produced under laboratory conditions to determine protein mass fraction in the starch, gelatinization temperature and viscosity of the starch gelatins on amylograph, the parameters of thermoplastic extrusion, and the structural and mechanical properties of the extrudates.

The transgenic maize grain complied with the specifications of Russian State Standards GOST 136-90 "Maize. Technical requirements".

The study resulted in the following conclusions:

- No differences or difficulties were observed in the technological process when producing starch from the transgenic maize line GA 21 or conventional maize;
- According to Russian State Standards GOST 7698-93 "Starch. Formal Acceptance and Analytical Methods", the quality of all examined samples was superior;
- By gelatinization temperature and rheological properties of 7% gelatins, the starches of transgenic maize did not differ from those derived from conventional maize;

- The structural and mechanical properties of the extrusive products derived from transgenic maize line GA 21 were virtually identical to those of the conventional maize.

## Conclusions

By all examined parameters, the data of the complex safety assessment of glyphosate-tolerant transgenic maize line GA 21 attest to the absence of any toxic, genotoxic, immune system modulating, or allergenic effects of this maize variety. Analysis of the biochemical composition of transgenic maize line GA 21 established its identity to the composition of conventional maize.

Based on the results of the studies, the State Sanitation Service of the Russian Federation (Department of State Sanitation and Epidemiological Inspectorate) granted the Registration Certificate which allows the transgenic maize line GA 21 to be used in the food industry and placed on the market without restrictions.

### 5.2.2 TRANSGENIC MAIZE LINE MON 810 RESISTANT TO EUROPEAN CORN BORER

#### Molecular Characteristics of Transgenic Maize Line MON 810

*Recipient Organism*

Maize (*Zea mays* L.) is characterized by a long history of safe use as human food and animal feed.

*Donor Organism*

The donor of the *cry 1Ab* gene imparting resistance to damage caused by larvae of the European cornborer *Ostrinia nubilalis* is a widespread gram-positive soil bacterium *Bacillus thuringiensis*, which produces proteins during sporogenesis that selectively affect particular groups of insects. The insecticidal proteins bind to the specific sites of the cells in the insect digestive tract and form ion-selective pores in plasmalemma resulting in lysis of the cells and death of the insects [25].

*Method of Genetic Transformation*

To insert the *cry 1Ab* gene, genetic modification of maize embryonic cells was performed by biolistic transformation using the PV-ZMBK07 plasmid. This plasmid incorporates the *cry 1Ab* sequence from *B. thuringiensis* subsp. HD-1, which was modified to increase the level of Cry1Ab protein responsible for resistance against European cornborer in transgenic maize line MON 810 [13,26]. The vector contains 35S promoter from cauliflower mosaic virus and NOS3′ terminator isolated from the Ti-plasmid of Agrobacterium. Gene *nptII*

### 5.2.2 Transgenic Maize Line MON 810 Resistant to European Corn Borer

**Table 5.131** Registration Status of Transgenic Maize Line MON 810 in Various Countries

| Country | Date of Approval | Scope |
|---|---|---|
| Australia | 2000 | Food |
| Argentina | 1998 | Food, feed, environmental release |
| EU | 1998 | Food, feed, environmental release |
| Canada | 1997 | Food, feed, environmental release |
| USA | 1995 | Environmental release |
|  | 1996 | Food, feed |
| Switzerland | 2000 | Food, feed |
| South Africa | 1997 | Food, feed, environmental release |
| Japan | 1996 | Environmental release |
|  | 1997 | Food, feed |

Note: The current registration status of the transgenic crop is on http://www.biotradestatus.com/

encoding neomycin phosphotransferase was employed as a selective bacterial marker (it is not present in the genome of maize line MON 810) [19].

## Global Registration Status of Maize Line MON 810
Table 5.131 shows the countries that had granted registration to use transgenic maize line MON 810 at the time of registration in Russia [19].

## Safety Assessment of Transgenic Maize Line MON 810 Conducted in the Russian Federation
The studies were conducted in accordance with the requirements of the Ministry of Health of the Russian Federation authorized for risk and safety assessment of food derived from GM sources [8]. PCR analysis of the maize test and control samples was performed to confirm the identity of the transformation event and its absence in the conventional control line.

### Biochemical Composition of Transgenic Maize Grain
The content of proteins and amino acid composition of grain of transgenic maize line MON 810 did not significantly differ from the corresponding values for conventional maize (Table 5.132).

Similarly, the content of carbohydrates in grain of transgenic maize line MON 810 did not significantly differ from the corresponding values for conventional maize (Table 5.133). The variations in the content of fructose remained within the range characteristic of maize: 0.01–0.35 g/100 g (data of the State Research Institute of Nutrition, RAMS). The content of lipids in the

**Table 5.132** Comparative Amino Acid Composition (g/100 g protein) and Protein Content in Maize Grain

| Ingredient | Conventional Maize | Transgenic Maize Line MON 810 |
|---|---|---|
| Protein, % | 9.06 | 8.42 |
| *Amino acids* | | |
| Lysine | 2.61 | 2.22 |
| Histidine | 2.31 | 1.94 |
| Arginine | 8.92 | 9.72 |
| Aspartic acid | 9.69 | 9.17 |
| Threonine | 3.07 | 3.05 |
| Serine | 4.46 | 4.16 |
| Glutamic acid | 21.50 | 20.55 |
| Proline | 11.23 | 12.22 |
| Glycine | 2.46 | 2.22 |
| Alanine | 5.69 | 5.28 |
| Cysteine | 2.61 | 3.33 |
| Valine | 2.92 | 2.78 |
| Methionine | 0.92 | 0.83 |
| Isoleucine | 2.00 | 1.94 |
| Leucine | 9.85 | 10.00 |
| Tyrosine | 4.77 | 5.28 |
| Phenylalanine | 4.92 | 5.28 |

**Table 5.133** Comparative Content of Carbohydrates (g/100 g product) in Maize Grain

| Carbohydrate | Conventional Maize | Transgenic Maize Line MON 810 |
|---|---|---|
| Fructose | 0.028 | 0.056 |
| Glucose | 0.027 | 0.032 |
| Sucrose | 1.330 | 1.600 |
| Starch | 54.300 | 55.600 |
| Cellulose | 1.960 | 1.730 |

grain of transgenic maize line MON 810 did not significantly differ from the corresponding value for conventional maize (Table 5.134).

The compositional similarity of transgenic and conventional maize was established in the content of vitamins as well (Table 5.135). Revealed variations in the content of vitamin $B_1$ did not surpass the physiological

### 5.2.2 Transgenic Maize Line MON 810 Resistant to European Corn Borer

**Table 5.134** Comparative Content of Lipids in Maize Grain

| Ingredient | Conventional Maize | Transgenic Maize Line MON 810 |
|---|---|---|
| Lipids, % | 4.7 | 4.9 |

**Table 5.135** Comparative Content of Vitamins (mg/100 g product) in Maize Grain

| Ingredient | Conventional Maize | Transgenic Maize Line MON 810 |
|---|---|---|
| Vitamin $B_1$ | 0.298 | 0.095 |
| Vitamin $B_2$ | 0.182 | 0.194 |
| Vitamin $B_6$ | 0.18 | 0.20 |
| Vitamin E ($\alpha$-tocopherol) | 8.0 | 6.0 |
| $\beta$-Carotene | 0.03 | 0.04 |
| Total carotenoids | 0.55 | 0.55 |

**Table 5.136** Analysis of Minerals (mg/kg) in Conventional and Transgenic Maize Grain

| Ingredient | Conventional Maize | Transgenic Maize Line MON 810 |
|---|---|---|
| Copper | 1.42 | 0.84 |
| Zinc | 28.1 | 21.3 |
| Iron | 37.8 | 26.8 |
| Sodium | 30.9 | 29.9 |
| Potassium | 3815 | 3752 |
| Calcium | 11.8 | 24.8 |
| Magnesium | 1322 | 1171 |

boundaries characteristic of maize according to the data of the State Research Institute of Nutrition, RAMS: 0.05–0.30 mg/100 g.

The content of minerals in conventional and transgenic maize MON 810 was also similar (Table 5.136). The changes in the content of calcium fell within the physiological range characteristic of maize (1.0–25.0 mg/kg) according to the data of the State Research Institute of Nutrition, RAMS.

The content of heavy metals and mycotoxins in the grains of conventional and transgenic maize line MON 810 did not surpass the acceptable limits according to the regulations valid in the Russian Federation (Tables 5.137 and 5.138) [5].

Thus, the above data showed that the grain of transgenic maize line MON 810 and conventional maize did not significantly differ by biochemical

**Table 5.137** Analysis of Mycotoxins in Maize Grain (mg/kg)

| Ingredient | Conventional Maize | Transgenic Maize Line MON 810 |
|---|---|---|
| Aflatoxin $B_1$ | Not detected | Not detected |
| Deoxynivalenol | Not detected | Not detected |
| Zearalenone | Not detected | Not detected |
| T2-toxin | Not detected | Not detected |
| Fumonisin $B_1$ | 0.1 | 0.1 |

**Table 5.138** Analysis of Heavy Metals in Maize Grain

| Ingredients | Conventional Maize | Transgenic Maize Line MON 810 |
|---|---|---|
| Lead, mg/kg | 0.048 | 0.014 |
| Cadmium, mg/kg | 0.008 | 0.006 |

composition. The revealed differences in the content of metals (calcium) and vitamins (vitamin $B_1$) remained within the range characteristic of maize [15,43,49–51]. The safety parameters of the grain of conventional and transgenic maize line MON 810 comply with the requirements of the regulations valid in the Russian Federation [5].

### Toxicological Assessment of Transgenic Maize Line MON 810

The chronic experiment (180 days) was carried out on male Wistar rats with an initial body weight of 65–80 g. In the control group, the rats were fed a daily diet including 3 g/day of grain derived from conventional maize. The test rats were fed a diet with an equivalent amount of grain from transgenic maize line MON 810 (composition of both diets is described in section 5.2.1).

#### Assessment of Proximate Parameters

During the entire duration of the experiment, no mortality was observed. General condition of the rats was satisfactory. The appearance of the rats in the control and test groups was similar. The difference in body weight was insignificant (Figure 5.7; Table 5.139).

Over the entire term of the experiment, the absolute weight of the internal organs of test rats fed diet with transgenic maize line MON 810 did not significantly differ from the corresponding values for the control rats fed conventional maize (Table 5.140). The changes in weight of liver (Day 30) and spleen (Day 180) remained within the physiological variations characteristic of rats of corresponding age (liver 8–15 g and 4–6 g/100 g; spleen 1.1–2.2 g and 0.3–0.5 g/100 g).

## 5.2.2 Transgenic Maize Line MON 810 Resistant to European Corn Borer

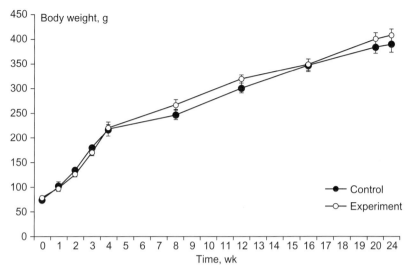

**FIGURE 5.7**
Comparative dynamics of body weight of rats fed diet with transgenic maize line MON 810 (test) or its conventional counterpart (control).

**Table 5.139** Body Weight (g) of Rats Fed Diet with Conventional Maize (Control) or Transgenic Maize Line MON 810 (Test) ($M \pm m$; $n = 25$)

| Duration of Experiment, Weeks | Control | Test |
| --- | --- | --- |
| 0 | 76.0 ± 2.2 | 78.3 ± 2.7 |
| 1 | 101.8 ± 10.2 | 96.8 ± 5.4 |
| 2 | 133.3 ± 6.2 | 126.0 ± 4.8 |
| 3 | 178.9 ± 7.2 | 170.9 ± 6.2 |
| 4 | 218.3 ± 14.0 | 220.3 ± 5.6 |
| 8 | 247.6 ± 9.9 | 267.2 ± 8.5 |
| 12 | 301.8 ± 9.3 | 320.5 ± 8.0 |
| 16 | 348.6 ± 14.3 | 348.8 ± 11.4 |
| 20 | 385.1 ± 13.5 | 400.0 ± 14.1 |
| 24 | 390.8 ± 16.5 | 407.5 ± 12.9 |

Prostate weight in test group rats was significantly higher than in control group rats: absolute by 43% (Day 30) and 73% (Day 180), relative by 12% and 66%, respectively. Analysis of data obtained in experiments performed by the State Research Institute of Nutrition, RAMS revealed the following variations of prostate weight in rats (data compiled from over 500 animals): on day 180 of the experiment the relative weight of the prostate

**Table 5.140** Absolute and Relative Weight of Internal Organs of Rats Fed Diet with Conventional Maize (Control) or Transgenic Maize Line MON 810 (Test) ($M \pm m$, $n = 6$–8)

| Organ | | 30 Days | | 180 Days | |
|---|---|---|---|---|---|
| | | Control | Test | Control | Test |
| Kidneys | Abs.[a], g | 1.61 ± 0.07 | 1.60 ± 0.08 | 2.3 ± 0.2 | 2.5 ± 0.2 |
| | Rel.[b], g/100 g | 0.74 ± 0.03 | 0.68 ± 0.03 | 0.59 ± 0.05 | 0.62 ± 0.02 |
| Liver | Abs., g | 12.0 ± 0.48 | 10.0 ± 0.40* | 12.7 ± 0.60 | 13.4 ± 0.96 |
| | Rel., g/100 g | 5.6 ± 0.34 | 4.3 ± 0.20* | 3.30 ± 0.10 | 3.30 ± 0.06 |
| Spleen | Abs., g | 1.30 ± 0.07 | 1.30 ± 0.20 | 1.63 ± 0.13 | 1.33 ± 0.08 |
| | Rel., g/100 g | 0.41 ± 0.1 | 0.55 ± 0.06 | 0.42 ± 0.03 | 0.33 ± 0.04* |
| Heart | Abs., g | 0.95 ± 0.04 | 0.93 ± 0.04 | 1.20 ± 0.09 | 1.20 ± 0.09 |
| | Rel., g/100 g | 0.44 ± 0.01 | 0.40 ± 0.02 | 0.30 ± 0.03 | 0.29 ± 0.02 |
| Testicles | Abs., g | 2.50 ± 0.25 | 2.70 ± 0.09 | 3.2 ± 0.16 | 3.1 ± 0.12 |
| | Rel., g/100 g | 1.19 ± 0.23 | 1.20 ± 0.04 | 0.83 ± 0.06 | 0.77 ± 0.05 |
| Hypophysis | Abs., mg | 7.6 ± 0.47 | 8.9 ± 1.80 | 8.4 ± 0.22 | 10.5 ± 1.3 |
| | Rel., mg/100 g | 3.56 ± 0.19 | 3.90 ± 0.80 | 2.1 ± 0.25 | 2.6 ± 0.28 |
| Adrenal glands | Abs., mg | 23.1 ± 2.8 | 32.0 ± 4.9 | 27.6 ± 3.4 | 23.2 ± 1.9 |
| | Rel., mg/100 g | 10.8 ± 1.9 | 13.8 ± 0.95 | 6.98 ± 1.0 | 5.70 ± 0.42 |
| Seminal vesicles | Abs., mg | 326.1 ± 17.4 | 379.3 ± 35.1 | 474.5 ± 35.6 | 513.5 ± 61.1 |
| | Rel., mg/100 g | 164.8 ± 12.4 | 160.8 ± 10.5 | 121.5 ± 8.6 | 123.5 ± 14.4 |
| Prostate | Abs., mg | 170.6 ± 9.4 | 244.5 ± 16.8* | 218.3 ± 21.6 | 378.6 ± 68.0* |
| | Rel., mg/100 g | 67.3 ± 8.9 | 75.7 ± 7.4 | 55.8 ± 5.4 | 92.5 ± 14.6* |

[a]Absolute weight of internal organs.
[b]Relative weight of internal organs (per 100 g body weight).
*$p < 0.05$ in comparison with control.

was 40–118 mg/100 g body weight. Survey microscopic morphological studies of rat prostate of control and experimental groups did not reveal differences between the groups. Unbiased, in the center of the tissue specimen the lumens of glands expanded in a greater or lesser degree or filled with eosinophilic contents; on the periphery, glands are smaller and mucosa folded.

There were no focal lesions. The structure was ordinary. Prostate parenchyma comprised numerous individual glands. The ducts, which collect the secretions of lobes, form tubulo-alveolar secretory units adapted both for formation and storage of the secretions, so they are able to expand greatly.

Thus, fluctuations in prostate weight within 200% may be considered typical of animals of the species and age.

The relative weight of internal organs of test rats fed diet with transgenic maize line MON 810 did not significantly differ from the corresponding values for the control rats fed conventional maize (Table 5.140). The changes in relative weight of liver (by 23.3%) observed on Day 30 and spleen

## 5.2.2 Transgenic Maize Line MON 810 Resistant to European Corn Borer

**Table 5.141** Biochemical Parameters of Blood Serum of Rats Fed Diet with Conventional Maize (Control) or Transgenic Maize Line MON 810 (Test) ($M \pm m$, $n = 6–8$)

| Parameter | 30 Days | | 180 Days | |
|---|---|---|---|---|
| | Control | Test | Control | Test |
| Total protein, g/L | $60.4 \pm 3.9$ | $63.2 \pm 4.2$ | $61.1 \pm 6.0$ | $64.0 \pm 2.4$ |
| Glucose, mM/L | $8.4 \pm 0.6$ | $8.1 \pm 0.4$ | $10.0 \pm 0.49$ | $10.9 \pm 0.99$ |
| Alanine aminotransferase, μcat/L | $0.74 \pm 0.05$ | $0.78 \pm 0.14$ | $0.78 \pm 0.04$ | $0.77 \pm 0.02$ |
| Aspartate aminotransferase, μcat/L | $0.31 \pm 0.03$ | $0.39 \pm 0.02$ | $0.46 \pm 0.05$ | $0.47 \pm 0.06$ |
| Alkaline phosphatase, μcat/L | $7.4 \pm 1.9$ | $8.7 \pm 1.5$ | $5.66 \pm 0.64$ | $5.58 \pm 0.60$ |

*Note: Here and in Table 5.142 the differences are not significant ($p > 0.05$).*

**Table 5.142** Urinary Biochemical Parameters of Control and Test Rats ($M \pm m$, $n = 6–8$)

| Parameter | 30 Days | | 180 Days | |
|---|---|---|---|---|
| | Control | Test | Control | Test |
| pH | 7.3 | 7.5 | 7.0 | 7.0 |
| Daily diuresis, mL | $6.2 \pm 0.7$ | $7.3 \pm 1.2$ | $9.1 \pm 1.2$ | $11.0 \pm 1.0$ |
| Relative density, g/mL | $1.01 \pm 0.06$ | $1.02 \pm 0.06$ | $1.07 \pm 0.01$ | $1.05 \pm 0.01$ |
| Creatinine, mg/mL | $0.72 \pm 0.11$ | $0.65 \pm 0.13$ | $1.60 \pm 0.23$ | $1.06 \pm 0.08$ |
| Creatinine, mg/day | $4.44 \pm 0.75$ | $4.03 \pm 0.31$ | $13.48 \pm 0.83$ | $11.24 \pm 0.54$ |

(by 27.2%) observed on Day 180 from the start of the experiment also remained within the physiological variations characteristic of rats of corresponding age, i.e., 4–6 g (liver) and 0.3–0.5 g (spleen).

### Assessment of Biochemical Parameters

During the entire experiment, the content of total protein, glucose, activity of alanine aminotransferase, aspartate aminotransferase, and alkaline phosphatase in blood serum of rats fed diet with transgenic maize line MON 810 (test group) did not significantly differ from the corresponding parameters of the control rats fed diet with conventional maize (Table 5.141).

There were no significant differences in daily diuresis, pH, relative density of urine, and urinary creatinine (Table 5.142).

### Assessment of Sensitive Biomarkers

On experimental Day 30, LPO intensity in blood of test group rats and control group rats did not significantly differ. A single exception was a significant (by 34%) decrease in the content of DC in erythrocytes of the test rats in comparison with the control. In 180 days, the concentrations of intermediate and final

**Table 5.143** Content of LPO Products in Blood and Liver of Rats Fed Diet with Conventional Maize (Control) or Transgenic Maize Line MON 810 (Test, $M \pm m$, $n = 6$–8)

| Parameter | 30 Days | | 180 Days | |
| --- | --- | --- | --- | --- |
| | Control | Test | Control | Test |
| *Erythrocytes* | | | | |
| DC, nM/mL | 6.118 ± 0.8 | 4.010 ± 0.4* | 4.365 ± 0.2 | 4.338 ± 0.3 |
| MDA, nM/mL | 5.272 ± 0.5 | 4.834 ± 0.5 | 6.256 ± 0.4 | 5.261 ± 0.6 |
| *Blood serum* | | | | |
| DC, nM/mL | 5.204 ± 0.4 | 4.416 ± 0.2 | 4.541 ± 0.4 | 4.436 ± 0.4 |
| MDA, nM/mL | 4.957 ± 1.5 | 4.455 ± 1.0 | 4.397 ± 0.5 | 5.320 ± 0.3 |
| *Liver* | | | | |
| DC, Unit | 1.070 ± 0.05 | 1.021 ± 0.05 | 0.958 ± 0.02 | 1.216 ± 0.04* |
| MDA, nM/g | 331.0 ± 9.9 | 328.0 ± 17.4 | 356.9 ± 12.2 | 408.3 ± 19.8* |

*$p < 0.05$ in comparison with control.

**Table 5.144** Activity of Enzymes of Antioxidant Protection System in Rats Fed Diet with Maize (Control) or Transgenic Maize Line MON 810 (Test) ($M \pm m$, $n = 6$–7)

| Parameter | 30 Days | | 180 Days | |
| --- | --- | --- | --- | --- |
| | Control | Test | Control | Test |
| Glutathione reductase, μmol/min/g Hb | 28.63 ± 1.6 | 29.36 ± 1.5 | 28.36 ± 1.1 | 30.37 ± 1.2 |
| Glutathione peroxidase, μmol/min/g Hb | 57.37 ± 2.1 | 54.59 ± 2.1 | 55.07 ± 2.4 | 51.97 ± 2.4 |
| Catalase, mmol/min/g Hb | 353.5 ± 20.9 | 369.4 ± 20.1 | 452.0 ± 10.7 | 424.7 ± 13.3 |
| Superoxide dismutase (SOD), U/min/g Hb | 1949.2 ± 43.8 | 1952.3 ± 39.4 | 2094.0 ± 63.3 | 2241.6 ± 66.5 |

LPO products increased in the liver of test rats: DC by 27% and MDA by 14% in comparison with the control group ($p < 0.05$); however the LPO intensity in blood of test and control group rats did not significantly differ (Table 5.143).

Similarly, the enzymatic activity of the antioxidant protection system did not significantly differ in both groups on experimental Days 30 and 180 (Table 5.144).

Thus, analysis of the parameters describing activity of LPO processes and activity of the enzymes in the antioxidant protection system showed that there was no pro-oxidant load in rats fed diet with transgenic maize line MON 810. Some variations of certain parameters result from individual peculiarities of oxidative lipid metabolism that do not affect the overall antioxidant status of the whole organism, which is corroborated by the absence

**Table 5.145** Activity of Enzymes Involved in Metabolism of Xenobiotics and Protein Content in Hepatic Microsomes of Rats Fed Diet with Maize (Control) or Transgenic Maize Line MON 810 (Test) ($M \pm m$, $n = 6–8$)

| Parameter | 30 Days | | 180 Days | |
|---|---|---|---|---|
| | Control | Test | Control | Test |
| Cytochrome P450, nM/mg protein | 0.75 ± 0.04 | 0.87 ± 0.04 | 0.79 ± 0.03 | 0.81 ± 0.02 |
| Cytochrome $b_5$, nM/mg protein | 0.61 ± 0.03 | 0.67 ± 0.03 | 0.78 ± 0.03 | 0.78 ± 0.02 |
| Aminopyrine N-demethylation, nM/min/mg protein | 10.2 ± 0.6 | 9.8 ± 0.5 | 7.62 ± 0.50 | 7.77 ± 0.13 |
| Benzpyrene hydroxylation, Fl-units/min/mg protein | 8.48 ± 0.54 | 8.36 ± 0.29 | 8.37 ± 0.61 | 8.97 ± 0.30 |
| Acetylesterase, µM/min/mg protein | 8.26 ± 0.35 | 8.78 ± 0.36 | 7.42 ± 0.32 | 7.62 ± 0.20 |
| Epoxide hydrolase, nM/min/mg protein | 8.83 ± 0.94 | 8.00 ± 0.40 | 7.98 ± 0.73 | 9.85 ± 1.03 |
| UDP-Glucuronosil transferase, nM/min/mg protein | 32.4 ± 1.6 | 26.5 ± 1.5* | 26.9 ± 1.2 | 26.7 ± 1.5 |
| CDNB-Glutathione transferase, µM/min/g tissue | 48.3 ± 1.8 | 48.3 ± 2.1 | 61.7 ± 2.1 | 63.5 ± 3.3 |
| Protein, mg/g | 12.3 ± 0.9 | 14.1 ± 0.5 | 12.3 ± 0.6 | 12.4 ± 0.3 |

*$p < 0.05$ in comparison with control.

of any changes in activity of the enzymes from the antioxidant protection system.

Activity of the hepatic enzymes involved in degradation of xenobiotics did not significantly differ in rats fed diet with conventional (control) and transgenic maize line MON 810 (test). In 30 days, activity of UDP-glucuronosil transferase in test rats was significantly lower than that in the control rats (by 18.3%). However, this decrease remained within the physiological boundaries and became insignificant by the end of the experiment (Table 5.145).

Activity of the hepatic lysosomal enzymes did not significantly differ between the groups of rats during the entire duration of the experiment (Table 5.146).

### Hematological Assessments

This part of the research examined the hematological parameters of peripheral blood drawn from the rats fed diet with transgenic maize line MON 810 (test group) and the control rats fed diet with conventional maize. Over the entire duration of the experiment (180 d), there were no significant differences between the two groups of rats in all examined hematological parameters: concentration of hemoglobin, total erythrocyte count, total leukocyte count, hematocrit, MCH, MCHC, and MCV (Table 5.147).

Similarly, there were no significant differences in all leukogram parameters between the control and test rats over the entire period of the experiment (Table 5.148).

**Table 5.146** Total and Non-sedimentable Activity of Hepatic Lysosomal Enzymes of Rats Fed Diet with Maize (Control) or Transgenic Maize Line MON 810 (Test) ($M \pm m$, $n = 6-8$)

| Parameters | 30 Days | | 180 Days | |
|---|---|---|---|---|
| | Control | Test | Control | Test |
| *Total activity, µM/min/g tissue* | | | | |
| Arylsulfatase A, B | 2.47 ± 0.05 | 2.43 ± 0.04 | 2.22 ± 0.04 | 2.10 ± 0.01 |
| β-Galactosidase | 2.60 ± 0.08 | 2.48 ± 0.08 | 2.39 ± 0.08 | 2.46 ± 0.06 |
| β-Glucuronidase | 2.17 ± 0.06 | 2.15 ± 0.04 | 2.16 ± 0.04 | 2.26 ± 0.04 |
| *Non-sedimentable activity, % total activity* | | | | |
| Arylsulfatase A, B | 3.51 ± 0.08 | 3.76 ± 0.06 | 3.35 ± 0.14 | 3.20 ± 0.11 |
| β-Galactosidase | 5.65 ± 0.15 | 5.65 ± 0.21 | 5.00 ± 0.25 | 4.84 ± 0.19 |
| β-Glucuronidase | 4.68 ± 0.25 | 4.88 ± 0.18 | 4.50 ± 0.24 | 4.12 ± 0.32 |

**Table 5.147** Hematological Parameters of Peripheral Blood Drawn from Rats Fed Diet with Maize (Control) or Transgenic Maize Line MON 810 (Test) ($M \pm m$, $n = 6-8$)

| Parameter | 30 Days | | 180 Days | |
|---|---|---|---|---|
| | Control | Test | Control | Test |
| Hemoglobin concentration, g/L | 160.1 ± 2.75 | 157.9 ± 5.36 | 163.6 ± 3.83 | 157.4 ± 6.36 |
| Total erythrocyte count, $\times 10^{12}$/L | 6.20 ± 0.18 | 5.88 ± 0.22 | 5.82 ± 0.15 | 5.90 ± 0.28 |
| Hematocrit, vol.% | 50.0 ± 0.57 | 49.4 ± 0.57 | 49.6 ± 0.57 | 49.0 ± 0.57 |
| MCH, pg | 25.82 ± 0.88 | 26.29 ± 0.75 | 28.23 ± 1.29 | 26.48 ± 0.88 |
| MCHC, % | 32.0 ± 0.57 | 31.95 ± 0.69 | 33.01 ± 1.01 | 32.06 ± 1.00 |
| MCV, µm$^3$ | 80.56 ± 1.44 | 84.4 ± 2.31 | 85.4 ± 1.86 | 82.58 ± 1.15 |
| Total leukocyte count, $\times 10^9$/L | 14.37 ± 1.49 | 12.65 ± 0.64 | 14.0 ± 1.25 | 16.15 ± 0.94 |

### Morphological Assessments

Over the entire duration of the experiment, there was no mortality in the control group of rats fed diet with conventional maize nor in the test group fed diet with transgenic maize line MON 810. Post-mortem examination revealed no alterations in the internal organs in both groups of rats. Similarly, histological examinations performed on experimental Days 30 and 180 revealed no significant differences in the internal organs of the rats between groups (Table 5.149). Therefore, the chronic toxicological experiment carried out during 180 days with biochemical, hematological, and morphological examinations revealed no adverse affects of transgenic maize line MON 810 on the animals.

**Table 5.148** Leukogram Parameters of Rats Fed Diet with Maize (Control) or Transgenic Maize Line MON 810 (Test) ($M \pm m$, $n = 6$–8)

| Parameter | 30 Days | | 180 Days | |
|---|---|---|---|---|
| | Control | Test | Control | Test |
| *Neutrophils* | | | | |
| rel., % | 11.6 ± 1.53 | 15.6 ± 3.0 | 14.2 ± 1.34 | 13.6 ± 2.3 |
| abs., ×10$^9$/L | 1.72 ± 0.34 | 1.99 ± 0.48 | 2.01 ± 0.35 | 2.12 ± 0.22 |
| *Eosinophils* | | | | |
| rel., % | 1.2 ± 0.38 | 0.8 ± 0.38 | 1.4 ± 0.38 | 1.2 ± 0.38 |
| abs., ×10$^9$/L | 0.15 ± 0.059 | 0.21 ± 0.05 | 0.19 ± 0.065 | 0.18 ± 0.07 |
| *Lymphocytes* | | | | |
| rel., % | 87.2 ± 1.14 | 83.6 ± 3.06 | 84.4 ± 1.34 | 85.2 ± 2.3 |
| abs., ×10$^9$/L | 12.49 ± 1.17 | 10.5 ± 0.61 | 11.79 ± 1.08 | 13.8 ± 1.13 |

**Table 5.149** Microscopic Examination of Internal Organs in Rats Fed Diet with Conventional Maize (Control) or Transgenic Maize Line MON 810 (Test) (Combined Data Obtained on Experimental Days 30 and 180)

| Organ | Control | Test |
|---|---|---|
| Liver | Clear trabecular structure; no alterations in hepatocytes or portal ducts | No differences from control |
| Kidneys | Usual appearance of cortical and medullar substance; no alterations in glomeruli or pelvis epithelium | No differences from control |
| Lung | Alveolar space is air-filled; no alterations in bronchi and blood vessels | No differences from control |
| Spleen | Large folliculi with wide clear marginal zones and reactive centers; splenic pulp is moderately plethoric | No differences from control |
| Small intestine | Preserved villous epithelium; usual infiltration in villous stroma | No differences from control |
| Testicle | Clearly definable spermatogenesis and age-related spermiogenesis | No differences from control |

## Assessment of Potential Impact of Transgenic Maize Line MON 810 on Immune System in Studies on Mice
### Potential Effect on Humoral Component of Immune System

The immunomodulating effect of transgenic maize line MON 810 on the humoral component of the immune system was examined on two oppositely reacting mice lines, C57Bl/6 and CBA, by determining the level of hemagglutination to sheep erythrocytes (SE). The experimental conditions are described in section 5.1.1, and the diet is shown in Table 5.114.

In control and test groups of CBA mice (high-sensitivity animals), the antibody titer was 1:2–1:8 at any time after immunization. In control and test groups of C57Bl/6 mice (low-sensitivity animals), the antibodies appeared on post-immunization Day 7 (1:10–1:20) and they can be detected on post-immunization Day 21. Therefore, the antibodies raised against SE appeared in both groups on post-immunization Day 7 and did not depend on maize variety in the diet.

### Potential Effect on Cellular Component of Immune System

The immunomodulating effect of transgenic maize was assessed by delayed hypersensitivity reaction to SE. The experimental conditions are described in section 5.1.1. Transgenic maize MON 810 did not elevate RI in either line of mice: in test and control CBA mice, the corresponding values were $17 \pm 4$ and $18 \pm 2$, while the respective values in test and control C57Bl/6 mice were $28 \pm 3$ and $28 \pm 3$. Thus, delayed hypersensitivity reaction to SE showed that the transgenic maize line MON 810 produced no effect on the cellular component of the immune system.

### Assessment of Potential Sensibilization Effect of Transgenic Maize MON 810

Assessment of possible sensitizing action of transgenic maize on the immune response to endogenous metabolic products was carried out by testing mouse sensitivity to histamine. The experimental conditions are described in section 5.1.1. The reaction was assessed by mortality (in percentage) counted at 1 h and 24 h after injection. There was no mortality and no differences in behavior of the test and control mice, so it was concluded that transgenic maize line MON 810 has no sensitizing agent.

### Potential Effect of Transgenic Maize on Susceptibility of Mice to Salmonella typhimurium

The potential effect of transgenic maize MON 810 on the susceptibility of mice to infection by salmonella of murine typhus was examined in experiments on C57Bl/6 mice injected intraperitoneally with various doses of Strain 415 *Salmonella typhimurium*. Four weeks prior to infection, the diet of control and test mice was supplemented with conventional and transgenic maize MON 810, respectively. The injected doses ranged from $10^2$ to $10^5$ microbial cells per mouse and varied on a 10-fold basis. The post-injection observation period was 21 days. The following data was obtained:

- In both groups, mortality of mice was first observed during post-injection week 1, and all mice died by post-injection Day 18;
- The mean lifetime of mice was approximately equal ($14.5 \pm 8.0$ day);
- $LD_{50}$ values in test and control groups were 301 and 84 microbial cells, correspondingly;

### 5.2.2 Transgenic Maize Line MON 810 Resistant to European Corn Borer

**Table 5.150** Severity of Active Anaphylactic Shock in Rats Fed Diet with Conventional Maize (Control) or Transgenic Maize Line MON 810 (Test)

| Group of Rats | AI | Severe Reactions, % | Mortality, % |
|---|---|---|---|
| Control ($n = 26$) | 3.27 | 77 | 73 |
| Test ($n = 24$) | 3.42 | 79 | 79 |

These data showed that *Salmonella typhimurium* produced a typical infection both in control and test mice, and the course of disease was similar in both groups. Moreover, the transgenic maize increased resistance of test mice: $LD_{50}$ was 3.5 times greater than the control value. Severity of the disease and the average lifetime were similar in both groups of mice. Overall, transgenic maize line MON 810 did not significantly modify resistance of mice against salmonella of murine typhus. Therefore, transgenic maize line MON 810 had no sensitizing potencies and produced no stimulating effect on humoral or cellular components of the immune system in the oppositely reacting mouse lines.

### Assessment of Potential Impact of Transgenic Maize Line MON 810 on Immune System in Studies on Rats

The study was carried out on male Wistar rats ($n = 50$) weighing initially $160 \pm 10$ g. After a 7-day adaptation period to standard vivarium diet, the rats for the next 28 days were fed diet supplemented with conventional maize (control group) or transgenic maize line MON 810 (test group). Both types of maize cereals were dissolved in boiled water to the consistency of dense curd and supplemented with sunflower-seed oil to improve intake. The feed was used instead of an equally caloric amount of oatmeal (composition of the diet is given in Table 5.114).

The model of generalized anaphylaxis was developed according to [8] as described in section 5.1.1.

During the entire experiment, the rats of both groups grew normally, which indicates nutritional adequacy of both diets. On experimental Day 29, the body weights of test and control rats were $257 \pm 7$ g and $248 \pm 6$ g, respectively ($p > 0.1$). Severity of active anaphylactic shock and mortality revealed no significant differences ($p > 0.1$; Table 5.150).

Table 5.151 shows the mean values of $D_{492}$, the concentration of antibodies, and common logarithm of this concentration in the control and test groups. The difference between test rats fed diet with transgenic maize line MON 810 and the control rats maintained on conventional maize was insignificant ($p > 0.1$). Thus, the intensity of humoral immune response was practically identical in both groups of animals.

**Table 5.151** Parameters of Humoral Immune Response (Level of Specific IgG Antibodies Raised Against Ovalbumin) in Rats Fed Diet with Conventional Maize or Transgenic Maize Line MON 810

| Group of Rats | $D_{492}$ | Concentration of Antibodies, mg/mL | Logarithm of Antibody Concentration |
|---|---|---|---|
| Control ($n = 26$) | $0.802 \pm 0.032$ | $8.8 \pm 1.5$ | $0.783 \pm 0.074$ |
| Test ($n = 25$) | $0.800 \pm 0.029$ | $8.3 \pm 1.6$ | $0.781 \pm 0.067$ |

On the whole, these data conclude that the degree of sensitization by ovalbumin in test rats fed diet with transgenic maize line MON 810 did not increase in comparison with that in the control rats fed diet with conventional maize.

The studies showed that transgenic maize MON 810 did not significantly change the allergenic reactivity and sensitization by a model allergen in comparison with the rats fed diet with conventional maize. Feeding rats with transgenic maize MON 810 did not enhance sensitization and allergic reactivity in comparison with the rats fed diet with conventional maize.

### Assessment of Potential Genotoxicity of Transgenic Maize MON 810

Genotoxic studies were carried out on male C57Bl/6 mice and hybrid female CBA mice fed diet shown in Table 5.114. During the experiment, the animals were fed diet composed of a soft feed with milled maize of examined varieties.

The cytogenetic analysis was carried out by metaphasic method; genetic alterations in the sex cells were examined by revealing the dominant lethal mutations in C57Bl/6 male mice [8]. The details of experiments are described in section 5.1.1.

In the test mice the number of cells with structural alterations was insignificant. Both control and test mice exhibited cells with chromosomal abnormalities (single segments and gaps), known as transient disturbances, that are eliminated in the subsequent nuclear divisions. The number of cells with polyploid chromosome set did not differ between control and test mice (Table 5.152).

At the stages of early and late spermatids or mature spermatozoa, the postimplantation embryonic mortality (the most reliable index of mutagenic activity of an examined agent) in the test group did not surpass the control level during the entire period of the experiment. During this period, there was no induced mortality, which attests to the absence of a mutagenic effect

### 5.2.2 Transgenic Maize Line MON 810 Resistant to European Corn Borer

**Table 5.152** Bone Marrow Cytogenetic Parameters in Mice Fed 45-Day Diet with Conventional Maize (Control) or Transgenic Maize Line MON 810 (Test) ($M \pm m$)

| Number of Cells, % | Control | Test |
|---|---|---|
| with chromosomal aberrations | 0.57 ± 0.40 | 0.52 ± 0.36 |
| with gaps | 1.14 ± 0.56 | 1.05 ± 0.52 |
| with polyploid chromosome set | 1.42 ± 0.63 | 1.32 ± 0.58 |

Note: Here and in Table 5.153 the differences are not significant ($p > 0.05$). The numbers of analyzed metaphases were 378 (Test) and 350 (control).

**Table 5.153** Dominant Lethal Mutations in Sex Cells of Female Mice Fed Diet of Conventional Maize (Control) or Transgenic Maize Line MON 810 (Test) ($M \pm m$)

| Parameter, % | Week 1, Mature Spermatozoa | | Week 2, Late Spermatids | | Week 3, Early Spermatids | |
|---|---|---|---|---|---|---|
| | Control | Test | Control | Test | Control | Test |
| Pre-implantation mortality | 8.30 | 6.17 | 7.0 | 5.82 | 7.90 | 5.23 |
| Post-implantation mortality | 3.30 | 4.19 | 4.79 | 3.37 | 4.66 | 3.06 |
| Survival rate | 88.70 | 89.8 | 88.53 | 92.0 | 87.70 | 91.86 |
| Induced mortality | 0 | 0 | 0 | 0 | 0 | 0 |

Note: The number of females was 90 (test group) and 78 (control group), 885 embryos and 960 corpus lutea were analyzed.

of long-term feeding with transgenic maize MON 810 on spermiogenesis in mice (Table 5.153). Therefore, transgenic maize MON 810 exhibited no mutagenic effects under the experimental conditions described.

### Assessment of Technological Parameters

The study of technological parameters was carried out in Moscow State University of Applied Biotechnology (Ministry of Science and Education of Russian Federation).

To characterize the samples of transgenic maize line MON 810 resistant to European corn borer and non-transgenic (conventional) maize, the moisture and ash contents were determined. Maize starch was produced under laboratory conditions to determine protein mass fraction in the starch, gelatinization temperature and viscosity of the starch gelatins on amylograph, the parameters of thermoplastic extrusion, and the structural and mechanical properties of the extrudates.

The transgenic maize grain complied with the specifications of Russian State Standards GOST 136-90 "Maize. Technical requirements".

The study resulted in the following conclusions:

- No differences or difficulties were observed in the technological process when producing starch from transgenic maize MON 810 or conventional maize;
- According to Russian State Standards GOST 7698-93 "Starch. Formal Acceptance and Analytical Methods", the quality of all examined samples was superior;
- By gelatinization temperature and rheological properties of 7% gelatins, the starches derived from transgenic maize did not differ from those obtained from conventional maize;
- The structural and mechanical properties of the extrusive products derived from transgenic maize line MON 810 were virtually identical to those obtained from the conventional maize.

### Conclusions

By all examined parameters, the data of complex safety assessment of transgenic maize line MON 810 resistant to European corn borer attest to the absence of any toxic, genotoxic, immune system modulating, or allergenic effects of this maize line. By chemical composition, transgenic maize line MON 810 was identical to conventional maize.

Based on the results of the studies, the State Sanitation Service of the Russian Federation (Department of State Sanitation and Epidemiological Inspectorate) granted the Registration Certificate which allows the transgenic maize line MON 810 to be used in the food industry and placed on the market without restrictions.

## 5.2.3 TRANSGENIC GLYPHOSATE-TOLERANT MAIZE LINE NK 603

### Molecular Characterization of Transgenic Maize Line NK 603

#### Recipient Organism
Maize (*Zea mays* L.) is characterized by a long history of safe use as human food and animal feed.

#### Donor Organism
Gene *cp4 epsps* was isolated from *Agrobacterium* sp., strain CP4.

#### Mechanism of Genetic Transformation
Transgenic maize line NK 603 was produced by incorporation of two expression cassettes, each containing one copy of the *cp4 epsps* coding sequence

### 5.2.3 Transgenic Glyphosate-Tolerant Maize Line NK 603

**Table 5.154** Registration Status of Transgenic Maize Line NK 603 in Various Countries

| Country | Date of Approval | Scope |
|---|---|---|
| Australia | 2002 | Food |
| Canada | 2001 | Food, feed, environmental release |
| Korea | 2002 | Food |
|  | 2004 | Feed |
| Mexico | 2002 | Food, feed |
| USA | 2000 | Food, feed, environmental release |
| South Africa | 2002 | Food, feed, environmental release |
| Japan | 2001 | Food, feed, environmental release |

*Note: The current registration status of the transgenic crop is on http://www.biotradestatus.com/*

from the PV-ZMG32 plasmid, into embryonic cells of maize line LH82 × B73 by biolistic transformation [21]. The first expression cassette contained the following sequences controlling expression of the *cp4 epsps* gene: promoter of rice actin gene, intron of rice actin, and NOS terminator isolated from the Ti-plasmid of *Agrobacterium*. The second expression cassette contained 35S promoter of the cauliflower mosaic virus, maize intron hsp 70, and NOS terminator, isolated also from the Ti-plasmid of *Agrobacterium*. In both cases, the post-translational translocation of CP4 EPSPS protein to the chloroplast (the site of synthesis of aromatic amino acids) was performed with transit CTP 2 peptide of *Arabidopsis thaliana* EPSPS.

## Global Registration Status of Maize Line NK 603
Table 5.154 shows registration status of transgenic maize line NK 603 at the time of registration in Russia [19].

## Safety Assessment of Transgenic Maize Line NK 603 Conducted in the Russian Federation
The studies were conducted in accordance with the requirements of the Ministry of Health of the Russian Federation authorized for risk and safety assessment of food derived from GM sources [8]. PCR analysis of the maize test and control samples was performed to confirm the identity of the transformation event and its absence in the conventional control line.

### *Biochemical Composition of Grain*
The content of proteins and amino acid composition in the grain of transgenic maize line NK 603 did not significantly differ from the corresponding values for conventional maize (Table 5.155).

**Table 5.155** Protein Content (%) and Amino Acid Composition (g/100 g protein) in Maize Grain

| Ingredient | Conventional Maize | GM Maize Line NK 603 |
|---|---|---|
| Protein,% | 8.86 | 9.56 |
| *Amino acid composition, g/100 g protein* | | |
| Lysine | 3.44 | 3.49 |
| Histidine | 3.87 | 3.77 |
| Arginine | 4.61 | 4.22 |
| Aspartic acid | 6.34 | 6.01 |
| Threonine | 2.91 | 2.82 |
| Serine | 4.02 | 3.80 |
| Glutamic acid | 18.03 | 19.84 |
| Proline | 9.72 | 10.39 |
| Glycine | 4.09 | 3.93 |
| Alanine | 6.89 | 6.23 |
| Cysteine | 1.44 | 1.47 |
| Valine | 4.28 | 4.40 |
| Methionine | 1.38 | 1.44 |
| Isoleucine | 3.50 | 3.31 |
| Leucine | 11.52 | 10.87 |
| Tyrosine | 3.89 | 3.74 |
| Phenylalanine | 4.59 | 4.46 |

**Table 5.156** Carbohydrates Content (g/100 g product) in Maize Grain

| Carbohydrate | Conventional Maize | Transgenic Maize Line NK 603 |
|---|---|---|
| Fructose | 0.34 | 0.21 |
| Glucose | 0.60 | 0.29 |
| Sucrose | 3.20 | 2.23 |
| Starch | 36.8 | 40.0 |
| Cellulose | 2.30 | 2.06 |

Similarly, the content of carbohydrates in the grain of transgenic maize line NK 603 did not significantly differ from the corresponding values for conventional maize. Variations in the content of glucose remained within the range characteristic of maize: 0.01–0.6 g/100 g (Table 5.156).

The content of lipids in the grain of transgenic maize line NK 603 did not significantly differ from the corresponding value for conventional maize (Table 5.157). Similarly, the content of fatty acids in the grain did not significantly differ between the maize varieties (Table 5.158).

### 5.2.3 Transgenic Glyphosate-Tolerant Maize Line NK 603

**Table 5.157** Content of Lipids (g/100 g product) in Maize Grain

| Ingredient | Conventional Maize | Transgenic Maize Line NK 603 |
|---|---|---|
| Lipids | 4.17 | 4.43 |

**Table 5.158** Content of Fatty Acids (Rel. %) in Maize Grain

| Fatty Acid | Conventional Maize | Transgenic Maize Line NK 603 |
|---|---|---|
| Lauric 12:0 | 0.01 | 0.01 |
| Myristic 14:0 | 0.02 | 0.11 |
| Pentadecanoic 15:0 | 0.02 | 0.06 |
| Palmitic 16:0 | 10.11 | 11.78 |
| Palmitoleic 16:1 | 0.17 | 0.18 |
| Margaric (heptadecanoic) 17:0 | 0.09 | 0.10 |
| Heptadecenoic 17:1 | 0.04 | 0.03 |
| Stearic 18:0 | 2.20 | 2.80 |
| cis-9-Oleic 18:1 | 25.24 | 27.91 |
| trans-11-Vaccenic 18:1 | 0.40 | 0.64 |
| Linoleic 18:2 | 59.86 | 54.67 |
| Linolenic 18:3 | 0.92 | 0.70 |
| Arachidic 20:0 | 0.43 | 0.55 |
| Gondoic 20:1 | 0.27 | 0.24 |
| Behenic 22:0 | 0.10 | 0.10 |
| Erucic 22:1 | 0.12 | 0.12 |

**Table 5.159** Content of Vitamins (mg/100 g product) in Maize Grain

| Ingredient | Conventional Maize | Transgenic Maize Line NK 603 |
|---|---|---|
| Vitamin $B_1$ | 0.279 | 0.290 |
| Vitamin $B_2$ | 0.189 | 0.247 |
| Vitamin $B_6$ | 0.43 | 0.51 |
| β-Carotene | 0.02 | 0.02 |

The content of vitamins and minerals in transgenic maize NK 603 and conventional maize did not significantly differ. The revealed variations in the content of vitamin $B_2$ and selenium did not surpass the physiological levels characteristic of maize according to the data of the State Research Institute of Nutrition, RAMS: 0.05–0.30 mg/100 g and 50–250 μg/100 g, respectively (Tables 5.159 and 5.160).

**Table 5.160** Mineral Composition in Maize Grain

| Ingredient | Conventional Maize | Transgenic Maize Line NK 603 |
|---|---|---|
| Sodium, mg/kg | 41.0 | 72.9 |
| Calcium, mg/kg | 1.88 | 1.85 |
| Magnesium, mg/kg | 1545 | 1546 |
| Iron, mg/kg | 21.4 | 20.0 |
| Potassium, mg/kg | 3170 | 3262 |
| Zinc, mg/kg | 21.0 | 17.6 |
| Copper, mg/kg | 1.14 | 1.06 |
| Selenium, µg/kg | 119 | 90 |

**Table 5.161** Analysis of Toxic Elements of Maize Grain (mg/kg)

| Ingredient | Conventional Maize | Transgenic Maize Line NK 603 |
|---|---|---|
| Deoxynivalenol | Not detected | Not detected |
| Zearalenone | Not detected | Not detected |
| Aflatoxin $B_1$ | Not detected | Not detected |
| Cadmium | 0.005 | 0.006 |
| Lead | 0.029 | 0.048 |

The content of heavy metals and mycotoxins in the grain of conventional and transgenic maize NK 603 did not surpass the acceptable limits according to the regulations valid in the Russian Federation [6] (Table 5.161).

Thus, the above-mentioned data showed that the grain of transgenic maize line NK 603 and conventional maize did not significantly differ by biochemical composition. The revealed differences in the content of metals and vitamins remained within the range characteristic of maize [15,43,49–52]. The safety parameters of the grain of conventional and transgenic maize NK 603 comply with the requirements of the regulations valid in Russian Federation [8].

### *Toxicological Assessment of Transgenic Maize Line NK 603*

The experiment was carried out on male Wistar rats with an initial body weight of 70–80 g. After admission to the vivarium of the State Research Institute of Nutrition, the rats were placed in quarantine for 10 days. At the onset of feeding the experimental diet, the body weight of rats was 85–95 g. During the entire experiment, the animals were fed a standard semi-synthetic diet with conventional (control rats) or transgenic (test rats) maize (Table 5.162).

### 5.2.3 Transgenic Glyphosate-Tolerant Maize Line NK 603

**Table 5.162** Standard Semi-synthetic Rat Diet with Conventional (Control Rats) or Transgenic NK 603 (Test Rats) Maize

| Ingredient | Mass, g/day |
|---|---|
| Casein | 18.6 |
| Starch | 32.6 |
| Vegetable oil | 2.7 |
| Lard | 5.0 |
| Salt mix[a] | 4.0 |
| Liposoluble vitamins[a] | 0.1 |
| Vitamin mix[a] | 1.0 |
| Maize | 36.0 |
| Total | 100 |

[a]See Tables 5.39–5.41 for the content of mixes.

**Table 5.163** Daily Intake (g/day) of Control and Test Maize ($M \pm m$, $n = 25$)

| Duration of Experiment, Weeks | Conventional (Control) Maize | Transgenic Maize Line NK 603 |
|---|---|---|
| 1 | 6.8 ± 0.1 | 6.7 ± 0.1 |
| 2 | 6.8 ± 0.5 | 7.0 ± 0.4 |
| 3 | 7.5 ± 0.2 | 7.4 ± 0.1 |
| 4 | 8.0 ± 0.0 | 8.0 ± 0.0 |

Note: Here and in Table 5.164 the differences are not significant ($p > 0.05$).

Throughout the entire duration of the experiment, the general condition of the rats was similar in the control and test groups. No mortality was observed in either group. During the first 3 weeks of the experiment, daily feed intake of maize did not significantly differ between groups (6.8–7.5 g/day, Table 5.163). To the end of experimental Month 1 and during the following 5 months, the animals of both groups consumed the maize grain completely (8 g/day). During the entire period of the experiment, the body weight of rats fed diet with transgenic maize line NK 603 did not significantly differ from that of the control rats fed diet with equivalent amount of conventional maize (Figure 5.8; Table 5.164).

The absolute and relative weight of internal organs of rats was measured by sacrificing the animals on experimental Days 30 and 180 (Tables 5.165).

During the entire experiment, the absolute and relative weights of internal organs did not significantly differ between the control and test groups.

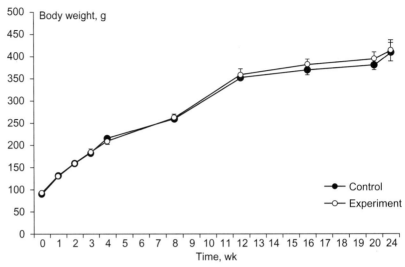

**FIGURE 5.8**
Comparative dynamics of body weight of rats fed diet with transgenic maize line NK 603 (test) or conventional maize (control).

**Table 5.164** Body Weight of Rats (g) Fed Diet with Conventional Maize (Control) or Transgenic Maize Line NK 603 (Test) ($M \pm m$; $n = 25$)

| Duration of Experiment, Weeks | Control | Test |
| --- | --- | --- |
| 0 | 89.2 ± 2.9 | 90.7 ± 2.8 |
| 1 | 131.2 ± 4.3 | 130.4 ± 4.3 |
| 2 | 159.2 ± 5.2 | 158.3 ± 6.0 |
| 3 | 182.8 ± 5.9 | 184.4 ± 6.9 |
| 4 | 214.2 ± 6.3 | 209.6 ± 8.1 |
| 8 | 259.4 ± 6.8 | 260.9 ± 9.0 |
| 12 | 351.8 ± 10.8 | 357.8 ± 12.3 |
| 16 | 368.7 ± 11.0 | 380.0 ± 12.1 |
| 20 | 380.0 ± 11.7 | 392.1 ± 17.1 |
| 24 | 409.3 ± 20.8 | 412.9 ± 24.2 |

An exception was a significant gain in absolute and relative weight of prostate in test rats fed diet with transgenic maize for 30 days, which nevertheless remained within the physiological variations characteristic of rats of corresponding age, i.e., 100–300 mg (absolute weight) or 50–100 mg/100 g (relative weight).

### 5.2.3 Transgenic Glyphosate-Tolerant Maize Line NK 603

**Table 5.165** Absolute and Relative Weight of Internal Organs of Rats Fed Diet with Conventional Maize or Transgenic Maize Line NK 603 ($M \pm m$, $n = 6$–$8$)

| Organ | | 30 Days | | 180 Days | |
|---|---|---|---|---|---|
| | | Control | Test | Control | Test |
| Kidneys | Abs.[a], g | 1.62 ± 0.12 | 1.73 ± 0.10 | 1.92 ± 0.22 | 2.39 ± 0.17 |
| | Rel.[b], g /100 g | 0.76 ± 0.02 | 0.77 ± 0.02 | 0.54 ± 0.04 | 0.57 ± 0.012 |
| Liver | Abs., g | 10.21 ± 0.55 | 10.69 ± 0.39 | 13.29 ± 0.92 | 13.08 ± 0.88 |
| | Rel., g /100 g | 4.75 ± 0.09 | 4.82 ± 0.31 | 3.18 ± 0.05 | 3.06 ± 0.06 |
| Spleen | Abs., g | 1.33 ± 0.12 | 1.45 ± 0.09 | 1.32 ± 0.16 | 1.37 ± 0.08 |
| | Rel., g /100 g | 0.62 ± 0.04 | 0.64 ± 0.03 | 0.32 ± 0.04 | 0.33 ± 0.02 |
| Heart | Abs., g | 0.83 ± 0.06 | 0.85 ± 0.04 | 1.21 ± 0.05 | 1.19 ± 0.08 |
| | Rel., g /100 g | 0.39 ± 0.01 | 0.37 ± 0.02 | 0.31 ± 0.28 | 0.29 ± 0.03 |
| Testicles | Abs., g | 2.45 ± 0.15 | 2.30 ± 0.25 | 2.60 ± 0.35 | 3.06 ± 0.19 |
| | Rel., g /100 g | 1.15 ± 0.11 | 1.08 ± 0.06 | 0.65 ± 0.09 | 0.76 ± 0.08 |
| Hypophysis | Abs., mg | 9.83 ± 0.87 | 8.80 ± 1.20 | 9.17 ± 0.94 | 10.50 ± 0.56 |
| | Rel., mg /100 g | 4.65 ± 0.74 | 3.99 ± 0.58 | 2.60 ± 0.24 | 2.53 ± 0.17 |
| Adrenal glands | Abs., mg | 23.00 ± 2.23 | 22.00 ± 1.37 | 30.00 ± 2.21 | 22.33 ± 1.39 |
| | Rel., mg /100 g | 10.75 ± 1.00 | 9.82 ± 0.67 | 6.20 ± 1.74 | 5.60 ± 0.66 |
| Seminal vesicles | Abs., mg | 314.2 ± 23.2 | 352.3 ± 28.7 | 616.0 ± 65.9 | 658.3 ± 76.96 |
| | Rel., mg /100 g | 147.2 ± 7.9 | 156.3 ± 6.9 | 149.6 ± 11.3 | 156.6 ± 14.3 |
| Prostate | Abs., mg | 125.8 ± 11.9 | 172.6 ± 14.5* | 280.0 ± 39.2 | 376.6 ± 56.37 |
| | Rel., mg /100 g | 58.60 ± 4.06 | 76.26 ± 6.36* | 68.08 ± 7.31 | 89.36 ± 19.5 |

[a]Absolute weight of internal organs.
[b]Relative weight of internal organs (per 100 g body weight).
*$p < 0.05$ in comparison with control.

### Biochemical Parameters

Throughout the entire duration of the experiment (180 d), the content of total protein and glucose, activity of aspartate aminotransferase, alanine aminotransferase, and alkaline phosphatase in blood serum of rats fed diet with transgenic maize line NK 603 did not significantly differ from the corresponding parameters of the control rats fed diet with conventional maize (Table 5.166).

The urinary biochemical parameters did not significantly differ between the groups of rats (Table 5.167).

### Assessment of Sensitive Biomarkers

The study examined the effect of transgenic maize line NK 603 on the content of LPO products in rats (Table 5.168). During the entire experiment, LPO level in blood and liver did not significantly differ between control and test rats. However, there were differences (by 4%) in the hepatic content of DC, that fell within the physiological variations of parameters characterizing

**Table 5.166** Biochemical Parameters of Blood Serum from Control Rats Fed Diet with Conventional Maize and Test Rats Fed Diet with Transgenic Maize Line NK 603 ($M \pm m$, $n = 6–8$)

| Parameter | 30 Days | | 180 Days | |
|---|---|---|---|---|
| | Control | Test | Control | Test |
| Total protein, g/L | 58.2 ± 1.8 | 63.2 ± 2.8 | 95.4 ± 9.6 | 99.3 ± 6.2 |
| Glucose, mM/L | 6.8 ± 0.38 | 7.3 ± 0.30 | 4.3 ± 0.44 | 5.7 ± 1.51 |
| Alanine aminotransferase, μcat/L | 0.68 ± 0.07 | 0.51 ± 0.04 | 0.57 ± 0.05 | 0.69 ± 0.07 |
| Aspartate aminotransferase, μcat/L | 0.47 ± 0.03 | 0.47 ± 0.03 | 0.58 ± 0.07 | 0.42 ± 0.04 |
| Alkaline phosphatase, μcat/L | 5.05 ± 0.61 | 5.18 ± 0.20 | 6.78 ± 0.42 | 5.89 ± 0.81 |

Note: Here and in Table 5.167 the differences are not significant ($p > 0.05$).

**Table 5.167** Urinary Biochemical Parameters of Rats Fed Diet with Conventional Maize (Control) or Transgenic Maize Line NK 603 (Test) ($M \pm m$, $n = 6–8$)

| Parameter | 30 Days | | 180 Days | |
|---|---|---|---|---|
| | Control | Test | Control | Test |
| pH | 6.0 | 6.0 | 7.0 | 7.0 |
| Daily diuresis, mL | 2.7 ± 0.2 | 2.5 ± 0.4 | 4.5 ± 0.5 | 5.3 ± 0.5 |
| Relative density, g/mL | 1.01 ± 0.04 | 1.06 ± 0.02 | 1.12 ± 0.02 | 1.13 ± 0.01 |
| Creatinine, mg/mL | 0.62 ± 0.05 | 0.82 ± 0.11 | 1.78 ± 0.11 | 1.44 ± 0.13 |
| Creatinine, mg/day | 1.83 ± 0.29 | 1.94 ± 0.31 | 7.68 ± 0.11 | 7.29 ± 0.35 |

**Table 5.168** Content of LPO Products in Blood and Liver of Rats Fed Diet with Conventional Maize (Control) or Transgenic Maize Line NK 603 (Test) ($M \pm m$, $n = 6–8$)

| Parameter | 30 Days | | 180 Days | |
|---|---|---|---|---|
| | Control | Test | Control | Test |
| *Erythrocytes* | | | | |
| DC, nM/mL | 6.206 ± 0.3 | 6.002 ± 0.3 | 4.241 ± 0.2 | 4.462 ± 0.4 |
| MDA, nM/mL | 3.149 ± 0.2 | 3.223 ± 0.1 | 2.770 ± 0.2 | 2.935 ± 0.3 |
| *Blood serum* | | | | |
| DC, nM/mL | 3.361 ± 0.1 | 3.348 ± 0.2 | 3.276 ± 0.2 | 3.341 ± 0.3 |
| MDA, nM/mL | 3.200 ± 0.2 | 3.118 ± 0.3 | 3.736 ± 0.1 | 3.572 ± 0.1 |
| *Liver* | | | | |
| DC, Unit | 1.018 ± 0.004 | 1.015 ± 0.006 | 1.153 ± 0.01 | 1.106 ± 0.01* |
| MDA, nM/g | 276.5 ± 6.0 | 281.0 ± 6.0 | 272.2 ± 4.7 | 278.6 ± 6.9 |

Note: Here and in Tables 5.169 and 5.170.
*$p < 0.05$ in comparison with control.

### 5.2.3 Transgenic Glyphosate-Tolerant Maize Line NK 603

**Table 5.169** Activity of Enzymes of Erythrocytic Antioxidant Protection System in Rats Fed Diet with Conventional Maize (Control) or Transgenic Maize Line NK 603 (Test) ($M \pm m$, $n = 6$–8)

| Parameter | 30 Days | | 180 Days | |
|---|---|---|---|---|
| | Control | Test | Control | Test |
| Glutathione reductase, μmol/min/g Hb | 36.71 ± 2.7 | 40.25 ± 2.9 | 38.59 ± 3.1 | 38.77 ± 3.5 |
| Glutathione peroxidase, μmol/min/g Hb | 55.23 ± 3.0 | 57.29 ± 1.3 | 55.29 ± 3.2 | 52.48 ± 1.4 |
| Catalase, mmol/min/g Hb | 428.1 ± 19.5 | 464.5 ± 13.1 | 439.4 ± 11.2 | 424.2 ± 18.6 |
| Superoxide dismutase, U/min/g Hb | 1991.9 ± 68.7 | 2036.0 ± 57.0 | 2205.1 ± 78.7 | 1908.6 ± 59.3* |

intensity of LPO processes in rats, which amount to 30%; therefore the revealed variations in DC content remained within the physiological norm.

The enzymatic activity of the antioxidant protection system did not significantly differ between groups on experimental Days 30 or 180. The significant increase of superoxide dismutase activity in the test group rats compared with the control group rats remained within the physiological variations caused by individual peculiarities in the intensity of protein synthesis (Table 5.169).

Thus, addition of glyphosate-tolerant transgenic maize line NK 603 to the diet produced no effect on the rat antioxidant status.

Table 5.170 shows activity of the hepatic enzymes involved in phase I and phase II xenobiotic degradation and lysosomal enzymes protecting the organism against exogenous and endogenous toxic agents.

On experimental Day 30, there were no significant differences in activity of the xenobiotic degradation enzymes. However, in 180 days the test group rats demonstrated a significant increase in activity of cytochrome P450, cytochrome $b_5$, acetyl esterase, and in the intensity of benzpyrene hydroxylation. Whereas detected changes occurred under conditions of the absence of complex stress manifestations of enzymes activity, it was suggested that prolonged intake of GM maize has no effect on the xenobiotics degradation system.

On Days 30 and 180, there were no significant differences in the total and non-sedimentable activity of hepatic lysosomal enzymes. Thus, addition of glyphosate-tolerant transgenic maize line NK 603 into the diet produced no significant effects on the metabolism of xenobiotics and lysosomal enzymes in rats (Table 5.171).

**Table 5.170** Activity of Enzymes Involved in Metabolism of Xenobiotics and Hepatic Protein Content in Liver of Rats Fed Diet with Conventional Maize (Control) or Transgenic Maize Line NK 603 (Test) ($M \pm m$, $n = 6$–$8$)

| Parameter | 30 Days | | 180 Days | |
|---|---|---|---|---|
| | Control | Test | Control | Test |
| Cytochrome P450, nM/mg protein | 0.56 ± 0.02 | 0.57 ± 0.05 | 0.77 ± 0.03 | 0.91 ± 0.03* |
| Cytochrome $b_5$, nM/mg protein | 0.50 ± 0.02 | 0.45 ± 0.03 | 0.67 ± 0.02 | 0.76 ± 0.01* |
| Aminopyrine N-demethylation, nM/min/mg protein | 7.33 ± 0.22 | 7.46 ± 0.16 | 10.00 ± 0.24 | 10.28 ± 0.57 |
| Benzpyrene hydroxylation, Fl-units/min/mg protein | 9.1 ± 0.6 | 8.6 ± 0.6 | 10.0 ± 0.7 | 12.2 ± 0.4* |
| Acetylesterase, μM/min/mg protein | 5.73 ± 0.10 | 5.78 ± 0.33 | 7.80 ± 0.38 | 9.33 ± 0.42* |
| Epoxide hydrolase, nM/min/mg protein | 5.55 ± 0.53 | 5.13 ± 0.38 | 4.20 ± 0.70 | 4.94 ± 0.57 |
| UDP-Glucuronosil transferase, nM/min/mg protein | 20.9 ± 0.7 | 20.5 ± 1.2 | 15.7 ± 1.6 | 18.2 ± 0.9 |
| CDNB-Glutathione transferase, μM/min/mg protein | 1.12 ± 0.03 | 1.17 ± 0.07 | 0.86 ± 0.03 | 0.97 ± 0.05 |
| Microsomal protein, mg/kg | 16.6 ± 0.6 | 17.0 ± 0.7 | 13.5 ± 0.3 | 14.5 ± 0.1 |
| Cytosolic protein, mg/kg | 74.0 ± 2.1 | 68.0 ± 1.8 | 84.3 ± 1.0 | 83.0 ± 2.1 |

**Table 5.171** Activity of Hepatic Lysosomal Enzymes of Rats Fed Diet with Conventional Maize (Control) or Transgenic Maize Line NK 603 (Test) ($M \pm m$, $n = 6$–$8$)

| Parameters | 30 Days | | 180 Days | |
|---|---|---|---|---|
| | Control | Test | Control | Test |
| *Total activity, μM/min/g tissue* | | | | |
| Arylsulfatase A, B | 2.32 ± 0.01 | 2.34 ± 0.02 | 2.14 ± 0.01 | 2.12 ± 0.02 |
| β-Galactosidase | 2.54 ± 0.05 | 2.52 ± 0.05 | 2.12 ± 0.03 | 2.16 ± 0.04 |
| β-Glucuronidase | 2.41 ± 0.05 | 2.36 ± 0.07 | 2.12 ± 0.02 | 2.17 ± 0.04 |
| *Non-sedimentable activity, % total activity* | | | | |
| Arylsulfatase A, B | 3.48 ± 0.18 | 3.52 ± 0.29 | 3.10 ± 0.03 | 3.16 ± 0.08 |
| β-Galactosidase | 5.98 ± 0.18 | 5.73 ± 0.14 | 5.79 ± 0.23 | 5.52 ± 0.23 |
| β-Glucuronidase | 5.79 ± 0.14 | 5.56 ± 0.19 | 5.06 ± 0.11 | 5.03 ± 0.15 |

Note: Here and in Tables 5.172 and 5.173, the differences are not significant ($p > 0.05$).

### Hematological Assessments

Hematological parameters of peripheral blood drawn from the test and control group rats were examined. Over the entire duration of the experiment, there were no significant differences between the two groups of rats in all examined hematological and leukogram parameters (Tables 5.172 and 5.173).

### 5.2.3 Transgenic Glyphosate-Tolerant Maize Line NK 603

**Table 5.172** Hematological Parameters of Peripheral Blood Drawn from Rats Fed Diet with Conventional Maize (Control) or Transgenic Maize Line NK 603 (Test) ($M \pm m$, $n = 6$–$8$)

| Parameter | 30 Days | | 180 Days | |
|---|---|---|---|---|
| | Control | Test | Control | Test |
| Hemoglobin concentration, g/L | 153.0 ± 5.35 | 147.0 ± 3.42 | 147.3 ± 3.25 | 155.2 ± 2.43 |
| Total erythrocyte count, ×10$^{12}$/L | 6.55 ± 0.08 | 6.5 ± 0.14 | 6.4 ± 0.16 | 6.4 ± 0.19 |
| Hematocrit, vol.% | 52.0 ± 0.0 | 51.83 ± 0.16 | 51.0 ± 0.48 | 51.1 ± 0.48 |
| MCH, pg | 23.37 ± 0.94 | 22.60 ± 0.57 | 23.18 ± 0.41 | 24.56 ± 0.62 |
| MCHC, % | 29.42 ± 1.0 | 28.33 ± 0.52 | 28.02 ± 0.7 | 30.6 ± 0.49 |
| MCV, µm$^3$ | 79.44 ± 0.79 | 79.81 ± 1.55 | 78.9 ± 1.22 | 80.16 ± 1.56 |
| Total leukocyte count, ×10$^9$/L | 12.43 ± 0.33 | 12.22 ± 0.3 | 12.23 ± 0.45 | 13.15 ± 0.45 |

**Table 5.173** Leukogram Parameters of Rats Fed Diet with Conventional Maize (Control) or Transgenic Maize Line NK 603 (Test) ($M \pm m$, $n = 6$–$8$)

| Parameter | 30 Days | | 180 Days | |
|---|---|---|---|---|
| | Control | Test | Control | Test |
| *Neutrophils* | | | | |
| rel., % | 19.5 ± 0.48 | 19.8 ± 0.32 | 30.3 ± 0.64 | 28.83 ± 0.32 |
| abs., ×10$^9$/L | 2.42 ± 0.12 | 2.41 ± 0.09 | 3.71 ± 0.19 | 3.78 ± 0.14 |
| *Eosinophils* | | | | |
| rel., % | 0.5 ± 0.16 | 0.5 ± 0.32 | 1.5 ± 0.48 | 1.67 ± 0.48 |
| abs., ×10$^9$/L | 0.06 ± 0.02 | 0.06 ± 0.039 | 0.176 ± 0.04 | 0.217 ± 0.06 |
| *Lymphocytes* | | | | |
| rel., % | 80.0 ± 0.48 | 79.7 ± 0.16 | 68.16 ± 0.97 | 69.5 ± 0.8 |
| abs., ×10$^9$/L | 9.94 ± 0.25 | 9.73 ± 0.26 | 8.34 ± 0.35 | 9.14 ± 0.37 |

**Morphological Assessments**

During the entire duration of the experiment, there was no mortality in either the control group of rats fed diet with conventional maize or in the group of test rats fed diet with transgenic maize line NK 603. Post-mortem examination revealed no alterations in the internal organs in either group of rats.

Similarly, histological examinations performed on experimental Days 30 and 180 revealed no significant differences between the groups in the internal organs of the rats (Tables 5.174).

Therefore, the chronic toxicological experiment carried out during 180 days with biochemical, hematological, and morphological examinations revealed no adverse affects of transgenic maize line NK 603 on the animals.

Table 5.174 Microscopic Examination of Internal Organs in Rats Fed Diet with Conventional Maize (Control) or Transgenic Maize Line NK 603 (Test) (Combined Data Obtained on Experimental Days 30 and 180)

| Organ | Control | Test |
|---|---|---|
| Liver | Clear trabecular structure; no alterations in hepatocytes or portal ducts | No differences from control |
| Kidneys | Usual appearance of cortical and medullar substance; no alterations in glomeruli or pelvis epithelium | No differences from control |
| Lung | Alveolar space is air-filled; no alterations in bronchi or blood vessels | No differences from control |
| Spleen | Large folliculi with wide clear marginal zones and reactive centers; splenic pulp is moderately plethoric | No differences from control |
| Small intestine | Preserved villous epithelium; usual infiltration in villous stroma | No differences from control |
| Testicle | Clearly definable spermatogenesis and age-related spermiogenesis | No differences from control |

### Assessment of Potential Impact of Transgenic Maize Line NK 603 on Immune System in Studies on Mice

#### Potential Effect on Humoral Component of Immune System

The immunomodulating effect of transgenic maize line NK 603 on the humoral component of the immune system was examined on two oppositely reacting mice lines, C57Bl/6 and CBA, by determining the level of hemagglutinins to sheep erythrocytes (SE). The transgenic maize line and its conventional counterpart were fed to test and control mice (respectively) in daily dose of 2.4 g/mouse for 21 days. The experimental conditions are described in section 5.1.1.

In control and test groups of CBA mice, the antibodies appeared on post-immunization Day 7 at a titer of 1:21–1:32, and then their titer decreased to 1:7–1:16 on Day 21. In test group of C57Bl/6 mice, the antibodies appeared on post-immunization Day 7 and remained to Day 21 at a titer of 1:128. In the control group of C57Bl/6 mice, there was insignificant elevation of antibody titer (1:2–1:8) during the experiment. Thus, the dynamics of antibody production in control and test C57Bl/6 mice was similar.

#### Potential Effect on Cellular Component of Immune System

The immunomodulating effect of transgenic maize was assessed by delayed hypersensitivity reaction to SE. The experimental conditions are described in section 5.1.1. The test and control CBA mice, fed diet with transgenic maize line NK 603 or its conventional counterpart correspondingly, demonstrated no significant difference in RI (respectively, $13 \pm 2$ and $18 \pm 2$). In test C57Bl/6 mice, RI was $9 \pm 3$, while in the control mice it rose to $17 \pm 2$ ($p < 0.001$). Thus, the transgenic maize line NK 603 exerted no immunostimulating effect, because RI did not drop markedly in high-sensitivity CBA mice nor did it elevate in low-sensitivity C57Bl/6 mice.

**Assessment of Potential Sensibilization Effect of Transgenic Maize**
Examination of possible sensitizing action of transgenic maize line NK 603 on the immune response to endogenous metabolic products was carried out in the test of mouse sensitivity to histamine. Grain from transgenic maize (test) and conventional maize (control) were fed to mice for 21 days; thereafter the mice of both groups were injected intraperitoneally with 2.5 mg histamine hydrochloride dissolved in 0.5 mL physiological solution. The reaction was assessed at 1 h and 24 h by the mortality of the mice. In this test, there was no mortality nor were there differences in behavior between test and control group, which attests to the absence of sensitization agent in transgenic maize line NK 603.

**Potential Effect of Transgenic Maize on Susceptibility of Mice to Salmonella typhimurium**
The effect of transgenic maize NK 603 on susceptibility of mice to infection by salmonella of murine typhus was examined in experiments on C57Bl/6 mice injected intraperitoneally with various doses of Strain 415 *Salmonella typhimurium*. Three weeks prior to infection, the diet of control and test mice was supplemented with conventional and transgenic maize NK 603, respectively. The injected doses ranged from $10^2$ to $10^5$ microbial cells per mouse and varied on a 10-fold basis. The post-injection observation period was 21 days. The following data were obtained:

- In both groups, the mortality was observed starting on post-injection week 1, and all mice died by post-injection Day 21;
- The mean lifetime of mice was approximately equal (16.6–17.6 day);
- $LD_{50}$ values in control and test groups were 3981 and 2042 microbial cells, correspondingly.

These data showed that *Salmonella typhimurium* produced a typical infection both in control and test mice, and the development of the disease was similar in both groups. Severity of the disease, $LD_{50}$ values, and the mean lifetime indicate a similar course of infectious disease in both groups of mice. Thus, transgenic maize line NK 603 did not modify resistance of mice against salmonella of murine typhus.

Therefore, transgenic maize line NK 603 has no sensitizing potencies and does not affect the resistance of mice to *S. typhimurium*. Both maize varieties produced no effect on the development of humoral and cellular components of the immune system in mice lines CBA and C57Bl/6.

### Assessment of Potential Impact of Transgenic Maize Line NK 603 on Immune System in Studies on Rats

The study was carried out on Wistar rats ($n = 48$) weighing initially $140 \pm 10$ g. After a 7-day adaptation period to standard vivarium diet, the rats for the next 28 days were fed diet supplemented with conventional maize (control group) or transgenic maize line NK 603 (test group). Maize flour of both test and control samples was dissolved in boiled water to the consistency of dense curd and supplemented with sunflower-seed oil to improve intake. The feed was used instead of an equally caloric amount of oatmeal (composition of the diet is given in Table 5.114).

The model of generalized anaphylaxis was developed according to [8] as described in section 5.1.1.

During the entire experiment, the rats of both groups grew normally, which indicates nutritional adequacy of both diets. Assessment of active anaphylactic shock showed that, by all examined parameters, the anaphylactic reaction was less severe in the test group of rats fed transgenic maize (Table 5.175). However, the difference in the severity of anaphylactic reaction was insignificant ($p > 0.1$).

By all parameters of humoral immune response, the difference between test rats fed diet with transgenic maize line NK 603 and the control rats maintained on conventional maize demonstrated the same trend as for anaphylactic reaction: the lowest parameters of sensitization were observed in the test group, while in the control group they were somewhat larger (Table 5.176; $p > 0.05$). However, the intensity of humoral immune response did not significantly differ between the two groups of rats ($p > 0.1$).

**Table 5.175** Comparative Severity of Active Anaphylactic Shock in Rats Fed Diet with Conventional Maize (Control) or Transgenic Maize Line NK 603 (Test)

| Group of Rats | AI | Severe Reactions, % | Mortality, % |
|---|---|---|---|
| Test ($n = 24$) | 2.83 | 54.2 | 54.2 |
| Control ($n = 24$) | 3.29 | 75.0 | 75.0 |

**Table 5.176** Comparative Intensity of Humoral Immune Response (Level of Specific IgG Antibodies Raised Against Ovalbumin) in Rats ($M \pm m$)

| Group of Rats | $D_{492}$ | Concentration of Antibodies, mg/mL | Logarithm of Antibody Concentration |
|---|---|---|---|
| Transgenic maize line NK 603 ($n = 24$) | $0.213 \pm 0.016$ | $3.7 \pm 0.8$ | $0.324 \pm 0.100$ |
| Conventional maize ($n = 24$) | $0.246 \pm 0.015$ | $5.1 \pm 1.0$ | $0.513 \pm 0.091$ |

### 5.2.3 Transgenic Glyphosate-Tolerant Maize Line NK 603

**Table 5.177** Bone Marrow Cytogenetic Parameters in Mice Fed Diet with Conventional Maize (Control) or Transgenic Maize Line NK 603 (Test) ($M \pm m$)

| Number of Cells, % | Control | Test |
|---|---|---|
| with chromosomal aberrations | 0.54 ± 0.38 | 0.27 ± 0.26 |
| with gaps | 1.08 ± 0.53 | 1.08 ± 0.53 |
| with polyploid chromosome set | 0.81 ± 0.51 | 0.54 ± 0.38 |

Note: The numbers of analyzed metaphases were: 370 (test) and 368 (control).

On the whole, these data support the conclusion that the degree of sensitization by ovalbumin in test rats fed diet with transgenic maize line NK 603 did not increase in comparison with that for the control rats fed diet with conventional maize. Addition of transgenic maize to the test diet insignificantly moderated severity of anaphylactic reaction and the response of specific antibodies in comparison with the control diet based on conventional maize. The studies showed that transgenic maize NK 603 did not enhance the allergic reactivity and degree of sensitization produced by a model allergen in comparison with the rats fed diet with conventional maize.

### Assessment of Potential Genotoxicicity of Transgenic Maize

The genotoxic studies were carried out on male C57Bl/6 mice and hybrid female CBA mice fed diet shown in Table 5.162. During the experiment, the animals were fed diet composed of a soft feed with milled maize of test or control variety.

The cytogenetic analysis was carried out by metaphasic method; genetic alterations in the sex cells were examined by revealing the dominant lethal mutations in C57Bl/6 male mice [8]. The details of the experiments are described in section 5.1.1.

The results of possible genotoxic effects of transgenic maize on mice are shown in Tables 5.177 and 5.178.

The major structural chromosomal abnormalities in control and test mice were single segments and gaps. There were no significant differences in the number of cells with gaps. In male C57Bl/6 mice, the number of observed structural chromosomal abnormalities was typical. Such mutations appear spontaneously; they are not stable and, as a rule, these mutations disappear in subsequent nuclear divisions. The pre-implantation mortality of unfertilized ovocytes, zygotes, and embryos in test mice did not surpass the corresponding values in the control group. The post-implantation embryonic mortality (the most reliable index of mutagenic activity of an examined agent) in the test group was somewhat lower

**Table 5.178** Dominant Lethal Mutations in Sex Cells of Mice Fed Diet with Conventional Maize (Control) or Transgenic Maize Line NK 603 (Test)

| Parameter, % | Week 1, Mature Spermatozoa | | Week 2, Late Spermatids | | Week 3, Early Spermatids | |
|---|---|---|---|---|---|---|
| | Control | Test | Control | Test | Control | Test |
| Pre-implantation mortality | 5.96 | 5.26 | 6.42 | 5.84 | 10.3 | 6.60 |
| Post-implantation mortality | 3.52 | 2.77 | 3.81 | 2.94 | 4.10 | 2.85 |
| Survival rate | 90.70 | 97.20 | 90.00 | 90.00 | 85.9 | 90.6 |
| Induced mortality | – | 0 | – | 0 | – | 0 |

*Note: In control group, 394 embryos and 426 corpus lutea from 72 females were analyzed. The corresponding values in test group (90 females) were 420 and 448.*

than in the control group, although the difference was insignificant ($p > 0.05$). There was no induced mortality, which attests to the absence of adverse affects of transgenic maize on spermiogenesis in mice. The data obtained support the conclusion that transgenic maize NK 603 produced no mutagenic effects.

### Assessment of Technological Parameters

Assessment of technological parameters was carried out in Moscow State University of Applied Biotechnology (Ministry of Science and Education of Russian Federation).

To characterize the samples of transgenic maize line NK 603, tolerant to glyphosate, and non-transgenic conventional maize, the moisture and ash contents were determined. Maize starch was produced under laboratory conditions to determine protein mass fraction in the starch, gelatinization temperature and viscosity of the starch gelatins on amylograph, the parameters of thermoplastic extrusion, and the structural and mechanical properties of the extrudates.

The transgenic maize grain complied with the specifications of Russian State Standards GOST 136-90 "Maize. Technical requirements". The study resulted in the following conclusions:

- No differences or difficulties were observed in the technological process when producing starch from transgenic maize NK 603 or conventional maize;
- According to Russian State Standards GOST 7698-93 "Starch. Formal Acceptance and Analytical Methods", the quality of all examined samples was superior;
- By gelatinization temperature and rheological properties of 7% gelatins, the starches derived from transgenic maize did not differ from those obtained from conventional maize;
- The structural and mechanical properties of the extrusive products derived from transgenic maize line NK 603 were virtually identical to those obtained from the conventional maize.

Thus, the study revealed no significant differences in the properties of examined grain samples of transgenic maize line NK 603 and its conventional counterpart.

### Conclusions
By all examined parameters, the data of complex safety assessment of transgenic maize line NK 603, tolerant to glyphosate, attest to the absence of any toxic, genotoxic, immune system modulating, or allergenic effects of this maize line. By chemical composition, transgenic maize line NK 603 was identical to conventional maize.

Based on the results of the studies, the State Sanitation Service of the Russian Federation (Department of State Sanitation and Epidemiological Inspectorate) granted the Registration Certificate which allows the transgenic maize line NK 603 to be used in the food industry and placed on the market without restrictions.

## 5.2.4 DIABROTICA-RESISTANT TRANSGENIC MAIZE LINE MON 863

### Molecular Characterization of Transgenic Maize Line MON 863
#### Recipient Organism
Maize (*Zea mays* L.) is characterized by a long history of safe use as human food and animal feed.

#### Donor Organism
The donor of the *cry3Bb1* gene is a gram-positive soil bacterium *Bacillus thuringiensis* (subspecies *kumamotoensis*), which produces Cry3Bb1 protein, that is active against corn rootworm (CRW, *Diabrotica* spp.).

#### Method of Genetic Transformation
The DNA-vector with *cry3Bb1* gene was incorporated into the embryonic cells of inbred maize line A634 by biolistic transformation. The insecticidal protein binds to specific sites in the cells of the CRW digestive system and forms ion-selective channels in the cell membrane, resulting in lysis of the cells and death of the pest larvae [25]. The incorporated DNA contained the following nucleotide sequences:

- *cry3Bb1* gene responsible for resistance against corn rootworm;
- *nptII* gene encoding tolerance to antibiotics (paromomycin), which was isolated from prokaryotic transposon Tn5 to be used as a selective marker gene;
- 35S promoter of the cauliflower mosaic virus;
- NOS 3′ terminator of nopaline synthase gene from *Agrobacterium tumefaciens*.

Table 5.179 Registration Status of Transgenic Maize Line MON 863 in Various Countries

| Country | Date of Approval | Scope |
| --- | --- | --- |
| Australia | 2003 | Food |
| Canada | 2003 | Food, feed, environmental release |
| Korea | 2003 | Food |
| Mexico | 2003 | Food, feed |
| USA | 2001 | Food, feed |
|  | 2003 | Environmental release |
| Taiwan | 2003 | Food |
| Philippines | 2003 | Food, feed |

Note: The current registration status of the transgenic crop is on http://www.biotradestatus.com/

## Global Registration Status of Maize Line MON 863

Table 5.179 shows the countries that had granted registration to use transgenic maize line MON 863 at the time of registration in Russia [19].

## Safety Assessment of Transgenic Maize Line MON 863 Conducted in the Russian Federation

Studies were conducted in accordance with the requirements of the Ministry of Health of Russian Federation authorized for risk and safety assessment of food derived from GM sources [8]. PCR analysis of the maize test and control samples was performed to confirm the identity of the transformation event and its absence in the conventional control line.

### *Biochemical Composition of Transgenic Maize Grain*

The content of proteins and amino acid composition in the grain of transgenic maize line MON 863 did not significantly differ from the corresponding values for conventional maize (Table 5.180).

The contents of carbohydrates and lipids in the grain of both maize cultivars were similar (Tables 5.181 and 5.182).

Similarly, the fatty acid and vitamin composition did not significantly differ in test and control maize samples (Tables 5.183 and 5.184).

The content of minerals in transgenic maize MON 863 and conventional non-transgenic maize did not significantly differ. The revealed variations in the content of selenium did not surpass the physiological values characteristic of this maize: 50–250 µg/100 g (Table 5.185).

The content of heavy metals and mycotoxins in the grains of conventional and transgenic maize line MON 863 did not surpass the acceptable limits according to the regulations of the Russian Federation (Table 5.186) [6].

### 5.2.4 Diabrotica-Resistant Transgenic Maize Line MON 863

**Table 5.180** Protein Content (%) and Amino Acid Composition (g/100 g protein) in Maize Grain

| Ingredient | Conventional Maize | GM Maize Line MON 863 |
|---|---|---|
| Protein | 7.10 | 8.12 |
| *Amino acids* | | |
| Lysine | 3.59 | 3.21 |
| Histidine | 4.45 | 4.88 |
| Arginine | 3.65 | 3.30 |
| Aspartic acid | 5.73 | 5.85 |
| Threonine | 3.42 | 3.26 |
| Serine | 4.78 | 4.78 |
| Glutamic acid | 19.30 | 19.20 |
| Proline | 7.00 | 7.33 |
| Glycine | 3.94 | 3.62 |
| Alanine | 6.46 | 6.66 |
| Cysteine | 1.41 | 1.36 |
| Valine | 5.32 | 5.12 |
| Methionine | 1.30 | 1.24 |
| Isoleucine | 3.86 | 3.62 |
| Leucine | 11.07 | 12.40 |
| Tyrosine | 4.00 | 3.79 |
| Phenylalanine | 5.00 | 4.83 |

**Table 5.181** Carbohydrates Content (g/100 g product) in Maize Grain

| Carbohydrate | Conventional Maize | Transgenic Maize Line MON 863 |
|---|---|---|
| Fructose | 0.021 | 0.02 |
| Glucose | 0.01 | 0.02 |
| Sucrose | 1.90 | 2.40 |
| Starch | 54.90 | 53.10 |
| Cellulose | 1.99 | 2.10 |

**Table 5.182** Lipids Content (%) in Maize Grain

| Conventional Maize | Transgenic Maize Line MON 863 |
|---|---|
| 4.15 | 4.20 |

Thus, the above data demonstrated that grain of transgenic maize line MON 863 and conventional maize did not significantly differ by biochemical composition. The revealed differences in the content of metals and vitamins remained within the range characteristic of maize [15,43,49–51].

**Table 5.183** Fatty Acids Content (Rel. %) in Maize Grain

| Fatty Acid | Conventional Maize | Transgenic Maize Line MON 863 |
|---|---|---|
| Lauric 12:0 | 0.01 | 0.01 |
| Myristic 14:0 | 0.05 | 0.06 |
| Pentadecanoic 15:0 | 0.01 | 0.01 |
| Palmitic 16:0 | 14.45 | 15.18 |
| Palmitoleic 16:1 | 0.11 | 0.10 |
| Margaric (heptadecanoic) 17:0 | 0.09 | 0.11 |
| Heptadecenoic 17:1 | 0.02 | 0.02 |
| Stearic 18:0 | 2.42 | 2.80 |
| cis-9-Oleic 18:1 | 20.17 | 19.85 |
| trans-11-Vaccenic 18:1 | 0.68 | 0.60 |
| Linoleic 18:2 | 59.82 | 58.25 |
| Linolenic 18:3 | 1.14 | 0.97 |
| Arachidic 20:0 | 0.47 | 0.59 |
| Gondoic 20:1 | 0.19 | 0.17 |
| Behenic 22:0 | 0.20 | 0.10 |
| Erucic 22:1 | 0.15 | 0.11 |

**Table 5.184** Vitamins Content (mg/100 g product) in Maize Grain

| Vitamin | Conventional Maize | Transgenic Maize Line MON 863 |
|---|---|---|
| Vitamin $B_1$ | 0.087 | 0.113 |
| Vitamin $B_2$ | 0.257 | 0.262 |
| Total carotenoids | 0.9 | 0.8 |
| β-Carotene | 0.1 | 0.1 |
| Total tocopherols | 5.6 | 4.5 |
| Vitamin E (α-tocopherol) | 0.7 | 0.8 |

**Table 5.185** Mineral Composition of Maize Grain

| Mineral | Conventional Maize | Transgenic Maize Line MON 863 |
|---|---|---|
| Sodium, mg/kg | 89.1 | 84.1 |
| Calcium, mg/kg | 5.87 | 4.64 |
| Magnesium, mg/kg | 1160 | 1148 |
| Iron, mg/kg | 42.6 | 37.9 |
| Potassium, mg/kg | 3993 | 4285 |
| Zinc, mg/kg | 16.4 | 16.2 |
| Copper, mg/kg | 1.73 | 1.72 |
| Selenium, μg/kg | 131 | 67 |

### 5.2.4 Diabrotica-Resistant Transgenic Maize Line MON 863

**Table 5.186** Analysis of Toxic Elements of Maize Grain (mg/kg)

| Ingredient | Conventional Maize | Transgenic Maize Line MON 863 |
|---|---|---|
| Deoxynivalenol | Not detected | Not detected |
| Zearalenone | Not detected | Not detected |
| Aflatoxin $B_1$ | Not detected | Not detected |
| Cadmium | 0.035 | 0.027 |
| Lead | 0.119 | 0.085 |

The safety parameters of the grain of conventional and transgenic maize line MON 863 comply with the requirements of the regulations valid in the Russian Federation [8].

### *Toxicological Assessment of Transgenic Maize Line MON 863*

The experiment was carried out on male Wistar rats with an initial body weight of 70–80 g. After admission to the vivarium of the State Research Institute of Nutrition, the rats were placed in quarantine for 10 days. At the onset of feeding the experimental diets, the body weight of rats was 85–95 g. During the entire experiment, the control rats were fed a standard semi-synthetic casein diet with conventional maize. The test rats were fed the diet with cereal derived from transgenic maize line MON 863. The experimental diet is described in Table 5.162. The biochemical, hematological, and morphological studies were conducted in accordance to the requirements of the Ministry of Health of the Russian Federation authorized for risk and safety assessment of food derived from GM sources [8].

Throughout the entire duration of the experiment, the general condition of the rats was similar in control and test groups. In both groups, no mortality was observed.

Throughout the entire duration of the experiment, the daily feed intake of maize did not significantly differ in control and test groups: 6.5 g/day/rat (the first two weeks), 7.6 g/day/rat (Week 3 and 4), 8.0–9.4 g/day/rat (Month 2–6).

The comparison of body weight of the control and test rats throughout the duration of the experiment is provided in Figure 5.9 and Table 5.187.

Starting from week 2, the body weight of test rats was greater than that of control animals by 10–15%. In 2, 3, and 4 weeks, the differences in body weight between groups became significant. However, these differences remained within the age-related physiological range.

The absolute and relative weight of internal organs of the rats was measured after sacrificing the animals on experimental Days 30 and 180 (Table 5.188). On Day 30, the weight of kidneys and testicles in test rats was lower than in

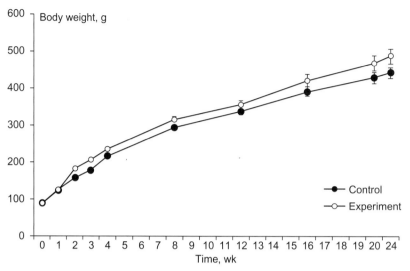

**FIGURE 5.9**
Comparative dynamics of body weight of rats fed diet with transgenic maize line MON 863 (test) or conventional maize (control).

**Table 5.187** Body Weight (g) of Rats Fed Diet with Conventional Maize (Control) or Transgenic Maize Line MON 863 (Test) ($M \pm m$; $n = 25$)

| Duration of Experiment, Weeks | Control | Test |
|---|---|---|
| 0 | 88.1 ± 2.1 | 87.9 ± 2.3 |
| 1 | 125.3 ± 3.4 | 126.9 ± 4.1 |
| 2 | 158.0 ± 2.7 | 181.0 ± 3.1* |
| 3 | 178.5 ± 3.1 | 205.5 ± 3.7* |
| 4 | 214.5 ± 5.6 | 236.0 ± 4.1* |
| 8 | 294.0 ± 8.8 | 314.5 ± 9.7 |
| 12 | 339.0 ± 10.2 | 356.5 ± 10.9 |
| 16 | 393.0 ± 12.3 | 422.0 ± 16.2 |
| 20 | 429.0 ± 14.6 | 469.0 ± 19.5 |
| 24 | 443.5 ± 14.9 | 488.0 ± 20.3 |

*$p < 0.05$ in comparison with control.

the controls: the absolute weight by 7% ($p > 0.05$) and 7% ($p < 0.05$), respectively, and the relative weight by 11% and 20%, respectively. In 180 days, there were no significant differences in the weight of internal organs. The revealed differences in the weight of internal organs remained within the age-related physiological range of these animals: 1–3 g (absolute weight of the

### 5.2.4 Diabrotica-Resistant Transgenic Maize Line MON 863

**Table 5.188** Absolute and Relative Weight of Internal Organs of Rats Fed Diet with Conventional Maize (Control) or Transgenic Maize Line MON 863 (Test) ($M \pm m$, $n = 8$)

| Organ | | 30 Days | | 180 Days | |
|---|---|---|---|---|---|
| | | Control | Test | Control | Test |
| Kidneys | Abs.[a], g | 2.08 ± 0.08 | 1.93 ± 0.08 | 2.39 ± 0.14 | 2.52 ± 0.06 |
| | Rel.[b], g/100 g | 0.64 ± 0.02 | 0.57 ± 0.01* | 0.62 ± 0.04 | 0.57 ± 0.02 |
| Liver | Abs., g | 13.43 ± 0.24 | 13.54 ± 1.20 | 12.61 ± 0.05 | 12.26 ± 0.55 |
| | Rel., g/100 g | 4.09 ± 0.11 | 3.95 ± 0.08 | 2.90 ± 0.05 | 2.70 ± 0.08 |
| Spleen | Abs., g | 1.64 ± 0.11 | 1.84 ± 0.31 | 1.39 ± 0.12 | 1.39 ± 0.07 |
| | Rel., g/100 g | 0.57 ± 0.02 | 0.51 ± 0.08 | 0.32 ± 0.06 | 0.31 ± 0.02 |
| Heart | Abs., g | 0.99 ± 0.07 | 1.00 ± 0.03 | 1.26 ± 0.12 | 1.28 ± 0.06 |
| | Rel., g/100 g | 0.30 ± 0.01 | 0.30 ± 0.01 | 0.29 ± 0.04 | 0.29 ± 0.01 |
| Testicles | Abs., g | 3.37 ± 0.06 | 3.14 ± 0.08* | 3.21 ± 0.18 | 3.36 ± 0.15 |
| | Rel., g/100 g | 1.16 ± 0.06 | 0.93 ± 0.03* | 0.74 ± 0.03 | 0.74 ± 0.02 |
| Hypophysis | Abs., mg | 10.30 ± 0.55 | 11.50 ± 0.32 | 7.60 ± 0.32 | 7.20 ± 0.79 |
| | Rel., mg/100 g | 3.15 ± 0.15 | 3.43 ± 0.19 | 1.80 ± 0.11 | 1.61 ± 0.20 |
| Adrenal glands | Abs., mg | 25.30 ± 1.15 | 27.50 ± 1.60 | 26.80 ± 2.40 | 21.20 ± 2.50 |
| | Rel., mg/100 g | 7.80 ± 0.24 | 8.08 ± 0.36 | 6.10 ± 0.69 | 4.10 ± 0.90 |
| Seminal vesicles | Abs., mg | 476.3 ± 27.5 | 492.5 ± 37.1 | 576.6 ± 49.5 | 577.6 ± 68.2 |
| | Rel., mg/100 g | 138.2 ± 10.8 | 146.7 ± 10.3 | 137.7 ± 8.4 | 128.7 ± 14.4 |
| Prostate | Abs., mg | 311.7 ± 24.4 | 290.3 ± 33.3 | 371.1 ± 46.6 | 348.3 ± 71.6 |
| | Rel., mg/100 g | 93.3 ± 6.2 | 88.2 ± 10.3 | 89.6 ± 14.3 | 73.9 ± 15.5 |

[a] Absolute weight of internal organs.
[b] Relative weight of internal organs (per 100 g body weight).
*$p < 0.05$ in comparison with control.

kidneys), 2–6 g (absolute weight of the testicles), 0.4–0.8 g/100 g body weight (relative weight of kidneys), and 0.6–1.5 g/100 g body weight (relative weight of testicles).

#### Assessment of Biochemical Parameters

Tables 5.189 and 5.190 show that biochemical parameters of blood serum and urine of the test rats fed diet with transgenic maize for 180 days did not significantly differ from the corresponding parameters of the control rats fed diet with conventional maize.

#### Assessment of Sensitive Biomarkers

Activity of enzymes involved in phase I and phase II xenobiotic degradation in test rats fed diet with transgenic maize line MON 863 did not significantly differ from the corresponding values for the control rats fed diet with conventional maize (Table 5.191).

**Table 5.189** Biochemical Parameters of Blood Serum of Rats Fed Diet with Conventional Maize (Control) or Transgenic Maize Line MON 863 (Test) ($M \pm m$, $n = 8$)

| Parameter | 30 Days | | 180 Days | |
|---|---|---|---|---|
| | Control | Test | Control | Test |
| Total protein, g/L | 75.8 ± 4.9 | 72.8 ± 3.3 | 69.4 ± 2.9 | 71.2 ± 5.4 |
| Glucose, mM/L | 5.04 ± 0.60 | 5.10 ± 0.45 | 6.00 ± 0.52 | 6.10 ± 0.61 |
| Alanine aminotransferase, μcat/L | 0.75 ± 0.07 | 0.76 ± 0.03 | 0.82 ± 0.03 | 0.86 ± 0.04 |
| Aspartate aminotransferase, μcat/L | 0.55 ± 0.06 | 0.62 ± 0.06 | 0.69 ± 0.12 | 0.49 ± 0.18 |
| Alkaline phosphatase, μcat/L | 7.55 ± 0.61 | 6.23 ± 0.97 | 6.86 ± 0.24 | 6.24 ± 1.10 |

Note: Here and in Tables 5.190 to 5.194 the differences are not significant ($p > 0.05$).

**Table 5.190** Urinary Biochemical Parameters of Rats Fed Diet with Conventional Maize (Control) or Transgenic Maize Line MON 863 (Test) ($M \pm m$, $n = 6$–$8$)

| Parameter | 30 Days | | 180 Days | |
|---|---|---|---|---|
| | Control | Test | Control | Test |
| pH | 7.0 | 7.0 | 7.0 | 7.0 |
| Daily diuresis, mL | 3.5 ± 0.3 | 3.0 ± 0.7 | 5.0 ± 0.6 | 4.6 ± 0.4 |
| Relative density, g/mL | 1.18 ± 0.05 | 1.23 ± 0.05 | 1.14 ± 0.02 | 1.12 ± 0.01 |
| Creatinine, mg/mL | 1.60 ± 0.11 | 2.13 ± 0.30 | 1.68 ± 0.18 | 1.75 ± 0.14 |
| Creatinine, mg/day | 5.69 ± 0.65 | 5.48 ± 0.46 | 8.15 ± 0.76 | 7.84 ± 0.61 |

**Table 5.191** Activity of Enzymes Involved in Metabolism of Xenobiotics in Rats Fed Diet with Conventional Maize (Control) or Transgenic Maize Line MON 863 (Test) ($M \pm m$, $n = 8$)

| Parameter | 180 Days | |
|---|---|---|
| | Control | Test |
| Cytochrome P450, nM/mg protein | 0.97 ± 0.08 | 1.00 ± 0.06 |
| Cytochrome $b_5$, nM/mg protein | 0.76 ± 0.03 | 0.80 ± 0.02 |
| Aminopyrine N-demethylation, nM/min/mg protein | 10.4 ± 0.4 | 10.5 ± 0.5 |
| Benzpyrene hydroxylation, Fl-units/min/mg protein | 8.10 ± 0.61 | 7.85 ± 0.48 |
| Acetylesterase, μM/min/mg protein | 8.67 ± 0.31 | 9.25 ± 0.28 |
| Epoxide hydrolase, nM/min/mg protein | 9.43 ± 1.05 | 9.72 ± 0.40 |
| UDP- glucuronosil transferase, nM/min/mg protein | 26.0 ± 2.3 | 27.8 ± 1.3 |
| CDNB-glutathione transferase, μM/min/mg protein | 58.5 ± 4.2 | 56.2 ± 3.4 |
| Protein, mg/g | 13.2 ± 0.4 | 13.4 ± 0.3 |

**Table 5.192** Activity of Hepatic Lysosomal Enzymes in Rats Fed Diet with Conventional Maize (Control) or Transgenic Maize Line MON 863 (Test) ($M \pm m$, $n = 8$)

| Parameters | 30 Days | | 180 Days | |
|---|---|---|---|---|
| | Control | Test | Control | Test |
| *Total activity, µM/min/g tissue* | | | | |
| Arylsulfatase A, B | 2.11 ± 0.02 | 2.16 ± 0.03 | 2.24 ± 0.03 | 2.22 ± 0.05 |
| β-Galactosidase | 2.12 ± 0.04 | 2.16 ± 0.05 | 2.35 ± 0.08 | 2.30 ± 0.06 |
| β-Glucuronidase | 2.32 ± 0.08 | 2.25 ± 0.04 | 2.39 ± 0.05 | 2.32 ± 0.04 |
| *Non-sedimentable activity, % total activity* | | | | |
| Arylsulfatase A, B | 3.14 ± 0.05 | 3.09 ± 0.06 | 3.59 ± 0.09 | 3.50 ± 0.05 |
| β-Galactosidase | 5.47 ± 0.21 | 5.33 ± 0.27 | 5.58 ± 0.26 | 5.29 ± 0.40 |
| β-Glucuronidase | 4.76 ± 0.22 | 4.87 ± 0.14 | 5.18 ± 0.20 | 5.28 ± 0.14 |

**Table 5.193** Content of LPO Products in Blood and Liver of Rats Fed Diet with Conventional Maize (Control) or Transgenic Maize Line MON 863 (Test) ($M \pm m$, $n = 6$–$8$)

| Parameter | 30 Days | | 180 Days | |
|---|---|---|---|---|
| | Control | Test | Control | Test |
| *Erythrocytes* | | | | |
| DC, nM/mL | 5.105 ± 0.420 | 5.300 ± 0.336 | 4.287 ± 0.398 | 4.190 ± 0.280 |
| MDA, nM/mL | 2.366 ± 0.074 | 2.423 ± 0.084 | 2.770 ± 0.092 | 3.017 ± 0.177 |
| *Blood serum* | | | | |
| DC, nM/mL | 4.069 ± 0.234 | 3.530 ± 0.226 | 3.172 ± 0.150 | 3.497 ± 0.147 |
| MDA, nM/mL | 6.676 ± 0.136 | 5.728 ± 0.161* | 6.126 ± 0.170 | 5.521 ± 0.127* |
| *Liver* | | | | |
| DC, Unit | 1.052 ± 0.018 | 1.037 ± 0.019 | 1.029 ± 0.020 | 1.002 ± 0.020 |
| MDA, nM/g | 394.3 ± 17.1 | 315.8 ± 10.0* | 386.4 ± 15.7 | 403.8 ± 21.9 |

Note: Here and in Table 5.194.
*$p < 0.05$ in comparison with control.

Activity of the hepatic lysosomal enzymes did not significantly differ between the two groups of rats over the entire duration of the experiment (Table 5.192).

Addition of transgenic maize to the rat diet for 180 days did not elevate the content of LPO products in blood serum and liver (Table 5.193).

Enzymatic activity of the antioxidant protection system did not significantly differ between groups in 30 and 180 days. A single exception was a small (by 12%) increase in the content of catalase in the test group on Day 30, which

**Table 5.194** Activity of Enzymes of the Antioxidant Protection System in Blood and Liver of Rats Fed Diet with Maize (Control) or Transgenic Maize Line MON 863 (Test) ($M \pm m$, $n = 8$)

| Parameter | 30 Days | | 180 Days | |
|---|---|---|---|---|
| | Control | Test | Control | Test |
| Glutathione reductase, μmol/min/g Hb | 39.09 ± 2.39 | 43.95 ± 1.35 | 34.66 ± 1.53 | 32.96 ± 1.47 |
| Glutathione peroxidase, μmol/min/g Hb | 60.78 ± 1.14 | 62.12 ± 2.90 | 56.02 ± 1.58 | 58.34 ± 1.98 |
| Catalase, mmol/min/g Hb | 405.1 ± 7.2 | 451.9 ± 16.1* | 456.9 ± 7.2 | 460.4 ± 11.7 |
| Superoxide dismutase, U/min/g Hb | 1446 ± 21 | 1510 ± 33 | 1908 ± 26 | 1934 ± 39 |

**Table 5.195** Hematological Parameters of Peripheral Blood Drawn from Rats Fed Diet with Maize (Control) or Transgenic Maize Line MON 863 (Test) ($M \pm m$, $n = 8$)

| Parameter | 30 Days | | 180 Days | |
|---|---|---|---|---|
| | Control | Test | Control | Test |
| Hemoglobin concentration, g/L | 158.2 ± 1.9 | 152.0 ± 5.0 | 144.8 ± 4.0 | 143.3 ± 5.6 |
| Total erythrocyte count, $\times 10^{12}$/L | 6.98 ± 0.28 | 6.92 ± 0.27 | 6.83 ± 0.23 | 6.85 ± 0.18 |
| Hematocrit, vol.% | 50.33 ± 0.52 | 50.50 ± 0.35 | 50.51 ± 0.48 | 50.70 ± 0.50 |
| MCH, pg | 22.78 ± 0.53 | 22.00 ± 0.65 | 21.00 ± 1.07 | 21.40 ± 0.41 |
| MCHC, % | 31.36 ± 0.58 | 30.10 ± 0.79 | 28.61 ± 1.10 | 28.89 ± 0.77 |
| MCV, μm$^3$ | 72.51 ± 1.92 | 73.30 ± 2.36 | 74.15 ± 1.74 | 74.00 ± 1.33 |
| Total leukocyte count, $\times 10^9$/L | 11.98 ± 0.39 | 12.32 ± 0.19 | 15.25 ± 1.36 | 14.29 ± 1.00 |

*Note: Here and in Table 5.196 the differences are not significant ($p > 0.05$).*

nevertheless did not surpass the age-related physiological variations in rats (Table 5.194).

### Hematological Assessments

Hematological parameters of peripheral blood drawn from the test and control group rats were examined. Over the entire period of the experiment, there were no significant differences between the groups of rats in all examined hematological and leukogram parameters (Tables 5.195 and 5.196).

### Morphological Assessments

Over the entire period of the experiment, there was no mortality in the control group of rats fed diet with conventional maize or in the group of test rats fed diet with transgenic maize line MON 863. Post-mortem examination revealed no alterations in the internal organs in both groups of rats. Similarly, histological examinations of internal organs performed on Days 30 and 180 did not revealed any statistically significant differences (Table 5.197).

Therefore, the chronic toxicological experiment carried out during 180 days with biochemical, hematological, and morphological examinations revealed no adverse affects of transgenic maize line MON 863 on the animals.

### 5.2.4 Diabrotica-Resistant Transgenic Maize Line MON 863

**Table 5.196** Leukogram Parameters of Control and Test Rats ($M \pm m$, $n = 8$)

| Parameter | 30 Days | | 180 Days | |
|---|---|---|---|---|
| | Control | Test | Control | Test |
| **Neutrophils** | | | | |
| rel., % | 20.51 ± 0.90 | 20.16 ± 1.23 | 20.17 ± 0.71 | 19.80 ± 1.23 |
| abs., ×10⁹/L | 2.44 ± 0.09 | 2.47 ± 0.14 | 3.12 ± 0.31 | 2.85 ± 0.34 |
| **Eosinophils** | | | | |
| rel., % | 0.79 ± 0.35 | 0.83 ± 0.18 | 0.70 ± 0.35 | 0.70 ± 0.35 |
| abs., ×10⁹/L | 0.11 ± 0.04 | 0.10 ± 0.02 | 0.11 ± 0.05 | 0.09 ± 0.05 |
| **Lymphocytes** | | | | |
| rel., % | 78.20 ± 0.88 | 79.00 ± 1.23 | 84.40 ± 1.34 | 85.20 ± 2.30 |
| abs., ×10⁹/L | 9.39 ± 0.40 | 9.74 ± 0.35 | 12.02 ± 1.00 | 11.35 ± 0.84 |

**Table 5.197** Microscopic Examination of Internal Organs in Rats Fed Diet with Conventional Maize (Control) or Transgenic Maize Line MON 863 (Test) (Combined Data Obtained on Experimental Days 30 and 180)

| Organ | Control | Test |
|---|---|---|
| Liver | Clear trabecular structure; no alterations in hepatocytes or portal ducts | No differences from control |
| Kidneys | Usual appearance of cortical and medullar substance; no alterations in glomeruli or pelvis epithelium | No differences from control |
| Lung | Alveolar space is air-filled; no alterations in bronchi or blood vessels | No differences from control |
| Spleen | Large folliculi with wide clear marginal zones and reactive centers; splenic pulp is moderately plethoric | No differences from control |
| Small intestine | Preserved villous epithelium; usual infiltration in villous stroma | No differences from control |
| Testicle | Clearly definable spermatogenesis and age-related spermiogenesis | No differences from control |

## Assessment of Potential Impact of Transgenic Maize Line MON 863 on Immune System in Studies on Mice
### Potential Effect on Humoral Component of Immune System

The immunomodulating effect of transgenic maize line MON 863 on the humoral component of the immune system was examined on two oppositely reacting mice lines, C57Bl/6 and CBA, by determining the level of hemagglutination to sheep erythrocytes (SE).

The transgenic maize and its conventional counterpart were fed to test and control mice (respectively) in daily dose of 2.4 g/mouse for 21 days. The experimental conditions are described in section 5.1.1, and the diet is shown in Table 5.114.

In test and control CBA mice fed diet with transgenic or conventional maize, respectively, the antibodies appeared on post-immunization Day 14 at the titers of 1:64 and 1:55, correspondingly, and remained to Day 21 at the titers of 1:85 and 1:32. There were no significant differences in the levels of antibodies in CBA mice. Under the same conditions, C57Bl/6 exhibited antibodies on post-immunization Day 14 at the titers of 1:128, which remained to Day 21 (1:85 ± 5). Thus, dynamics of antibody production were identical in control and test C57Bl/6 mice.

### Potential Effect on Cellular Component of Immune System

The immunomodulating effect of transgenic maize was assessed by delayed hypersensitivity reaction to SE. The experimental conditions are described in section 5.1.1. In CBA mice, RI did not differ between the test and control animals fed transgenic or conventional maize (41.7 ± 9.2 and 47.2 ± 10.1, respectively). In test and control C57Bl/6 mice, the RI values were 41.6 ± 10.3 and 54.3 ± 13.6, respectively, which were not statistically significantly different from each other.

### Assessment of Potential Sensibilization Effect of Transgenic Maize

Assessment of possible sensitizing action of transgenic maize line MON 863 on the immune response to endogenous metabolic products was conducted in the test of mouse sensitivity to histamine. The experiments were carried out on female C57Bl/6 mice. Diets containing transgenic maize (test) and conventional maize (control) were fed to mice for 21 days; thereafter the mice of both groups were injected intraperitoneally with 2.5 mg histamine hydrochloride dissolved in 0.5 mL physiological solution. The reaction was assessed in 1 h and 24 h by the mortality of the mice. In this test, there was no mortality or differences in behavior between test and control mice, which attests to the absence of any extra sensitization agent in transgenic maize line MON 863 in comparison with conventional maize.

### Potential Effect of Transgenic Maize on Susceptibility of Mice to Salmonella typhimurium

The effect of transgenic maize MON 863 on susceptibility of mice to infection by salmonella of murine typhus was examined in the experiments on C57Bl/6 mice injected intraperitoneally with various doses of Strain 415 *Salmonella typhimurium*. Three weeks prior to infection, the diet of control and test mice was supplemented with conventional and transgenic maize MON 863, respectively. The injected doses ranged from $10^2$ to $10^5$ microbial cells per mouse and varied on a 10-fold basis. The post-injection observation period was 21 days. The following data were obtained:

- In both groups, mortality was observed in post-injection week 1, and all mice died by post-injection Day 21;
- The mean lifetime of mice was approximately equal (14–16 days);

- $LD_{50}$ values in control and test groups were 127 and 162 microbial cells, correspondingly.

These data showed that *Salmonella typhimurium* produced a typical infection both in control and test mice. The course of murine typhus was similar in both groups. Therefore, transgenic maize line MON 863 had no sensitizing potencies; it did not affect the resistance of mice to *S. typhimurium* or the development of humoral and cellular immunity in mice.

### Assessment of Potential Impact of Transgenic Maize Line MON 863 on Immune System in Studies on Rats

The study was carried out on Wistar rats ($n = 49$) weighing initially $140 \pm 10$ g. After a 7-day adaptation period to standard vivarium diet, in the following 28 days the rats were fed diet supplemented with conventional maize (control group) or transgenic maize line MON 863 (test group). Maize flour of both types was dissolved in boiled water to the consistency of dense curd and supplemented with sunflower-seed oil to improve intake. The feed was used instead of an equally caloric amount of oatmeal.

The model of generalized anaphylaxis was developed according to [8] as described in section 5.1.1.

On Day 29, the body weights of test and control rats were practically identical ($275 \pm 8$ g and $277 \pm 8$ g, correspondingly, $p > 0.05$). The growth of rats in both groups was somewhat slower than that of the rats fed standard vivarium diet, which probably indicates low nutritional value of the protein in the examined maize samples.

Table 5.198 shows the data on severity of anaphylactic reaction in control and test rats. By all examined parameters, there was no difference in severity of anaphylactic reactions ($p > 0.1$).

Table 5.199 shows the mean values of $D_{492}$, concentration of antibodies, and common logarithm of this concentration in the control and test groups. By all parameters, the difference between test rats fed diet with transgenic maize Vline MON 863 and the control rats maintained on conventional maize diet

**Table 5.198** Comparative Severity of Active Anaphylactic Shock in Rats Fed Diet with Conventional Maize (Control) or Transgenic Maize Line MON 863 (Test)

| Group of Rats | AI | Severe Reactions, % | Mortality, % |
|---|---|---|---|
| Control ($n = 23$) | 2.00 | 26.1 | 26.1 |
| Test ($n = 264$) | 2.31 | 38.5 | 38.5 |

**Table 5.199** Comparative Intensity of Humoral Immune Response Assessed by the Level of Specific IgG Antibodies Raised Against Ovalbumin in Rats ($M \pm m$)

| Group of Rats | $D_{492}$ | Concentration of Antibodies, mg/mL | Logarithm of Antibody Concentration |
|---|---|---|---|
| Transgenic Maize Line MON 863 ($n = 23$) | $0.622 \pm 0.036$ | $4.8 \pm 0.4$ | $0.624 \pm 0.046$ |
| Conventional maize ($n = 23$) | $0.661 \pm 0.049$ | $5.4 \pm 0.60$ | $0.634 \pm 0.079$ |

was insignificant ($p > 0.1$). Thus, the intensity of humoral immune response was practically identical in both groups of animals.

On the whole, these data support the conclusion that the degree of sensitization by ovalbumin in test rats fed diet with transgenic maize line MON 863 did not increase in comparison with the control rats fed diet with conventional maize. The studies showed that transgenic maize MON 863 did not significantly change the allergic reactivity and degree of sensitization produced by a model allergen in comparison with these factors in the rats fed diet with conventional maize.

### Assessment of Potential Genotoxicity of Transgenic Maize

The genotoxic studies were carried out on male C57Bl/6 mice and hybrid female CBA mice fed diet shown in Table 5.162. During the experiment, the animals were fed diet composed of a soft feed with milled maize of test or control variety.

The cytogenetic analysis was carried out by metaphasic method; genetic alterations in the sex cells were examined by revealing the dominant lethal mutations in C57Bl/6 male mice [8]. The details of the experiments are described in section 5.1.1.

The results of possible genotoxic effects of transgenic maize on mice are shown in Tables 5.200 and 5.201.

The major structural chromosomal abnormalities in control and test mice were single segments and gaps. There were no significant differences in the numbers of cells with gaps. In male C57Bl/6 mice, the number of observed structural chromosomal abnormalities was typical. Such mutations appear spontaneously; they are not stable and usually disappear in the subsequent nuclear divisions.

The pre-implantation mortality of unfertilized ovocytes, zygotes, and embryos in test mice did not surpass the corresponding values in the control group. The post-implantation embryonic mortality (the most reliable

### 5.2.4 Diabrotica-Resistant Transgenic Maize Line MON 863

**Table 5.200** Cytogenetic Values of Bone Marrow in Mice Fed Diet with Conventional Maize (Control) or Transgenic Maize Line MON 863 (Test) ($M \pm m$)

| Number of Cells, % | Control | Test |
|---|---|---|
| with chromosomal aberrations | $0.83 \pm 0.47$ | $0.84 \pm 0.48$ |
| with gaps | $1.11 \pm 0.54$ | $0.56 \pm 0.40$ |
| with polyploid chromosome set | $0.55 \pm 0.38$ | $0.56 \pm 0.40$ |

*Note: The numbers of analyzed metaphases were: 355 test and 360 control. The differences are not significant ($p > 0.05$).*

**Table 5.201** Dominant Lethal Mutations in Sex Cells of Mice Fed Diet with Conventional Maize (Control) or Transgenic Maize Line MON 863 (Test)

| Parameter, % | Week 1, Mature Spermatozoa | | Week 2, Late Spermatids | | Week 3, Early Spermatids | |
|---|---|---|---|---|---|---|
| | Control | Test | Control | Test | Control | Test |
| Pre-implantation mortality | 5.55 | 2.63 | 4.70 | 5.12 | 6.14 | 4.50 |
| Post-implantation mortality | 2.35 | 2.16 | 3.70 | 1.62 | 3.57 | 1.88 |
| Survival rate | 92.22 | 95.26 | 91.76 | 93.33 | 90.50 | 94.14 |
| Induced mortality | – | 0 | – | 0 | – | 0 |

*Note: In the control group, 500 embryos and 529 corpus lutea from 66 control females were analyzed. The corresponding values in the test group (70 females) were 582 and 607.*

index of mutagenic activity of an examined agent) did not significantly differ between groups. There was no induced mortality, which attests to the absence of adverse effects of transgenic maize on spermiogenesis in mice. The data obtained support the conclusion that transgenic maize MON 863 produces no mutagenic effects in animals.

#### Assessment of Technological Parameters

Assessment of technological parameters was carried out in Moscow State University of Applied Biotechnology (Ministry of Science and Education of Russian Federation).

To characterize the samples of transgenic maize line MON 863 resistant to corn rootworm and non-transgenic conventional maize, the moisture and ash contents were determined. Maize starch was produced under laboratory conditions to determine protein mass fraction in the starch, gelatinization temperature and viscosity of the starch gelatins on amylograph, the parameters of thermoplastic extrusion, and the structural and mechanical properties of the extrudates.

The transgenic maize grain complied with the specifications of Russian State Standards GOST 136-90 "Maize. Technical requirements". The study resulted in the following conclusions:

- No difficulties or differences were observed in the technological process when producing starch from transgenic maize MON 863 or conventional maize;
- According to Russian State Standards GOST 7698-93 "Starch. Formal Acceptance and Analytical Methods", the quality of all examined samples was superior;
- By gelatinization temperature and rheological properties of 7% gelatins the starches derived from transgenic maize did not differ from those obtained from conventional maize;
- The structural and mechanical properties of the extrusive products derived from transgenic maize line MON 863 were virtually identical to those obtained from the conventional maize.

Thus, the study revealed no significant differences in the properties of examined grain samples of transgenic maize line MON 863 and its conventional counterpart.

### Conclusions

By all examined parameters, the data of complex safety assessment of transgenic maize line MON 863 resistant to corn rootworm attest to the absence of any toxic, genotoxic, immune system modulating, or allergenic effects of this maize line. By chemical composition, transgenic maize line MON 863 was identical to conventional maize.

Based on the results of the studies, the State Sanitation Service of the Russian Federation (Department of State Sanitation and Epidemiological Inspectorate) granted the Registration Certificate which allows the transgenic maize line MON 863 to be used in the food industry and placed on the market without restrictions.

## 5.2.5 TRANSGENIC MAIZE LINE BT11 RESISTANT TO EUROPEAN CORN BORER AND TOLERANT TO GLUFOSINATE AMMONIUM

### Molecular Characteristics of Transgenic Maize Line Bt11
*Recipient Organism*

Maize (*Zea mays* L.) is characterized by a long history of safe use as human food and animal feed.

## 5.2.5 Transgenic Maize Line Bt11 Resistant to European Corn

### *Donor Organism*
The donor of the *cry 1Ab* gene responsible for resistance to damage by the European corn borer *Ostrinia nubilalis* is a widespread gram-positive soil bacterium *B. thuringiensis*, which during sporification produces proteins with selective action against a narrow group of insects, European corn borer included. The insecticidal Bt proteins bind to specific sites in the cells of the insect digestive system and form ion-selective channels in the cell membrane, resulting in lysis of the cells and death of the insects [25].

The donor of the *pat* gene *Streptomyces viridochromogenes* is a gram-positive spore-forming soil bacterium. The *pat* gene encodes the synthesis of phosphinothricin acetyltransferase (PAT) [32,47].

The herbicide glufosinate ammonium inhibits glutamine synthetase, which plays an important role in nitrogen metabolism in plants. Inhibition of this enzyme results in accumulation of ammonia and death of the plant cells. Phosphinothricin acetyltransferase (PAT) acetylates the free $NH_2$-group of glufosinate ammonium and prevents accumulation of ammonia.

### *Method of Genetic Transformation*
The plasmid DNA was incorporated into the protoplasts of inbred maize line H8540 by direct transformation of DNA during incubation in chemical solution and regeneration in selective medium [19]. The transformation plasmid vector pZO1502 was used.

The incorporated DNA contained the following major nucleotide sequences:

- *cry1Ab* gene from *B. thuringiensis* conferring resistance against European cornborer;
- 35S promoter of cauliflower mosaic virus controlling expression of *cry1Ab* gene;
- IVS6 intron of maize 1S alcohol dehydrogenase gene, which controls expression of plant genes;
- NOS 3′ terminator of nopaline synthase gene from *A. tumefaciens*;
- *pat* gene from *Streptomyces viridochromogenes* that confers tolerance to ammonium glufosinate;
- 35S promoter of the Figwort mosaic virus, that controls expression of IVS2 gene of maize 1S alcohol dehydrogenase, which controls expression of plant genes;
- The *bla* gene isolated from *E. coli*, which encodes synthesis of β-lactamase to provide tolerance to some antibiotics, is not expressed in the plant [19].

### Global Registration Status of Maize Line Bt11
Table 5.202 shows the regulatory status of Bt11 in various countries at the time of registration in Russia [19].

**Table 5.202** Regulatory Status of Transgenic Maize Line Bt11 in Various Countries

| Country | Date of Approval | Application |
| --- | --- | --- |
| Australia | 2001 | Food, feed |
| Argentina | 2001 | Food, feed, environmental release |
| UK | 1998 | Food, feed |
| Canada | 1996 | Food, feed, environmental release |
| Korea | 2003 | Food |
| USA | 1996 | Food, feed, environmental release |
| Philippines | 2003 | Food, feed |
| South Africa | 2002 | Food, feed |
| | 2003 | Environmental release |
| Japan | 1996 | Food, feed, environmental release |

*Note: The current registration status of the transgenic crop is on http://www.biotradestatus.com/*

## Safety Assessment of Transgenic Maize Line Bt11 Conducted in the Russian Federation

Studies were conducted in accordance with the requirements of the Ministry of Health of the Russian Federation authorized for risk and safety assessment of food derived from GM sources [8]. PCR analysis of the maize test and control samples was performed to confirm the identity of the transformation event and its absence in the conventional control line.

### *Biochemical Composition of Transgenic Maize Grain*

The content of proteins and amino acid composition in the grain of transgenic maize line Bt11 did not significantly differ from the corresponding values for the conventional maize (Table 5.203).

The contents of carbohydrates in the grain of both maize cultivars were similar (Table 5.204). The variations in the content of fructose and sucrose remained within the range characteristic of maize: 0.010–0.350 and 1.00–3.50 g/100 g, correspondingly (data from the State Research Institute of Nutrition, RAMS). There was no significant difference in the content of lipids in the grain of the two maize varieties (Table 5.205).

The vitamin composition was virtually the same in test and control maize samples (Table 5.206). Differences were found only in the contents of carotenoids and vitamin $B_2$, although they were within physiological values characteristic of maize according to the RAMS data: 0.2–1.5 (total carotenoids) and 0.05–0.30 mg/100 g (vitamin $B_2$).

The contents of minerals in transgenic maize Bt11 and conventional non-transgenic maize were similar (Table 5.207). The revealed variations in the content of

### 5.2.5 Transgenic Maize Line Bt11 Resistant to European Corn

**Table 5.203** Protein Content (%) and Amino Acid Composition (g/100 g protein) in Maize Grain

| Ingredient | Conventional Maize | Transgenic Maize Line Bt11 |
|---|---|---|
| Protein, % | 8.46 | 9.33 |
| *Amino acids* | | |
| Lysine | 2.64 | 3.48 |
| Histidine | 2.81 | 2.65 |
| Arginine | 9.25 | 10.31 |
| Aspartic acid | 7.11 | 10.58 |
| Threonine | 2.64 | 2.78 |
| Serine | 3.97 | 4.04 |
| Glutamic acid | 20.66 | 18.52 |
| Proline | 11.73 | 10.44 |
| Glycine | 2.15 | 2.37 |
| Alanine | 5.62 | 5.71 |
| Cysteine | 2.15 | 1.25 |
| Valine | 2.81 | 3.06 |
| Methionine | 1.82 | 0.83 |
| Isoleucine | 2.15 | 2.37 |
| Leucine | 11.57 | 11.14 |
| Tyrosine | 5.29 | 5.01 |
| Phenylalanine | 5.62 | 5.43 |

**Table 5.204** Content of Carbohydrates (g/100 g product) in Maize Grain

| Carbohydrate | Conventional Maize | Transgenic Maize Line Bt11 |
|---|---|---|
| Cellulose | 1.74 | 1.88 |
| Starch | 55.9 | 54.9 |
| Fructose | 0.013 | 0.11 |
| Glucose | 0.03 | 0.04 |
| Sucrose | 1.13 | 2.01 |

**Table 5.205** Content of Lipids (%) in Maize Grain

| Conventional Maize | Transgenic Maize Line Bt11 |
|---|---|
| 4.4 | 5.2 |

potassium, calcium, magnesium, and selenium were within physiological values characteristic for maize according to the data from the State Research Institute of Nutrition, RAMS: 3000–4800 mg/kg (potassium), 1.0–25.0 mg/kg (calcium), 1000–2000 mg/kg (magnesium), and 50.0–250.0 µg/100 g (selenium).

### Table 5.206 Content of Vitamins (mg/100 g product) in Maize Grain

| Ingredient | Conventional Maize | Transgenic Maize Line Bt11 |
|---|---|---|
| Vitamin $B_1$ | 0.164 | 0.202 |
| Vitamin $B_2$ | 0.112 | 0.062 |
| Vitamin $B_6$ | 0.216 | 0.240 |
| Vitamin E ($\alpha$-tocopherol) | 5.3 | 5.8 |
| $\beta$-Carotene | 0.10 | 0.07 |
| Total carotenoids | 0.83 | 0.46 |

### Table 5.207 Mineral Composition in Maize Grain

| Ingredient | Conventional Maize | Transgenic Maize Line Bt11 |
|---|---|---|
| Copper, mg/kg | 0.986 | 2.04 |
| Zinc, mg/kg | 21.5 | 18.5 |
| Iron, mg/kg | 65.5 | 45.8 |
| Sodium, mg/kg | 44.2 | 34.2 |
| Potassium, mg/kg | 3162 | 4523 |
| Calcium, mg/kg | 4.86 | 1.6 |
| Magnesium, mg/kg | 1166 | 1459 |
| Selenium, µg/kg | 62 | 141 |

### Table 5.208 Analysis of Toxic Elements of Maize Grain (mg/kg)

| Ingredient | Conventional Maize | Transgenic Maize Line Bt11 |
|---|---|---|
| Aflatoxin $B_1$ | Not detected | Not detected |
| Zearalenone | Not detected | Not detected |
| T2-toxin | Not detected | Not detected |
| Lead | 0.094 | 0.058 |
| Cadmium | <0.001 | 0 |

The content of mycotoxins in the grain of conventional and transgenic maize line Bt11 did not surpass the acceptable limits according to the regulations valid in the Russian Federation (Table 5.208) [6].

Thus, the above data showed that the grain of transgenic maize line Bt11 and conventional maize did not significantly differ by biochemical composition. The revealed differences in the content of calcium, potassium, magnesium, selenium, vitamin $B_1$, and carotenoids remained within the range characteristic of maize [15,43,49–51]. The safety parameters of the grain of conventional and transgenic maize line Bt11 comply with the requirements of the regulations valid in the Russian Federation [6].

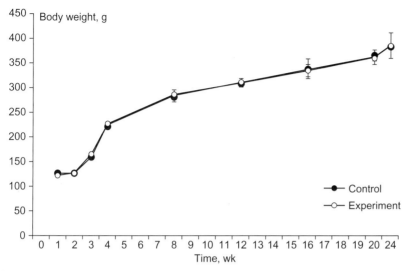

**FIGURE 5.10**
Comparative dynamics of body weight of rats fed diet with transgenic maize line Bt11 (test) or conventional maize (control).

### Toxicological Assessment of Transgenic Maize Line Bt11

The experiments were carried out on male Wistar rats with an initial body weight of 75–90 g. The control rats were fed the diet with conventional maize grain (3 g/day), while the test rats were fed diet with equal amount of cereals derived from transgenic maize line Bt11 (composition of the diet is given in Table 5.114).

The biochemical, hematological, and morphological studies were conducted in accordance with the requirements of the Ministry of Health of the Russian Federation authorized for risk and safety assessment of food derived from GM sources [6]. Duration of the chronic experiment was 180 days.

Throughout the entire duration of the experiment, the general condition of the rats was similar and satisfactory in the control and test groups. No mortality was observed in either group. The body weight of the control and test rats did not significantly differ over the entire term of the experiment (Figure 5.10 and Table 5.209).

The absolute weight of internal organs in control and test rats did not significantly differ (Table 5.210); slightly different results were obtained for the relative weight of organs. Increases in the relative weight of heart and hypophysis were observed on Day 30 in the test group. However, these changes remained within the age-related physiological range (heart: 0.3–0.5 g/100 g; hypophysis: 2.0–6.0 mg/100 g) and leveled by the end of the experiment.

## CHAPTER 5: Human and Animal Health Safety Assessment

**Table 5.209** Comparative Body Weight (g) of Rats Fed Diet with Conventional Maize (Control) or Transgenic Maize Line Bt11 ($M \pm m$; $n = 25$)

| Duration of Test, Weeks | Control | Test |
|---|---|---|
| 1 | 124.2 ± 2.1 | 120.5 ± 1.9 |
| 2 | 126.7 ± 2.8 | 126.3 ± 1.7 |
| 3 | 159.5 ± 2.9 | 163.5 ± 1.9 |
| 4 | 223.9 ± 5.5 | 226.5 ± 4.7 |
| 8 | 282.6 ± 11.4 | 285.8 ± 9.7 |
| 12 | 308.0 ± 7.5 | 310.5 ± 8.0 |
| 16 | 336.8 ± 20.6 | 332.8 ± 12.7 |
| 20 | 362.9 ± 8.9 | 360.9 ± 14.5 |
| 24 | 382.0 ± 7.3 | 384.2 ± 26.0 |

Note: The differences are not significant ($p > 0.05$).

**Table 5.210** Absolute and Relative Weight of Internal Organs of Rats Fed Diet with Conventional Maize (Control) or Transgenic Maize Line Bt11 (Test) ($M \pm m$, $n = 6\text{--}8$)

| Organ | | 30 Days | | 180 Days | |
|---|---|---|---|---|---|
| | | Control | Test | Control | Test |
| Kidneys | Abs.[a], g | 1.68 ± 0.09 | 1.62 ± 0.11 | 2.65 ± 0.15 | 2.42 ± 0.17 |
| | Rel.[b], g /100 g | 0.70 ± 0.02 | 0.73 ± 0.04 | 0.70 ± 0.04 | 0.67 ± 0.02 |
| Liver | Abs., g | 9.38 ± 0.34 | 8.77 ± 0.31 | 11.88 ± 0.20 | 10.30 ± 0.69 |
| | Rel., g /100 g | 3.90 ± 0.09 | 3.93 ± 0.06 | 3.12 ± 0.15 | 2.80 ± 0.09 |
| Spleen | Abs., g | 1.25 ± 0.06 | 1.27 ± 0.09 | 1.48 ± 0.15 | 1.40 ± 0.14 |
| | Rel., g /100 g | 0.52 ± 0.02 | 0.57 ± 0.04 | 0.39 ± 0.06 | 0.38 ± 0.03 |
| Heart | Abs., g | 0.87 ± 0.02 | 0.90 ± 0.03 | 1.25 ± 0.02 | 1.22 ± 0.66 |
| | Rel., g /100 g | 0.36 ± 0.01 | 0.41 ± 0.02* | 0.33 ± 0.01 | 0.33 ± 0.01 |
| Testicles | Abs., g | 2.95 ± 0.15 | 2.50 ± 0.34 | 3.42 ± 0.20 | 3.10 ± 0.31 |
| | Rel., g /100 g | 1.23 ± 0.09 | 1.12 ± 0.15 | 0.90 ± 0.04 | 0.86 ± 0.08 |
| Hypophysis | Abs., mg | 7.60 ± 0.68 | 8.60 ± 0.22 | 11.80 ± 1.02 | 10.30 ± 1.30 |
| | Rel., mg /100 g | 3.13 ± 0.17 | 3.83 ± 0.12* | 2.44 ± 0.26 | 2.40 ± 0.26 |
| Adrenal glands | Abs., mg | 28.17 ± 3.44 | 26.00 ± 2.54 | 26.67 ± 2.60 | 29.70 ± 2.90 |
| | Rel., mg /100 g | 11.75 ± 1.51 | 11.66 ± 0.96 | 6.90 ± 0.60 | 6.50 ± 0.88 |
| Seminal vesicles | Abs., mg | 307.3 ± 34.2 | 300.5 ± 47.4 | 576.8 ± 41.8 | 486.3 ± 95.0 |
| | Rel., mg /100 g | 129.0 ± 11.9 | 132.9 ± 18.2 | 151.2 ± 10.6 | 198.2 ± 54.8 |
| Prostate | Abs., mg | 168.8 ± 14.3 | 138.5 ± 20.1 | 444.8 ± 49.3 | 298.2 ± 54.8 |
| | Rel., mg /100 g | 71.20 ± 7.60 | 62.67 ± 8.87 | 114.7 ± 9.5 | 86.4 ± 3.4* |

[a]Absolute weight of internal organs.
[b]Relative weight of internal organs (per 100 g body weight).
*$p < 0.05$ in comparison with control.

The decrease in relative weight of prostate revealed in test rats on Day 180 did not surpass the age- and species-related physiological variations. The data obtained on more than 500 rats in the Department of Novel Food Sources of the State Research Institute of Nutrition, RAMS revealed the following variations in the relative prostate weight of rats maintained on the standard vivarium diet for 30 and 180 days: 50–140 and 50–125 mg/100 g body weight, respectively.

Analysis of the data obtained in experiments carried out at the State Research Institute of Nutrition, RAMS revealed the following variations of prostate weight in rats (data compiled from more than 500 animals): on day 180 of the experiment the relative weight of the prostate was 50–125 mg/100 g body weight. Survey microscopic morphological studies of rat prostate of the control and experimental groups did not reveal differences between the groups. Unbiased, in the center of the tissue specimen the lumens of glands expanded in a greater or lesser degree or filled with eosinophilic contents; on the periphery, glands are smaller and mucosa folded.

There were no focal lesions. The structure was ordinary. Prostate parenchyma comprises numerous individual glands. The ducts, which collect the secretions of lobes, form tubulo-alveolar secretory units adapted both for formation and storage of the secretions, so they are able to expand greatly.

Thus, fluctuations in prostate weight within 200% may be considered typical of animals of the species and age.

### Assessment of Biochemical Parameters
Over the entire period of the experiment, the biochemical parameters of blood serum and urine did not significantly differ between the groups of control rats fed diet with conventional maize and test rats whose diet was supplemented with transgenic maize Bt11 (Tables 5.211 and 5.212).

### Assessment of Sensitive Biomarkers
Activity of hepatic enzymes involved in xenobiotic degradation in test rats fed diet with transgenic maize line Bt11 did not significantly differ from the corresponding values for the control rats fed diet with conventional maize (Table 5.213). Similarly, there were no significant intergroup differences in activity of the hepatic lysosomal enzymes (Table 5.214).

### Hematological Assessments
Over the entire duration of the experiment, there were no significant differences in hematological parameters between the control and test groups of rats (Table 5.215). Similarly, there were no significant differences in the leukogram parameters (Table 5.216).

**Table 5.211** Biochemical Parameters of Blood Serum of Rats Fed Diet with Conventional Maize (Control) or Transgenic Maize Line Bt11 (Test) ($M \pm m$, $n = 6–8$)

| Parameter | 30 Days | | 180 Days | |
|---|---|---|---|---|
| | Control | Test | Control | Test |
| Total protein, g/L | 64.6 ± 3.7 | 73.4 ± 3.5 | 61.6 ± 5.6 | 55.5 ± 3.3 |
| Glucose, mM/L | 7.40 ± 0.36 | 8.30 ± 0.26 | 6.50 ± 0.50 | 7.00 ± 0.70 |
| Aspartate aminotransferase, µcat/L | 0.40 ± 0.04 | 0.44 ± 0.02 | 0.70 ± 0.02 | 0.77 ± 0.05 |
| Alanine aminotransferase, µcat/L | 0.28 ± 0.04 | 0.29 ± 0.08 | 0.20 ± 0.03 | 0.22 ± 0.01 |
| Alkaline phosphatase, µcat/L | 10.7 ± 2.6 | 13.7 ± 0.8 | 3.53 ± 0.55 | 3.40 ± 0.55 |

*Note: Here and in Tables 5.212 to 5.216 the differences are not significant ($p > 0.05$).*

**Table 5.212** Urinary Biochemical Parameters of Rats Fed Diet with Conventional Maize (Control) or Transgenic Maize Line Bt11 (Test) ($M \pm m$, $n = 6–8$)

| Parameter | 30 Days | | 180 Days | |
|---|---|---|---|---|
| | Control | Test | Control | Test |
| pH | 7.0 | 7.0 | 7.0 | 7.0 |
| Daily diuresis, mL | 10.6 ± 1.4 | 10.9 ± 1.3 | 5.7 ± 1.1 | 6.8 ± 1.0 |
| Relative density, g/mL | 1.04 ± 0.01 | 1.03 ± 0.02 | 1.00 ± 0.01 | 1.00 ± 0.01 |
| Creatinine, mg/mL | 0.97 ± 0.16 | 0.73 ± 0.07 | 1.46 ± 0.16 | 1.02 ± 0.16 |
| Creatinine, mg/day | 9.15 ± 0.37 | 7.73 ± 1.08 | 7.60 ± 0.70 | 6.31 ± 0.65 |

**Table 5.213** Activity of Enzymes Involved in Metabolism of Xenobiotics in Rats Fed Diet with Conventional Maize (Control) or Transgenic Maize Line Bt11 (Test) ($M \pm m$, $n = 6–8$)

| Parameter | 180 Days | |
|---|---|---|
| | Control | Test |
| Cytochrome P450, nM/mg protein | 0.97 ± 0.08 | 1.00 ± 0.06 |
| Cytochrome $b_5$, nM/mg protein | 0.76 ± 0.03 | 0.80 ± 0.02 |
| Aminopyrine N-demethylation, nM/min/mg protein | 10.4 ± 0.4 | 10.5 ± 0.5 |
| Benzpyrene hydroxylation, Fl-units/min/mg protein | 8.10 ± 0.61 | 7.85 ± 0.48 |
| Acetylesterase, µM/min/mg protein | 8.67 ± 0.31 | 9.25 ± 0.28 |
| Epoxide hydrolase, nM/min/mg protein | 9.43 ± 1.05 | 9.72 ± 0.40 |
| UDP-Glucuronosil transferase, nM/min/mg protein | 26.0 ± 2.3 | 27.8 ± 1.3 |
| CDNB-Glutathione transferase, µM/min/mg protein | 58.5 ± 4.2 | 56.2 ± 3.4 |
| Protein, mg/g | 13.2 ± 0.4 | 13.4 ± 0.3 |

## 5.2.5 Transgenic Maize Line Bt11 Resistant to European Corn

**Table 5.214** Activity of Hepatic Lysosomal Enzymes of Rats Fed Diet with Conventional Maize (Control) or Transgenic Maize Line Bt11 (Test) ($M \pm m$, $n = 6$–$8$)

| Parameters | 180 Days | |
|---|---|---|
| | Control | Test |
| *Total activity, μM/min/g tissue* | | |
| Arylsulfatase A, B | 2.15 ± 0.06 | 2.13 ± 0.03 |
| β-Galactosidase | 2.35 ± 0.06 | 2.34 ± 0.11 |
| β-Glucuronidase | 2.26 ± 0.07 | 2.35 ± 0.02 |
| *Non-sedimentable activity, % total activity* | | |
| Arylsulfatase A, B | 3.51 ± 0.20 | 3.77 ± 0.16 |
| β-Galactosidase | 5.21 ± 0.26 | 5.09 ± 0.32 |
| β-Glucuronidase | 5.06 ± 0.17 | 5.04 ± 0.19 |

**Table 5.215** Hematological Parameters of Peripheral Blood Drawn from Rats Fed Diet with Conventional Maize (Control) or Transgenic Maize Line Bt11 (Test) ($M \pm m$, $n = 6$–$8$)

| Parameter | 30 Days | | 180 Days | |
|---|---|---|---|---|
| | Control | Test | Control | Test |
| Hemoglobin concentration, g/L | 158.9 ± 4.8 | 162.6 ± 9.3 | 162.5 ± 2.8 | 166.4 ± 3.1 |
| Total erythrocyte count, ×$10^{12}$/L | 6.08 ± 0.05 | 6.18 ± 0.21 | 6.40 ± 0.15 | 6.36 ± 0.15 |
| Hematocrit, vol.% | 49.2 ± 0.6 | 50.0 ± 0.6 | 51.0 ± 0.4 | 50.8 ± 0.4 |
| MCH, pg | 27.8 ± 1.8 | 26.3 ± 0.7 | 25.1 ± 0.6 | 26.3 ± 0.8 |
| MCHC, % | 32.3 ± 0.6 | 32.5 ± 1.4 | 30.2 ± 2.2 | 32.7 ± 0.3 |
| MCV, μm$^3$ | 81.6 ± 2.1 | 81.3 ± 1.5 | 78.8 ± 1.3 | 80.0 ± 1.3 |
| Total leukocyte count, ×$10^9$/L | 13.6 ± 1.0 | 13.5 ± 1.2 | 15.8 ± 0.9 | 15.5 ± 0.6 |

**Table 5.216** Leukogram Parameters of Rats Fed Diet with Conventional Maize (Control) or Transgenic Maize Line Bt11 (Test) ($M \pm m$, $n = 6$–$8$)

| Parameter | 30 Days | | 180 Days | |
|---|---|---|---|---|
| | Control | Test | Control | Test |
| *Neutrophils* | | | | |
| rel., % | 14.80 ± 2.68 | 13.60 ± 0.77 | 14.00 ± 1.53 | 12.00 ± 1.90 |
| abs., ×$10^9$/L | 2.07 ± 0.50 | 1.83 ± 0.17 | 2.26 ± 0.33 | 1.86 ± 0.32 |
| *Eosinophils* | | | | |
| rel., % | 1.00 ± 0.38 | 1.20 ± 0.38 | 1.00 ± 0.38 | 1.40 ± 0.19 |
| abs., ×$10^9$/L | 0.126 ± 0.050 | 0.148 ± 0.048 | 0.148 ± 0.050 | 0.214 ± 0.059 |
| *Lymphocytes* | | | | |
| rel., % | 84.20 ± 2.30 | 85.20 ± 0.57 | 85.00 ± 1.34 | 87.00 ± 1.72 |
| abs., ×$10^9$/L | 11.36 ± 0.25 | 11.47 ± 0.48 | 13.40 ± 0.69 | 13.57 ± 0.45 |

**Table 5.217** Microscopic Examination of Internal Organs in Rats Fed Diet with Conventional Maize (Control) or Transgenic Maize Line Bt11 (Test) (Combined Data Obtained on Experimental Days 30 and 180)

| Organ | Control | Test |
|---|---|---|
| Liver | Clear trabecular structure; no alterations in hepatocytes or portal ducts | No differences from control |
| Kidneys | Usual appearance of cortical and medullar substance; no alterations in glomeruli or pelvis epithelium | No differences from control |
| Lung | Alveolar space is air-filled; no alterations in bronchi or blood vessels | No differences from control |
| Spleen | Large folliculi with wide clear marginal zones and reactive centers; splenic pulp is moderately plethoric | No differences from control |
| Small intestine | Preserved villous epithelium; usual infiltration in villous stroma | No differences from control |
| Testicle | Clearly definable spermatogenesis and age-related spermiogenesis | No differences from control |

### Morphological Assessments

Over the entire duration of the experiment, there was no mortality in the control group of rats fed diet with conventional maize or in the group of test rats fed diet with transgenic maize line Bt11. Visual post-mortem examination revealed no alterations in the internal organs in either group of rats. Similarly, histological examinations performed on experimental Days 30 and 180 found no significant differences between groups in the internal organs of the rats (Table 5.217).

Therefore, the chronic toxicological experiment carried out during 180 days with biochemical, hematological, and morphological examinations revealed no adverse affects of transgenic maize line Bt11 on the animals.

### *Assessment of Potential Impact of Transgenic Maize Line Bt11 on Immune System in Studies on Mice*
#### Potential Effect on Humoral Component of Immune System

The immunomodulating effect of transgenic maize line Bt11 on the humoral component of the immune system was examined on two oppositely reacting mice lines, CBA and C57Bl/6, by determining the level of hemagglutination to sheep erythrocytes (SE). The experimental conditions are described in section 5.1.1. In control and test groups of CBA mice (high-sensitivity animals), the antibodies appeared at the titers of 1:10–1:30 only on post-immunization Day 7. In both groups of low-sensitivity C57Bl/6 mice, the antibodies appeared on post-immunization Day 21 at the titers of 1:10–1:15. Therefore, the antibodies raised against SE appeared in both groups of rats at low titers irrespective of maize variety.

### Potential Effect on Cellular Component of Immune System

The immunomodulating effect of transgenic maize was assessed by delayed hypersensitivity reaction to SE. The experimental conditions are described in section 5.1.1. In CBA mice, RI did not differ between the test and control animals fed diet with transgenic or conventional maize ($13 \pm 2$ and $15 \pm 3$, respectively). Consumption of transgenic maize did not change RI: in test and control C57Bl/6 mice, the RI values were $38 \pm 4$ and $38 \pm 3$, respectively, while in the group of intact control (fed maize-free diet) mice it was $17 \pm 3$. Thus, transgenic maize Bt11 produced no effect on the cellular component of the immune system.

### Assessment of Potential Sensibilization Effect of Transgenic Maize

Examination of possible sensitizing action of transgenic maize line Bt11 on the immune response to endogenous metabolic products was carried out in the test of mouse sensitivity to histamine. Transgenic or conventional maize was supplemented to the diets of test and control mice, respectively, during 21 days. Thereafter the mice of both groups were injected intraperitoneally with 2.5 mg histamine hydrochloride dissolved in 0.5 mL physiological solution. The reaction was assessed in 1 h and 24 h by mortality of the mice. In this test, there was no mortality, nor were there differences in behavior of test or control mice, which attest to the absence of any sensitization agent in transgenic maize line Bt11.

### Potential Effect of Transgenic Maize on Susceptibility of Mice to Salmonella typhimurium

The effect of transgenic maize Bt11 on the susceptibility of mice to infection by salmonella of murine typhus was examined in experiments on C57Bl/6 mice injected intraperitoneally with various doses of Strain 415 *Salmonella typhimurium*. For 4 weeks, the diets of test and control mice comprised transgenic or conventional maize, respectively. The injected doses ranged from $10^2$ to $10^5$ microbial cells per mouse and varied on a 10-fold basis. The post-injection observation period was 21 days.

The following data were obtained:

- In both groups, the first mortality was observed on Day 1, and all mice died by post-injection Day 21;
- The mean lifetime was approximately equal in test and control groups: 13.3 and 16.8 day, respectively;
- $LD_{50}$ values in test and control groups were 1258 and 5011 microbial cells, respectively.

These data showed that *Salmonella typhimurium* produced a typical infection both in test and control mice. The course of murine typhus was similar in both groups. However, there was some increase in resistance in control mice, which was above the limits of physiological variations and increased $LD_{50}$ value by 4 times in comparison with the test group.

Therefore, transgenic maize line Bt11 had no sensitizing potencies; it did not affect the susceptibility of mice to *S. typhimurium* and it did not modulate the immune system according to the tests performed in this study.

### Assessment of Potential Impact of Transgenic Maize Line Bt11 on Immune System in Studies on Rats

The study was carried out on male Wistar rats ($n = 52$) weighing initially $200 \pm 10$ g. After a 7-day adaptation period to standard vivarium diet, in the following 28 days the rats were fed diet supplemented with conventional maize (control group) or transgenic maize line Bt11 (test group). Milled maize was dissolved in boiled water to the consistency of dense curd and supplemented with sunflower-seed oil to improve intake. The feed was used instead of an equally caloric amount of oatmeal (composition of the diet is given in Table 5.114).

The model of generalized anaphylaxis was developed according to [8] as described in section 5.1.1.

Throughout the entire duration of the experiment, the rats of both groups grew normally, which attests to adequate nutritional value of both diets. On Day 29, the body weights of test and control rats were practically identical ($313 \pm 7$ and $314 \pm 6$ g, correspondingly, $p > 0.1$).

Table 5.218 shows the data on severity of the anaphylactic reaction in control and test rats. By all examined parameters, there was no difference in severity of anaphylactic reactions ($p > 0.1$).

Table 5.219 shows the mean parameters of $D_{492}$, concentration of antibodies, and common logarithm of this concentration in the control and test groups. By all parameters, the difference between test rats fed diet with transgenic maize line Bt11 and the control rats maintained on conventional maize was insignificant ($p > 0.1$). Thus, the intensity of humoral immune response was practically identical in both groups of animals.

On the whole, these data support the conclusion that the degree of sensitization by ovalbumin in test group rats did not increase in comparison with that for the control group rats.

The studies showed that transgenic maize Bt11 did not enhance the allergic reactivity and sensitization produced by a model allergen in comparison with that for the rats fed diet with conventional maize. Thus, the use of transgenic maize does not enhance sensitization and allergic reactivity in laboratory animals in comparison with conventional maize.

### 5.2.5 Transgenic Maize Line Bt11 Resistant to European Corn

**Table 5.218** Severity of Active Anaphylactic Shock in Rats Fed Diet with Conventional Maize (Control) or Transgenic Maize Line Bt11 (Test)

| Group of Rats | AI | Severe Reactions, % | Mortality, % |
|---|---|---|---|
| Control ($n = 26$) | 3.154 | 69.2 | 69.2 |
| Test ($n = 26$) | 3.231 | 61.5 | 61.5 |

**Table 5.219** Parameters of Humoral Immune Response (Level of Specific IgG Antibodies Raised Against Ovalbumin) in Rats Fed Diet with Conventional Maize (Control) or Transgenic Maize Line Bt11 (Test)

| Group of Rats | $D_{492}$ | Concentration of Antibodies, mg/mL | Logarithm of Antibody Concentration |
|---|---|---|---|
| Control ($n = 24$) | $1.029 \pm 0.030$ | $7.5 \pm 1.1$ | $0.753 \pm 0.074$ |
| Test ($n = 26$) | $1.013 \pm 0.026$ | $6.5 \pm 0.9$ | $0.703 \pm 0.065$ |

### Assessment of Potential Genotoxicicity of Transgenic Maize

The genotoxic studies were carried out on male C57Bl/6 mice and hybrid female CBA mice fed diet shown in Table 5.114. During the experiment, the animals were fed a diet composed of a soft feed with milled maize of test or control variety.

The cytogenetic analysis was carried out by metaphasic method; genetic alterations in the sex cells were examined by revealing the dominant lethal mutations in C57Bl/6 male mice [8]. The details of experiments are described in section 5.1.1.

The results of possible genotoxic effects of transgenic maize on mice are shown in Tables 5.220 and 5.221.

Major chromosomal abnormalities in test and control mice were single segments and gaps. The number of cells with gaps did not significantly differ in control and test mice (Table 5.220). It should be noted that such a number of chromosomal aberrations is typical of male C57Bl/6 mice. Such mutations appear spontaneously; they are not stable and usually disappear in the subsequent nuclear divisions.

The pre-implantation mortality of unfertilized ovocytes, zygotes, and embryos in test mice did not surpass the corresponding values in the control group. The post-implantation embryonic mortality (the most reliable index of mutagenic activity of an examined agent) did not significantly differ between groups. There was no induced mortality, which attests to the absence of adverse affects of transgenic maize on spermiogenesis in mice (Table 5.221).

The data obtained showed that transgenic maize line Bt11 exerts no mutagenic effects.

**Table 5.220** Cytogenetic Parameters of Bone Marrow in Mice Fed Diet with Conventional Maize (Control) or Transgenic Maize Line Bt11 (Test) ($M \pm m$)

| Number of Cells, % | Control | Test |
|---|---|---|
| with chromosomal aberrations | 0.81 ± 0.46 | 0.76 ± 0.43 |
| with gaps | 1.63 ± 0.66 | 1.27 ± 0.56 |
| with polyploid chromosome set | 1.90 ± 0.70 | 1.27 ± 0.56 |

*Note: The numbers of analyzed metaphases were: 393 (test) and 367 (control).*

**Table 5.221** Dominant Lethal Mutations in Sex Cells of Mice Fed Diet with Conventional Maize (Control) or Transgenic Maize Line Bt11 (Test)

| Parameter, % | Week 1, Mature Spermatozoa | | Week 2, Late Spermatids | | Week 3, Early Spermatids | |
|---|---|---|---|---|---|---|
| | Control | Test | Control | Test | Control | Test |
| Pre-implantation mortality | 7.14 | 4.60 | 7.10 | 9.31 | 11.72 | 7.38 |
| Post-implantation mortality | 2.85 | 1.60 | 4.58 | 4.79 | 4.68 | 3.05 |
| Survival rate | 90.00 | 93.36 | 88.65 | 92.54 | 84.13 | 97.72 |
| Induced mortality | 0 | 0 | 0 | 0 | 0 | 0 |

*Note: In the control group, 399 embryos and 437 corpus lutea from 78 females were analyzed. The corresponding values in the test group (90 females) were 519 and 533.*

### Assessment of Technological Parameters

Assessment of technological parameters was carried out in Moscow State University of Applied Biotechnology (Ministry of Science and Education of Russian Federation).

To characterize the grain of transgenic maize line Bt11 resistant to European corn borer and tolerant to glufosinate ammonium in comparison with non-transgenic conventional maize, the moisture and ash contents were determined. Maize starch was produced under laboratory conditions to determine protein mass fraction in the starch, gelatinization temperature and viscosity of the starch gelatins on amylograph, the parameters of thermoplastic extrusion, and the structural and mechanical properties of the extrudates.

The transgenic maize grain complied with the requirements of Russian State Standards GOST 136-90 "Maize. Technical requirements". The study resulted in the following conclusions:

- No difficulties or differences were observed in the technological process when producing starch from transgenic maize Bt11 or conventional maize;

- According to Russian State Standards GOST 7698-93 "Starch. Formal Acceptance and Analytical Methods", the quality of all examined samples was superior;
- By gelatinization temperature and rheological properties, the starches derived from transgenic maize line Bt11 did not differ from those obtained from conventional maize;
- The structural and mechanical properties of the extrusive products derived from transgenic maize line Bt11 were practically identical to those obtained from the conventional maize.

Thus, the study revealed no significant differences in the properties of examined grain samples of transgenic maize line Bt11 and its conventional counterpart.

### Conclusions
By all examined parameters, the data of complex safety assessment of transgenic maize line Bt11, resistant to damage by European corn borer and tolerant to glufosinate ammonium, attest to the absence of any toxic, genotoxic, immune system modulating, or allergenic effects of this maize line. By chemical composition, transgenic maize line Bt11 was identical to conventional maize.

Based on the results of the studies, the State Sanitation Service of the Russian Federation (Department of State Sanitation and Epidemiological Inspectorate) granted the Registration Certificate which allows the transgenic maize line Bt11 to be used in the food industry and placed on the market without restrictions.

## 5.2.6 TRANSGENIC MAIZE LINE T25 TOLERANT TO GLUFOSINATE AMMONIUM

### Molecular Characteristics of Transgenic Maize Line T25
#### Recipient Organism
Maize (*Zea mays L.*) is characterized by a long history of safe use as human food and animal feed.

#### Donor Organism
The donor of the *pat* gene conferring tolerance to ammonium glufosinate, *Streptomyces viridochromogenes* strain Tu 494, is a gram-positive spore-forming soil bacterium which produces bialafos (phosphinothricin), a tripeptide composed of two molecules of L-alanine and an analog of L-glutamine acid. The *pat* gene codes synthesis of phosphinothricin acetyltransferase [32,47].

The herbicide glufosinate ammonium inhibits glutamine synthetase, an enzyme playing an important role in nitrogen metabolism in plants. Inhibition of this enzyme results in an accumulation of ammonia and death of the plant cells. Phosphinothricin acetyltransferase acetylates the free $NH_2$-group of glufosinate ammonium, thereby preventing accumulation of ammonia.

### Method of Genetic Transformation

The plasmid DNA was incorporated into maize protoplasts by means of direct transformation of DNA during incubation in chemical solution [19].

The transformation plasmid vector pUC/Ac used to produce maize line T25 has the following basic genetic elements: synthetic *pat* gene (a sequence from *Streptomyces viridochromogenes*)[44], 35S promoter of cauliflower mosaic virus, and 35S terminator of cauliflower mosaic virus. The *bla* gene, isolated from *E. coli*, which encodes synthesis of β-lactamase providing tolerance to some antibiotics, was used as selective bacterial marker [36]. β-Lactamase is not expressed in T25 maize [19].

## Global Registration Status of Maize Line T25

Table 5.222 shows the countries that had granted registration for use of transgenic maize line T25 at the time of registration in Russia [19].

## Safety Assessment of Transgenic Maize Line T25 Conducted in the Russian Federation

The studies were conducted in accordance with the requirements of the Ministry of Health of the Russian Federation authorized for risk and safety assessment of food derived from GM sources [8]. PCR analysis of the maize test and control samples was performed to confirm the identity of the transformation event and its absence in the conventional control line.

**Table 5.222** Registration Status of Transgenic Maize Line T25 in Various Countries

| Country | Date of Approval | Scope |
|---|---|---|
| Argentina | 1998 | Food, feed, environmental release |
| EU | 1998 | Food, feed, environmental release |
| Canada | 1996 | Feed, environmental release |
|  | 1997 | Food |
| USA | 1995 | Food, feed, environmental release |
| Japan | 1997 | Food, feed |

*Note:* The current registration status of the transgenic crop is on http://www.biotradestatus.com/

### Biochemical Composition of Transgenic Maize Grain

The content of proteins and amino acid composition in the grain of transgenic maize line T25 did not significantly differ from the corresponding values for the conventional maize (Table 5.223). The content of carbohydrates in the grain of both maize cultivars were similar (Table 5.224). There were no significant differences in the content of lipids and fatty acids in the grain of the two maize varieties (Tables 5.225, 5.226).

**Table 5.223** Protein Content (%) and Amino Acid Composition (g/100 g protein) in Maize Grain

| Ingredient | Conventional Maize | Transgenic Maize Line T25 |
|---|---|---|
| Protein, % | 8.95 | 10.03 |
| *Amino acids* | | |
| Lysine | 1.62 | 1.74 |
| Histidine | 1.39 | 1.16 |
| Arginine | 8.00 | 7.08 |
| Aspartic acid | 10.09 | 8.59 |
| Threonine | 3.25 | 3.14 |
| Serine | 4.52 | 4.41 |
| Glutamic acid | 22.39 | 23.00 |
| Proline | 11.95 | 13.03 |
| Glycine | 2.09 | 1.97 |
| Alanine | 5.22 | 5.23 |
| Cysteine | 0.46 | 1.51 |
| Valine | 3.60 | 3.48 |
| Methionine | 1.51 | 1.51 |
| Isoleucine | 2.90 | 2.79 |
| Leucine | 11.37 | 11.96 |
| Tyrosine | 4.64 | 4.53 |
| Phenylalanine | 4.99 | 4.88 |

**Table 5.224** Comparative Content of Carbohydrates (g/100 g product) in Transgenic and Conventional Maize

| Carbohydrate | Conventional Maize | Transgenic Maize Line T25 |
|---|---|---|
| Sucrose | 1.32 | 1.02 |
| Starch | 57.4 | 56.8 |
| Cellulose | 1.56 | 2.63 |

**Table 5.225** Comparative Content of Lipids (%) in Maize Grain

| Conventional Maize | Transgenic Maize Line T25 |
|---|---|
| 4.26 | 5.01 |

**Table 5.226** Content of Fatty Acids (Rel. %) in Maize Grain

| Fatty Acid | Conventional Maize | Transgenic Maize Line T25 |
|---|---|---|
| Lauric 12:0 | 0.04 | 0.03 |
| Myristic 14:0 | 0.09 | 0.08 |
| Pentadecanoic 15:0 | 0.05 | 0.03 |
| Palmitic 16:0 | 14.09 | 17.80 |
| Palmitoleic 16:1 | 0.22 | 0.18 |
| Margaric (heptadecanoic) 17:0 | 0.08 | 0.09 |
| Heptadecenoic 17:1 | 0.03 | 0.04 |
| Stearic 18:0 | 1.81 | 2.29 |
| cis-9-Oleic 18:1 | 24.66 | 22.86 |
| trans-11-Vaccenic 18:1 | 0.52 | 0.55 |
| Linoleic 18:2 | 56.30 | 53.28 |
| Linolenic 18:3 | 1.33 | 1.98 |
| Arachidic 20:0 | 0.32 | 0.40 |
| Gondoic 20:1 | 0.19 | 0.22 |
| Behenic 22:0 | 0.09 | 0.07 |
| Erucic 22:1 | 0.08 | 0.10 |

**Table 5.227** Comparative Content of Vitamins (mg/100 g product) in Maize Grain

| Vitamin | Conventional Maize | Transgenic Maize Line T25 |
|---|---|---|
| Vitamin $B_1$ | 0.179 | 0.181 |
| Vitamin $B_2$ | 0.259 | 0.194 |
| Vitamin $B_6$ | 0.19 | 0.21 |
| β-Carotene | 0.08 | 0.08 |

The vitamin composition was also similar in test and control maize samples. Differences were observed only in the content of vitamin $B_2$, although they were within physiological boundaries characteristic of this maize: 0.05–0.30 mg/100 g (Table 5.227). The mineral composition of transgenic maize T25 did not significantly differ from that of conventional non-transgenic maize (Table 5.228).

The content of heavy metals and mycotoxins in the grains of conventional and transgenic maize line T25 did not surpass the acceptable limits according to the regulations valid in the Russian Federation (Table 5.229) [8].

Thus, the grain of transgenic maize line T25 and conventional maize did not significantly differ by biochemical composition. The revealed differences in

### 5.2.6 Transgenic Maize Line T25 Tolerant to Glufosinate Ammonium

**Table 5.228** Comparative Mineral Composition in Maize Grain

| Ingredient | Conventional Maize | Transgenic Maize Line T25 |
|---|---|---|
| Sodium, mg/kg | 67.8 | 46.0 |
| Calcium, mg/kg | 4.90 | 3.27 |
| Magnesium, mg/kg | 1688 | 1892 |
| Iron, mg/kg | 16.6 | 15.9 |
| Potassium, mg/kg | 3412 | 3601 |
| Zinc, mg/kg | 19.3 | 21.8 |
| Copper, mg/kg | 0.885 | 0.736 |
| Selenium, µg/kg | 93 | 92 |

**Table 5.229** Analysis of Toxic Elements of Conventional and Transgenic Maize Grain (mg/kg)

| Ingredient | Conventional Maize | Transgenic Maize Line T25 |
|---|---|---|
| Deoxynivalenol | Not detected | Not detected |
| Zearalenone | Not detected | Not detected |
| Aflatoxin $B_1$ | Not detected | Not detected |
| Cadmium | <0.001 | <0.001 |
| Lead | <0.001 | <0.001 |

the content of vitamin $B_1$ remained within the range characteristic for maize variety [15,43,49–51]. The safety parameters of the grain of conventional and transgenic maize line T25 complied with the requirements of the regulations valid in the Russian Federation [8].

#### Toxicological Assessment of Transgenic Maize Line T25

The experiment was carried out on male Wistar rats with an initial body weight of 70–90 g. After admission to the vivarium of the State Research Institute of Nutrition, the rats were placed in quarantine for 10 days. At the onset of feeding experimental diet, the body weight of rats was 105–115 g. During the entire experiment, the animals were fed a standard semi-synthetic casein diet with conventional (control group) or transgenic (test group) maize. Composition of the diets is given in Table 5.162. The biochemical, hematological, and morphological studies were conducted in accordance with the requirements of the Ministry of Health of the Russian Federation authorized for risk and safety assessment of food derived from GM sources [6].

Throughout the entire duration of the experiment, the general condition of the rats was similar and satisfactory in control and test groups. No animal mortality was observed in either group. Daily intake of maize was approximately equal in both groups (Table 5.230).

**Table 5.230** Comparative Daily Intake (g/day) of Conventional and Transgenic Maize Line T25 ($M \pm m$, $n = 25$)

| Duration, Week | Control Maize | Transgenic Maize T25 |
|---|---|---|
| 1 | 6.6 ± 0.3 | 6.3 ± 0.3 |
| 2 | 7.3 ± 0.3 | 7.4 ± 0.3 |
| 3 | 7.8 ± 0.2 | 7.7 ± 0.3 |
| 4 | 7.8 ± 0.2 | 7.8 ± 0.3 |

*Note: Here and in Table 5.231 the differences are not significant ($p > 0.05$).*

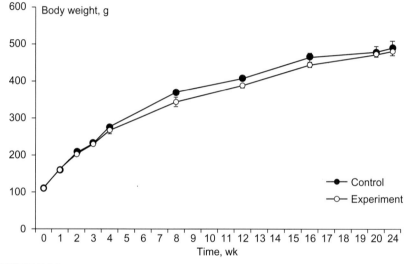

**FIGURE 5.11**

Comparative dynamics of body weight of rats fed diet with transgenic maize line T25 (test) or conventional maize (control).

During the first month of the experiment, the daily maize intake was 6.3–7.8 g without significant intergroup difference. In the following 5 months, the rats of both groups consumed maize completely. The body weight of the control and test rats did not significantly differ (Figure 5.11 and Table 5.231).

The absolute and relative weight of internal organs in control and test rats were determined at 30 and 180 days after the onset of the experiment (Table 5.232).

On Day 30, there were no differences in absolute weights except for a statistically significant decrease in absolute weight of kidney by 30% in the test group, which remained within the age-related physiological range (1.0–3.0 g) and disappeared by the end of the experiment. The weight loss of liver in

### 5.2.6 Transgenic Maize Line T25 Tolerant to Glufosinate Ammonium

**Table 5.231** Body Weight of Rats (g) Fed Diet with Conventional Maize (Control) or Transgenic Maize Line T25 (Test) ($M \pm m$; $n = 6-8$)

| Duration of Experiment, Weeks | Control Maize | Transgenic Maize T25 |
|---|---|---|
| 0  | 110.6 ± 3.2  | 109.0 ± 2.9  |
| 1  | 160.4 ± 4.7  | 160.2 ± 3.6  |
| 2  | 206.8 ± 5.5  | 203.1 ± 4.8  |
| 3  | 229.8 ± 3.1  | 229.6 ± 3.2  |
| 4  | 275.7 ± 9.2  | 267.9 ± 10.6 |
| 8  | 370.3 ± 8.6  | 344.3 ± 13.4 |
| 12 | 408.5 ± 8.6  | 390.4 ± 8.4  |
| 16 | 465.8 ± 11.1 | 445.0 ± 8.0  |
| 20 | 479.6 ± 13.9 | 474.1 ± 10.7 |
| 24 | 489.3 ± 19.5 | 481.4 ± 9.5  |

**Table 5.232** Absolute and Relative Weight of Internal Organs of Rats Fed Diet with Conventional Maize or Transgenic Maize Line T25 ($M \pm m$, $n = 6-8$)

| Organ | | 30 Days | | 180 Days | |
|---|---|---|---|---|---|
| | | Control | Test | Control | Test |
| Kidneys | Abs.[a], g | 2.10 ± 0.08 | 1.48 ± 0.09* | 2.66 ± 0.15 | 2.46 ± 0.07 |
|  | Rel.[b], g/100g | 0.73 ± 0.02 | 0.68 ± 0.02 | 0.54 ± 0.03 | 0.56 ± 0.02 |
| Liver | Abs., g | 13.20 ± 0.17 | 12.70 ± 0.58 | 14.51 ± 0.80 | 12.52 ± 0.25* |
|  | Rel., g/100g | 4.60 ± 0.13 | 4.80 ± 0.20 | 2.90 ± 0.10 | 2.92 ± 0.05 |
| Spleen | Abs., g | 1.48 ± 0.09 | 1.97 ± 0.24 | 1.53 ± 0.30 | 1.53 ± 0.15 |
|  | Rel., g/100g | 0.52 ± 0.10 | 0.73 ± 0.09 | 0.31 ± 0.02 | 0.35 ± 0.03 |
| Heart | Abs., g | 1.02 ± 0.02 | 1.08 ± 0.05 | 1.33 ± 0.06 | 1.25 ± 0.05 |
|  | Rel., g/100g | 0.36 ± 0.04 | 0.39 ± 0.01 | 0.27 ± 0.01 | 0.28 ± 0.02 |
| Testicles | Abs., g | 2.82 ± 0.09 | 2.68 ± 0.07 | 3.20 ± 0.11 | 3.33 ± 0.14 |
|  | Rel., g/100g | 1.00 ± 0.04 | 1.24 ± 0.21 | 0.64 ± 0.05 | 0.74 ± 0.07 |
| Hypophysis | Abs., mg | 8.7 ± 0.9 | 9.0 ± 0.9 | 9.8 ± 0.7 | 11.6 ± 0.8 |
|  | Rel., mg/100g | 3.06 ± 0.33 | 3.40 ± 0.38 | 1.95 ± 0.13 | 2.67 ± 0.14* |
| Adrenal glands | Abs., mg | 23.5 ± 2.3 | 24.8 ± 3.4 | 28.6 ± 2.2 | 27.5 ± 1.2 |
|  | Rel., mg/100g | 8.30 ± 1.14 | 9.25 ± 1.25 | 5.77 ± 0.27 | 6.30 ± 1.25 |
| Seminal vesicles | Abs., mg | 420.0 ± 17.5 | 433.3 ± 30.9 | 470.0 ± 53.7 | 515.0 ± 69.8 |
|  | Rel., mg/100g | 152.0 ± 9.2 | 158.2 ± 7.6 | 94.6 ± 8.8 | 118.2 ± 16.5 |
| Prostate | Abs., mg | 218.6 ± 8.7 | 186.6 ± 30.5 | 276.6 ± 51.4 | 338.3 ± 54.8 |
|  | Rel., mg/100g | 77.2 ± 3.4 | 70.5 ± 13.1 | 65.72 ± 14.18 | 80.02 ± 7.4 |

[a]Absolute weight of internal organs.
[b]Relative weight of internal organs (per 100g body weight).
*$p < 0.05$ in comparison with control.

**Table 5.233** Biochemical Parameters of Blood Serum of Rats Fed Diet with Conventional Maize (Control) or Transgenic Maize Line T25 (Test) ($M \pm m$, $n = 6–8$)

|  | 30 Days | | 180 Days | |
| --- | --- | --- | --- | --- |
| Parameter | Control | Test | Control | Test |
| Total protein, g/L | 60.8 ± 1.7 | 54.5 ± 1.5* | 63.1 ± 0.10 | 64.9 ± 0.10* |
| Glucose, mM/L | 9.6 ± 0.32 | 10.2 ± 0.49 | 9.19 ± 0.28 | 9.32 ± 0.32 |
| Alanine aminotransferase, µcat/L | 0.73 ± 0.08 | 0.71 ± 0.08 | 0.73 ± 0.04 | 0.63 ± 0.04 |
| Aspartate aminotransferase, µcat/L | 0.48 ± 0.01 | 0.49 ± 0.01 | 0.78 ± 0.05 | 0.70 ± 0.05 |
| Alkaline phosphatase, µcat/L | 3.61 ± 0.52 | 3.86 ± 0.31 | 4.00 ± 0.28 | 4.77 ± 0.37 |

*($p < 0.05$) here and in Tables 5.236 to 5.242.

**Table 5.234** Urinary Biochemical Parameters of Rats Fed Diet with Conventional Maize (Control) or Transgenic Maize Line T25 (Test) ($M \pm m$, $n = 6–8$)

|  | 30 Days | | 180 Days | |
| --- | --- | --- | --- | --- |
| Parameter | Control | Test | Control | Test |
| pH | 6.0 | 6.0 | 6.5 | 6.5 |
| Daily diuresis, mL | 4.1 ± 0.8 | 3.7 ± 0.4 | 6.4 ± 0.7 | 5.0 ± 0.3 |
| Relative density, g/mL | 1.16 ± 0.02 | 1.17 ± 0.03 | 1.14 ± 0.02 | 1.12 ± 0.01 |
| Creatinine, mg/mL | 0.66 ± 0.10 | 0.57 ± 0.12 | 2.43 ± 0.43 | 2.62 ± 0.18 |
| Creatinine, mg/day | 2.34 ± 0.10 | 1.93 ± 0.39 | 14.19 ± 1.10 | 12.93 ± 0.35 |

test rats observed on Day 180 remained within the age-related physiological range (8.0–15.0 g).

The relative weight of internal organs in control and test rats did not significantly differ. The significant gain in the relative weight of hypophysis in test rats remained within the age-related physiological range (1.0–3.0 mg/100 g).

### Assessment of Biochemical Parameters

The content of glucose and activity of aspartate aminotransferase, alanine aminotransferase, and alkaline phosphatase in the test group of rats did not significantly differ from the corresponding parameters of the control group of rats (Table 5.233). The slight variations of total protein in test group (decrease on Day 30 and increase on Day 180) were within the physiological range for rats (45.0–100.0 g/L).

Over the entire period of the experiment, the urinary biochemical parameters did not significantly differ between the groups of rats (Table 5.234).

### 5.2.6 Transgenic Maize Line T25 Tolerant to Glufosinate Ammonium

**Table 5.235** Activity of Enzymes Involved in Metabolism of Xenobiotics in Rats Fed Diet with Conventional Maize (Control) or Transgenic Maize Line T25 (Test) ($M \pm m$, $n = 6$–$8$)

| Parameter | 30 Days | | 180 Days | |
|---|---|---|---|---|
| | Control | Test | Control | Test |
| Cytochrome P450, nM/mg protein | 0.61 ± 0.07 | 0.67 ± 0.07 | 0.79 ± 0.04 | 0.76 ± 0.05 |
| Cytochrome b5, nM/mg protein | 0.57 ± 0.04 | 0.60 ± 0.04 | 0.71 ± 0.02 | 0.72 ± 0.03 |
| Aminopyrine N-demethylation, nM/min/mg protein | 8.47 ± 0.49 | 8.90 ± 0.21 | 7.72 ± 0.41 | 7.65 ± 0.34 |
| Benzpyrene hydroxylation, Fl-units/min/mg protein | 6.87 ± 0.39 | 7.04 ± 0.83 | 11.4 ± 1.2 | 10.9 ± 0.5 |
| Acetylesterase, µM/min/mg protein | 6.13 ± 0.30 | 6.42 ± 0.24 | 6.19 ± 0.16 | 7.05 ± 0.21 |
| Epoxide hydrolase, nM/min/mg protein | 4.08 ± 0.39 | 4.16 ± 0.22 | 7.13 ± 0.53 | 6.00 ± 0.57 |
| UDP-Glucuronosil transferase, nM/min/mg protein | 20.1 ± 1.2 | 21.4 ± 1.4 | 20.4 ± 0.7 | 19.0 ± 1.6 |
| CDNB-Glutathione transferase, µM/min/mg protein | 1.02 ± 0.04 | 1.08 ± 0.06 | 0.85 ± 0.03 | 0.79 ± 0.03 |
| Microsomal protein, mg/g | 13.5 ± 0.3 | 13.5 ± 0.5 | 14.0 ± 0.4 | 15.8 ± 0.4 |
| Cytosolic protein, mg/g | 66.5 ± 3.1 | 73.2 ± 3.1 | 79.1 ± 1.8 | 82.2 ± 1.3 |

**Table 5.236** Activity of Hepatic Lysosomal Enzymes in Control and Test in Rats Fed Diet with Conventional Maize (Control) or Transgenic Maize Line T25 (Test) ($M \pm m$, $n = 6$–$8$)

| Parameters | 30 Days | | 180 Days | |
|---|---|---|---|---|
| | Control | Test | Control | Test |
| *Total activity, µM/min/g tissue* | | | | |
| Arylsulfatase A, B | 2.31 ± 0.01 | 2.31 ± 0.02 | 2.13 ± 0.03 | 2.09 ± 0.02 |
| β-Galactosidase | 2.16 ± 0.03 | 2.35 ± 0.02* | 2.27 ± 0.05 | 2.29 ± 0.06 |
| β-Glucuronidase | 2.13 ± 0.03 | 2.20 ± 0.05 | 2.24 ± 0.03 | 2.28 ± 0.02 |
| *Non-sedimentable activity, % total activity* | | | | |
| Arylsulfatase A, B | 3.27 ± 0.06 | 3.12 ± 0.07 | 3.13 ± 0.06 | 3.14 ± 0.03 |
| β-Galactosidase | 5.51 ± 0.42 | 5.43 ± 0.43 | 5.31 ± 0.22 | 5.39 ± 0.24 |
| β-Glucuronidase | 5.29 ± 0.15 | 4.96 ± 0.21 | 4.65 ± 0.16 | 4.68 ± 0.14 |

*$p < 0.05$ in comparison with control.

### Assessment of Sensitive Biomarkers

Activity of hepatic enzymes involved in xenobiotic degradation in test rats fed diet with transgenic maize line T25 did not significantly differ from the corresponding values for the control rats fed diet with conventional maize (Table 5.235). Similarly, there were no significant differences in activity of the hepatic lysosomal enzymes (Table 5.236). An insignificant elevation of total

**Table 5.237** Content of LPO Products in Blood of Rats Fed Diet with Conventional Maize (Control) or Transgenic Maize Line T25 (Test) ($M \pm m$, $n = 6$–$8$)

| Parameter | 30 Days | | 180 Days | |
|---|---|---|---|---|
| | Control | Test | Control | Test |
| **Erythrocytes** | | | | |
| DC, nM/mL | 5.811 ± 0.4 | 5.820 ± 0.3 | 3.594 ± 0.1 | 5.952 ± 0.3* |
| MDA, nM/mL | 5.226 ± 0.3 | 5.078 ± 0.3 | 4.386 ± 0.1 | 4.812 ± 0.1* |
| **Blood serum** | | | | |
| DC, nM/mL | 3.211 ± 0.2 | 3.120 ± 0.2 | 4.187 ± 0.2 | 3.501 ± 0.2* |
| MDA, nM/mL | 3.008 ± 0.2 | 2.995 ± 0.3 | 3.736 ± 0.2 | 3.818 ± 0.1 |
| **Liver** | | | | |
| DC, Unit | 0.987 ± 0.006 | 0.996 ± 0.003 | 0.869 ± 0.01 | 0.912 ± 0.01* |
| MDA, nM/g | 336.5 ± 9.1 | 339.3 ± 12.5 | 274.8 ± 16.6 | 290.2 ± 9.6 |

activity of β-galactosidase by 9% observed on Day 30 in test rats in comparison with the control values remained within the physiological boundaries of this parameter and disappeared on Day 180.

The study examined the effect of transgenic maize line T25 on the content of LPO products in rats. On Day 30, LPO level in blood and liver did not significantly differ between control and test rats. By the end of the experiment, the content of LPO products in test rats was higher than the corresponding values in the control group. Specifically, the erythrocytic levels of DC and MDA, and the level of hepatic DC were higher in test group rats. At the same time, DC level in the blood serum of test rats was lower in comparison with the control group (Table 5.237).

The antioxidant status of the rats in both groups was in dynamic equilibrium, which is indicated by the content of LPO products in blood on Day 180: in comparison with the control group, the erythrocytic levels of DC and MDA were higher, while the serum levels of DC were lower. This 'opposed directivity' in the intensity of LPO processes in erythrocytes and serum indicates compensation and stability of the integral antioxidant status.

The enzymatic activity of the erythrocytic antioxidant protection system did not significantly differ between groups during the entire period of the experiment (Table 5.238). The above-mentioned variations in intensity of LPO resulted from individual peculiarities of oxidative lipid metabolism that do not affect the overall antioxidant status of the whole organism, which conclusion is supported by the absence of any changes in activity of the enzymes from the antioxidant protection system.

### 5.2.6 Transgenic Maize Line T25 Tolerant to Glufosinate Ammonium

**Table 5.238** Activity of Enzymes in Erythrocytic Antioxidant Protection System in Rats Fed Diet with Conventional Maize (Control) or Transgenic Maize Line T25 (Test) ($M \pm m$, $n = 6$–$8$)

| Parameter | 30 Days | | 180 Days | |
|---|---|---|---|---|
| | Control | Test | Control | Test |
| Glutathione reductase, µmol/min/g Hb | 31.91 ± 1.2 | 32.73 ± 1.4 | 32.06 ± 1.9 | 38.24 ± 3.5 |
| Glutathione peroxidase, µmol/min/g Hb | 52.12 ± 2.1 | 51.10 ± 1.9 | 53.34 ± 2.5 | 53.55 ± 2.0 |
| Catalase, mmol/min/g Hb | 396.4 ± 10.9 | 412.6 ± 11.8 | 403.5 ± 9.9 | 416.2 ± 23.9 |
| Superoxide dismutase, U/min/g Hb | 2034.3 ± 52.7 | 2058.9 ± 45.8 | 1806.2 ± 69.2 | 1957.2 ± 75.5 |

**Table 5.239** Hematological Parameters of Peripheral Blood Drawn from Rats Fed Diet with Conventional Maize (Control) or Transgenic Maize Line T25 (Test) ($M \pm m$, $n = 6$–$8$)

| Parameter | 30 Days | | 180 Days | |
|---|---|---|---|---|
| | Control | Test | Control | Test |
| Hemoglobin concentration, g/L | 162.17 ± 1.40 | 158.90 ± 1.26 | 161.48 ± 5.20 | 169.30 ± 6.45 |
| Total erythrocyte count, ×$10^{12}$/L | 6.39 ± 0.13 | 6.32 ± 0.12 | 6.50 ± 0.10 | 6.50 ± 0.08 |
| Hematocrit, vol.% | 50.8 ± 0.32 | 51.00 ± 0.32 | 51.30 ± 0.35 | 51.70 ± 0.35 |
| MCH, pg | 25.54 ± 0.45 | 25.25 ± 0.33 | 24.75 ± 0.74 | 25.82 ± 0.87 |
| MCHC, % | 32.05 ± 0.17 | 31.25 ± 0.22 | 31.43 ± 1.07 | 32.73 ± 0.95 |
| MCV, µm$^3$ | 79.67 ± 1.39 | 80.80 ± 1.10 | 78.81 ± 0.73 | 78.92 ± 0.53 |
| Total leukocyte count, ×$10^9$/L | 12.46 ± 0.59 | 12.05 ± 0.73 | 14.01 ± 1.19 | 13.38 ± 1.18 |

Thus, analysis of parameters describing activity of LPO processes and activity of the enzymes in the antioxidant protection system showed no pro-oxidant load in the test group rats.

#### Hematological Assessments
Hematological parameters of peripheral blood drawn from the test and control group rats were examined. Over the entire duration of the experiment, there were no significant differences between the groups of rats in all examined hematological and leukogram parameters (Tables 5.239 and 5.240).

#### Morphological Assessments
Over the entire duration of the experiment, there was no mortality in the control group of rats fed diet with conventional maize or in the group of test rats fed diet with transgenic maize line T25. Visual post-mortem examination revealed no alterations in internal organs in either group of rats. Similarly, histological examinations performed on experimental Days 30 and 180 revealed no significant differences between groups in the internal organs of the rats (Table 5.241)

**Table 5.240** Leukogram Parameters of Rats Fed Diet with Conventional Maize (Control) or Transgenic Maize Line T25 (Test) ($M \pm m$, $n = 8$)

| Parameter | 30 Days | | 180 Days | |
|---|---|---|---|---|
| | Control | Test | Control | Test |
| **Neutrophils** | | | | |
| rel., % | 19.17 ± 0.80 | 19.30 ± 0.64 | 24.00 ± 0.70 | 22.30 ± 1.06 |
| abs., ×10$^9$/L | 2.40 ± 0.20 | 2.36 ± 0.22 | 3.38 ± 0.35 | 3.10 ± 0.39 |
| **Eosinophils** | | | | |
| rel., % | 0.50 ± 0.16 | 0.50 ± 0.32 | 1.50 ± 0.97 | 1.50 ± 0.97 |
| abs., ×10$^9$/L | 0.036 ± 0.016 | 0.050 ± 0.032 | 0.19 ± 0.065 | 0.27 ± 0.17 |
| **Lymphocytes** | | | | |
| rel., % | 80.50 ± 0.64 | 80.20 ± 0.64 | 74.17 ± 0.16 | 75.75 ± 0.70 |
| abs., ×10$^9$/L | 10.02 ± 1.07 | 9.65 ± 0.54 | 10.38 ± 0.90 | 10.00 ± 0.80 |

**Table 5.241** Microscopic Examination of Internal Organs in Rats Fed Diet with Conventional Maize (Control) or Transgenic Maize Line T25 (Test) (Combined Data Obtained on Experimental Days 30 and 180)

| Organ | Control | Test |
|---|---|---|
| Liver | Clear trabecular structure; no alterations in hepatocytes or portal ducts | No differences from control |
| Kidneys | Usual appearance of cortical and medullar substance; no alterations in glomeruli or pelvis epithelium | No differences from control |
| Lung | Alveolar space is air-filled; no alterations in bronchi and blood vessels | No differences from control |
| Spleen | Large folliculi with wide clear marginal zones and reactive centers; splenic pulp is moderately plethoric | No differences from control |
| Small intestine | Preserved villous epithelium; usual infiltration in villous stroma | No differences from control |
| Testicle | Clearly definable spermatogenesis and age-related spermiogenesis | No differences from control |

Therefore, the chronic toxicological study carried out during 180 days with biochemical, hematological, and morphological examinations revealed no adverse affects of transgenic maize line T25 on the animals.

### Assessment of Potential Impact of Transgenic Maize Line T25 on Immune System in Studies on Mice
#### Potential Effect on Humoral Component of Immune System

The immunomodulating effect of transgenic maize line T25 on the humoral component of the immune system was examined on two oppositely reacting mice lines, CBA and C57Bl/6, by determining the level of hemagglutination to sheep erythrocytes (SE). The experimental conditions are described in section 5.1.1.

The antibodies appeared at equal titers on post-immunization days 7 and 21. In test C57Bl/6 mice, the antibodies appeared on post-immunization Day 7 at high titers of 1:107–1:96, which decreased to Day 14 and remained at low level to Day 21. Therefore, the antibodies raised against SE appeared in both groups of rats with equal titers irrespective of the maize variety.

**Potential Effect on Cellular Component of Immune System**
The immunomodulating effect of transgenic maize was assessed by delayed hypersensitivity reaction to SE. The experimental conditions are described in section 5.1.1. In CBA and C57Bl/6 mice, RI increased equally in the test mice (36 ± 11 and 22 ± 4) and in the control mice (27 ± 8 and 26 ± 4). Thus, transgenic maize T25 produced no extra effect on the cellular component of the immune system and modified it to the same degree as the conventional maize.

**Assessment of Sensibilization Effect of Transgenic Maize**
Assessment of possible sensitizing action of transgenic maize line T25 on the immune response to endogenous metabolic products was carried out in the test of mouse sensitivity to histamine. Transgenic maize T25 or conventional maize was supplemented to the diets of, correspondingly, test and control mice for 21 days. Thereafter mice of both groups were injected intraperitoneally with 2.5 mg histamine hydrochloride dissolved in 0.5 mL physiological solution. The reaction was assessed at 1 h and 24 h by mortality of the mice. In this test, there was no mortality nor differences in behavior between the test or control mice, which attest to the absence of any sensitization agent in transgenic maize line T25.

**Potential Effect of Transgenic Maize on Susceptibility of Mice to *Salmonella typhimurium***
The effect of transgenic maize on the susceptibility of mice to infection by salmonella of murine typhus was examined in experiments on C57Bl/6 mice weighing 18–20 g. For 21 days, the diets of test and control mice were supplemented with transgenic or conventional maize, respectively. The mice of both groups were infected with Strain 415 *Salmonella typhimurium* in doses ranging from $10^2$ to $10^5$ microbial cells per mouse, which differed on a 10-fold basis. On post-infection Day 20, the mean lifetime and $LD_{50}$ were estimated.

In both groups of mice, the values of $LD_{50}$ were roughly equal. The peak of mortality was observed on post-injection Day 4, and all mice died by post-injection Days 17–18. No effect of transgenic maize on lifetime was revealed: in both groups, the lifetime ranged from 13 to 15 days. These data showed that *Salmonella typhimurium* produced a typical infection of murine typhus in both test and control mice, which developed identically in both groups.

Therefore, transgenic maize line T25 had no sensitizing or immune modulating potencies.

### Assessment of Potential Impact of Transgenic Maize Line T25 on Immune System in Studies on Rats

The study was carried out on male Wistar rats ($n = 48$) weighing initially $140 \pm 10$ g. After a 7-day adaptation period to standard vivarium diet, the rats for the following 28 days were fed diet supplemented with conventional maize (control group) or transgenic maize line T25 (test group). Both types of milled maize were dissolved in warm boiled water to the consistency of dense curd and supplemented with sunflower-seed oil to improve intake. The feed was used instead of an equally caloric amount of oatmeal.

The model of generalized anaphylaxis was developed according to [8] as described in section 5.1.1.

Throughout the entire duration of the experiment, the rats of both groups grew normally, which attests to adequate nutritional value of both diets. On Day 29, the body weights of test and control rats were practically identical ($273 \pm 5$ and $284 \pm 8$ g, respectively, $p > 0.1$).

Over the entire duration of the experiment, the rats of both groups developed normally, which attests to adequate nutritional value of both diets. Table 5.242 shows the data on severity of the anaphylactic reaction in control and test rats. By all examined parameters, there were no differences in the severity of anaphylactic reaction.

The intensity of humoral immune response was practically identical in both groups of animals ($p > 0.05$; Table 5.243).

Taken together, these data support the conclusion that the degree of sensitization by ovalbumin in test rats fed diet with transgenic maize line T25 did not increase in comparison with the control rats fed diet with conventional maize.

**Table 5.242** Severity of Active Anaphylactic Shock in Rats Fed Diet with Conventional Maize (Control) or Transgenic Maize Line T25 (Test)

| Group of Rats | AI | Severe Reactions, % | Mortality, % |
|---|---|---|---|
| Control ($n = 24$) | 3.00 | 62 | 58 |
| Test ($n = 24$) | 2.38 | 50 | 46 |

### 5.2.6 Transgenic Maize Line T25 Tolerant to Glufosinate Ammonium

**Table 5.243** Parameters of Humoral Immune Response (Level of Specific IgG Antibodies Raised Against Ovalbumin) in Rats Fed Diet with Conventional Maize (Control) or Transgenic Maize Line T25 (Test) ($M \pm m$)

| Group of Rats | $D_{492}$ | Concentration of Antibodies, mg/mL | Logarithm of Antibody Concentration |
|---|---|---|---|
| Control ($n = 24$) | $0.324 \pm 0.015$ | $11.3 \pm 2.0$ | $0.854 \pm 0.094$ |
| Test ($n = 24$) | $0.279 \pm 0.018$ | $7.5 \pm 1.4$ | $0.588 \pm 0.121$ |

**Table 5.244** Cytogenetic Parameters of Bone Marrow in Mice Fed Diet with Conventional Maize (Control) or Transgenic Maize Line T25 (Test) ($M \pm m$)

| Number of Cells, % | Control | Test |
|---|---|---|
| with chromosomal aberrations | $0.78 \pm 0.45$ | $0.78 \pm 0.43$ |
| with gaps | $1.30 \pm 0.58$ | $1.78 \pm 0.66$ |
| with polyploid chromosome set | $1.04 \pm 0.51$ | $1.01 \pm 0.50$ |

*Note: The numbers of analyzed metaphases were 393 (test) and 382 (control).*

The studies on laboratory animals showed that transgenic maize T25 did not significantly modify the allergic reactivity and degree of sensitization by a model allergen in comparison with conventional maize.

### Assessment of Potential Genotoxicicity of Transgenic Maize

Genotoxic studies were carried out on male C57Bl/6 mice and hybrid female CBA mice fed the diet shown in Table 5.114. During the experiment, the animals were fed the diet composed of a soft feed with milled maize of test or control variety.

The cytogenetic analysis was carried out by metaphasic method; genetic alterations in the sex cells were examined by revealing the dominant lethal mutations in C57Bl/6 male mice [8]. The details of experiments are described in section 5.1.1.

The results of possible genotoxic effects of transgenic maize on mice are shown in Tables 5.244 and 5.245.

The major chromosomal abnormalities in test and control mice were single or paired segments and gaps (Table 5.244). The number of cells with gaps did not significantly differ between control and test mice. The observed number of chromosomal aberrations is typical of male C57Bl/6 mice. Such mutations

**Table 5.245** Dominant Lethal Mutations in Sex Cells of Mice Fed Diet with Conventional Maize (Control) or Transgenic Maize Line T25 (Test)

| Parameter, % | Week 1, Mature Spermatozoa | | Week 2, Late Spermatids | | Week 3, Early Spermatids | |
|---|---|---|---|---|---|---|
| | Control | Test | Control | Test | Control | Test |
| Pre-implantation mortality | 10.0 | 9.7 | 7.40 | 8.18 | 7.48 | 7.84 |
| Post-implantation mortality | 4.80 | 3.30 | 6.00 | 4.95 | 3.67 | 2.83 |
| Survival rate | 85.60 | 87.01 | 86.00 | 87.00 | 88.43 | 89.54 |
| Induced mortality | – | 0 | – | 0 | – | 0 |

Note: In the control group, 359 embryos and 394 corpus lutea from 72 females were analyzed. The corresponding values in the test group (90 females) were 373 and 397.

appear spontaneously; they are not stable and usually disappear in the subsequent nuclear divisions.

The pre-implantation mortality of unfertilized ovocytes, zygotes, and embryos in test mice did not surpass the corresponding values in the control group. The post-implantation embryonic mortality (the most reliable index of mutagenic activity of an examined agent) did not significantly differ between groups. There was no induced mortality, which attests to the absence of adverse affects of transgenic maize on spermiogenesis in mice (Table 5.245). The data obtained support the conclusion that transgenic maize line T25 has no mutagenic properties.

### Assessment of Technological Parameters

The study of technological parameters was carried out in Moscow State University of Applied Biotechnology (Ministry of Science and Education of Russian Federation).

To characterize the grain of transgenic maize line T25 tolerant to glufosinate ammonium in comparison with non-transgenic conventional maize, the moisture and ash contents were determined. Maize starch was produced under laboratory conditions to determine protein mass fraction in the starch, gelatinization temperature and viscosity of the starch gelatins on amylograph, the parameters of thermoplastic extrusion, and the structural and mechanical properties of the extrudates. The transgenic maize grain complied with the specifications of Russian State Standards GOST 136-90 "Maize. Technical requirements". The study resulted in the following conclusions:

- No differences of difficulties were observed in the technological process when producing starch from transgenic maize T25 or conventional maize;

- According to Russian National Standards GOST 7698-93 "Starch. Formal Acceptance and Analytical Methods", the quality of all examined samples was superior;
- By gelatinization temperature and rheological properties, the starch derived from transgenic maize line T25 did not differ from those obtained from conventional maize;
- The structural and mechanical properties of the extrusive products derived from transgenic maize line T25 were practically identical to those obtained from the conventional maize.

Thus, this study revealed no significant differences in the properties of examined grain samples of transgenic maize line T25 and its conventional counterpart.

## Conclusions

By all examined parameters, the data of complex safety assessment of transgenic maize line T25 tolerant to glufosinate ammonium attest to the absence of any toxic, genotoxic, immune system modulating, or allergenic effects of this maize line. By chemical composition, transgenic maize line T25 was identical to conventional maize.

Based on the results of the studies, the State Sanitation Service of the Russian Federation (Department of State Sanitation and Epidemiological Inspectorate) granted the Registration Certificate which allows the transgenic maize line T25 to be used in the food industry and placed on the market without restrictions.

## 5.2.7 DIABROTICA-RESISTANT AND GLYPHOSATE-TOLERANT TRANSGENIC MAIZE LINE MON 88017

### Molecular Characteristics of Transgenic Maize Line MON 88017

*Recipient Organism*

Maize (*Zea mays* L.) is characterized by a long history of safe use as human food as animal feed.

### Method of Genetic Transformation

Transgenic maize line MON 88017 was derived by agrobacterial transformation of maize genome with binary plasmid vector PV-ZMIR39 incorporating expression cassette of the gene of 5-enolpyruvilshikimate-3-phosphate synthase (*cp4 epsps*) and expression cassette of δ-endotoxin gene *Cry3Bb1*. The expression cassette of the *cp4 epsps* gene contained the following genetic elements:

- 5′-terminal region of rice actin gene, containing the promoter and the first intron;
- synthetic N-terminal sequence of chloroplast transit peptide ribulose-1.5-bisphosphate-carboxylase/oxygenase (CTP) responsible for translocation of CP4 EPSPS protein into the chloroplasts, where synthesis of the aromatic amino acid takes place;
- *cp4 epsps* gene isolated from *Agrobacterium* sp. CP-4 line;
- NOS 3′ terminator (the non-translated fragment of nopaline synthase gene) from the Ti-plasmid of *Agrobacterium tumefaciens*.

The expression cassette of the *Cry3Bb1* gene contained the following genetic elements:

- amplified *P-e35S* promoter of cauliflower mosaic virus;
- the leading sequence wtCAB leader 5′ isolated from the non-translated fragment of wheat chlorophyll a/b-binding protein;
- Ract1- intron of rice actin gene;
- *Cry3Bb1* gene isolated from *Bacillus thuringiensis* subsp. *kumamotoensis*, EG4691 line, which is responsible for synthesis of δ-endotoxin Cry3Bb1;
- Tahsp17 3′ terminator isolated from the non-translated fragment of the gene encoding wheat heat shock protein.

Tolerance of transgenic maize to glyphosate results from the presence of the *cp4 epsps* gene in the plant genome encoding synthesis of CP4 EPSPS protein ($M_r \sim 47.6\,kDa$), which confers tolerance to glyphosate. Resistance against corn rootworm pests results from the presence of the *cry3Bb1* gene encoding the corresponding protein ($M_r \sim 74.6\,kDa$) which is toxic for *Diabrotica* spp. [19].

## Global Registration Status of Maize Line MON 88017

Table 5.246 shows the countries that had granted registration for use of transgenic maize line MON 88017 at the time of registration in Russia [19].

## Safety Assessment of Transgenic Maize Line MON 88017 Conducted in the Russian Federation

The studies were conducted in accordance with Methodical Guidelines of the Ministry of Health of the Russian Federation MUK 2.3.2.970-00 and MU 2.3.2.2306-07 [8,9]. PCR analysis of the maize test and control samples was performed to confirm the identity of the transformation event and its absence in the conventional control line.

### Biochemical Composition of Transgenic Maize Grain

The content of proteins, fatty acids, lipids, and carbohydrates in the grain of transgenic maize line MON 88017 did not significantly differ from the corresponding values for the conventional maize (Table 5.247). The contents

### 5.2.7 Diabrotica-Resistant and Glyphosate-Tolerant Transgenic

**Table 5.246** Registration Status of Transgenic Maize Line MON 88017 in Various Countries

| Country | Date of Approval | Scope |
| --- | --- | --- |
| Australia | 2006 | Food |
| Canada | 2006 | Food, feed, environmental release |
| China | 2007 | Food, feed |
| Korea | 2006 | Food, feed |
| Mexico | 2006 | Food, feed |
| USA | 2005 | Food, feed, environmental release |
| Taiwan | 2006 | Food |
| Philippines | 2003 | Food, feed |
| Japan | 2006 | Food, feed, environmental release |

*Note: The current registration status of the transgenic crop is on http://www.biotradestatus.com/*

of vitamins and minerals in the examined samples (Table 5.248) varied within the limits characteristic of maize *Zea mays* L. [1,10,15]. The contents of toxic elements in transgenic and conventional maize samples (Table 5.249) did not surpass the acceptable levels specified by Sanitary Norms SanPiN 2.3.2.1078-01 (section 1.4.1).

Thus, the grain of transgenic maize line MON 88017 and isogenic conventional maize did not significantly differ by biochemical composition. The safety parameters of the grain of conventional and transgenic maize line MON 88017 complied with the requirements of the regulations valid in the Russian Federation [6].

#### Toxicological Assessment of Transgenic Maize Line MON 88017

The chronic 180-day experiment was carried out on male Wistar rats with an initial body weight of 80–90 g. During the entire experiment, the animals were fed the standard semi-synthetic casein diet that contained either conventional (control group) or transgenic (test group) maize. Composition of the diets is described in Table 5.162. Milled maize grain was included into the feed, replacing the diet ingredients with due account for the contents of proteins, fat, and carbohydrates, and adherence to the isocaloric principle. The toxicological studies were conducted in accordance with Methodical Guidelines of the Ministry of Health of the Russian Federation MUK 2.3.2.970-00 [8]. Throughout the entire duration of the experiment, daily feed intake and general condition of the animals were documented. Body weight was measured weekly. Tissue specimens were taken on experimental Days 30 and 180.

Table 5.247 Content of Proteins, Fatty Acids, Lipids, and Carbohydrates in Transgenic and Conventional Maize ($M \pm m$; $n = 6-8$)

| Parameter | Conventional Maize | Transgenic Maize Line MON 88017 |
|---|---|---|
| Total protein, g/100 g | 8.163 ± 0.206 | 8.913 ± 0.483 |
| *Carbohydrates, g/100 g* | | |
| Fructose | 0.057 ± 0.027 | 0.096 ± 0.035 |
| Glucose | 0.051 ± 0.000 | 0.052 ± 0.001 |
| Sucrose | 1.323 ± 0.053 | 1.282 ± 0.110 |
| *Dietary fibers, g/100 g* | | |
| Total | 12.36 ± 1.19 | 15.24 ± 0.48 |
| Insoluble | 9.89 ± 0.40 | 10.82 ± 0.12 |
| Soluble | 2.47 ± 0.80 | 4.42 ± 0.45 |
| *Lipids* | | |
| Whole lipids, g/100 g | 5.220 ± 0.046 | 5.193 ± 0.023 |
| *Fatty acid composition of whole lipids and fractions, rel. %* | | |
| Total PFA | 53.59 ± 0.50 | 54.51 ± 0.95 |
| Lauric 12:0 | 0.050 ± 0.011 | 0.068 ± 0.028 |
| Myristic 14:0 | 0.115 ± 0.035 | 0.198 ± 0.083 |
| Pentadecanoic Σ 15:0 | 0.225 ± 0.099 | 0.188 ± 0.135 |
| Palmitic 16:0 | 11.73 ± 0.87 | 11.95 ± 0.43 |
| Palmitoleic 16:1 | 0.330 ± 0.032 | 0.260 ± 0.030 |
| Margaric Σ 17:0 | 0.400 ± 0.221 | 0.360 ± 0.118 |
| Stearic 18:0 | 3.783 ± 0.777 | 3.068 ± 0.141 |
| Oleic Σ 18:1 cys | 27.52 ± 0.79 | 27.29 ± 1.22 |
| Linoleic 18:2 | 51.77 ± 0.66 | 52.73 ± 0.77 |
| γ-Linolenic 18:3 ω-6 | 0.303 ± 0.243 | 0.173 ± 0.126 |
| α-Linolenic 18:3 ω-3 | 1.520 ± 0.028 | 1.610 ± 0.068 |
| Arachidic 20:0 | 0.595 ± 0.055 | 0.658 ± 0.017 |
| Eicosenoic Σ 20:1 | 0.758 ± 0.365 | 0.493 ± 0.090 |
| Docosanoic 22:0 | 0.538 ± 0.210 | 0.623 ± 0.250 |
| Docosenoic Σ 22:1 | 0.565 ± 0.221 | 0.423 ± 0.133 |

Throughout the entire duration of the experiment, the general condition of the rats was similar and satisfactory in control and test groups. No mortality was observed in any group. By appearance, condition of coat, behavior, and growth rate the test rats did not differ from the control animals. Daily intake of feed was 20–22 g/day. Body weight did not significantly differ in control and test groups (Figure 5.12; Table 5.250).

### 5.2.7 Diabrotica-Resistant and Glyphosate-Tolerant Transgenic

**Table 5.248** Comparative Content of Vitamins and Minerals in Transgenic and Conventional Maize Grain ($M \pm m$; $n = 6–8$)

| Parameter | Conventional Maize | Transgenic Maize Line MON 88017 |
|---|---|---|
| Vitamin $B_1$, mg/100 g | $0.220 \pm 0.009$ | $0.225 \pm 0.008$ |
| Vitamin $B_2$, mg/100 g | $0.228 \pm 0.015$ | $0.238 \pm 0.032$ |
| Vitamin $B_6$, mg/100 g | $0.333 \pm 0.020$ | $0.340 \pm 0.012$ |
| Carotenoids (total), mg/100 g | $1.10 \pm 0.10$ | $1.20 \pm 0.10$ |
| Vitamin E ($\beta+\gamma$), mg/100 g | $6.90 \pm 0.70$ | $8.00 \pm 0.90$ |
| Sodium, mg/kg | $59.00 \pm 3.15$ | $48.44 \pm 3.29$ |
| Calcium, mg/kg | $8.74 \pm 1.19$ | $8.66 \pm 1.08$ |
| Magnesium, mg/kg | $1883 \pm 124$ | $1799 \pm 151$ |
| Iron, mg/kg | $11.45 \pm 0.16$ | $9.72 \pm 0.66$ |
| Potassium, mg/kg | $3452 \pm 89$ | $3321 \pm 26$ |
| Zinc, mg/kg | $14.97 \pm 1.45$ | $16.03 \pm 0.71$ |
| Copper, mg/kg | $1.573 \pm 0.079$ | $1.487 \pm 0.061$ |
| Selenium, µg/kg | $203.0 \pm 4.0$ | $185.0 \pm 3.5$ |
| Phosphorus, %$P_2O_5$ | $0.673 \pm 0.032$ | $0.683 \pm 0.020$ |

**Table 5.249** Sanitary and Hygienic Safety Parameters of Maize Grain

| Parameter | Conventional Maize | Transgenic Maize Line MON 88017 | SanPiN 2.3.2.1078-01 |
|---|---|---|---|
| *Toxic elements, mg/kg* | | | |
| Lead | $\leq 0.001 \pm 0.000$ | $0.003 \pm 0.002$ | 0.5 |
| Arsenic | $\leq 0.10 \pm 0.00$ | $\leq 0.10 \pm 0.00$ | 0.2 |
| Cadmium | $0.006 \pm 0.005$ | $\leq 0.001 \pm 0.000$ | 0.1 |
| Mercury | not detected | not detected | 0.03 |
| *Pesticides, mg/kg* | | | |
| Hexachlorocyclohexane | not detected | not detected | 0.5 |
| DDT and its metabolites | not detected | not detected | 0.02 |
| Benzpyrene | not detected | not detected | 0.001 |
| *Mycotoxins, mg/kg* | | | |
| Aflatoxin $B_1$ | not detected | not detected | 0.005 |
| Deoxynivalenol | not detected | not detected | 0.7 |
| T2-toxin | not detected | not detected | 1.0 |
| Zearalenone | not detected | not detected | 1.0 |
| Fumonisin $B_1 + B_2$ | $0.073 \pm 0.009$ | $0.020 \pm 0.000$ | – |

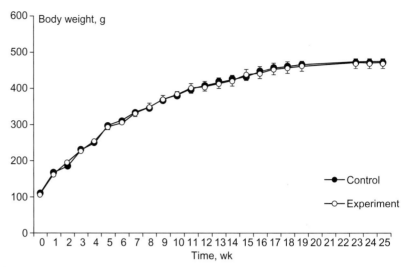

**FIGURE 5.12**
Comparative dynamics of body weight of rats fed diet with transgenic maize line MON 88017 (test) or its conventional counterpart (control).

The absolute and relative weight of internal organs in test rats did not differ from the corresponding values for control animals (Table 5.251).

Post-mortem examination revealed no alterations in the internal organs in both groups of rats. Similarly, histological examinations of internal organs performed on Days 30 and 180 did not reveal any statistically significant differences (Table 5.252).

The morphological examination of liver, kidney, lungs, spleen, small intestine, testicles, and prostate as well as structural morphometric analysis of small intestine, liver, kidneys, and spleen revealed no differences (Tables 5.253–5.255).

During the entire experiment, the hematological parameters of peripheral blood of control and test rats remained within the physiological range (Table 5.256). The biochemical parameters of blood serum (Table 5.257) and urine (Table 5.258) in the test group did not significantly differ from the analogous parameters of control rats. The increase of lipase activity by 50% in blood serum of test rats revealed on Day 30 and elevation of cholesterol by 25% observed on Day 180 remained within the physiological range of these parameters [28,45,48].

Analysis of antioxidant status and activity of the enzymes involved in protective and adaptive processes, which are the system biomarkers reflecting the level of adaptation of the entire organism to the environment, revealed no adverse effects of the transgenic maize on rats (Tables 5.259 and 5.260).

### 5.2.7 Diabrotica-Resistant and Glyphosate-Tolerant Transgenic

**Table 5.250** Body Weight (g) of Rats Fed Diet with Conventional Maize (Control) or Transgenic Maize Line MON 88017 (Test) ($M \pm m$, $n = 30$)

| Duration of Experiment, Day | Control | Test |
|---|---|---|
| 0 | 108.3 ± 1.1 | 108.3 ± 1.1 |
| 7 | 166.1 ± 3.2 | 162.3 ± 3.4 |
| 14 | 186.0 ± 4.7 | 196.3 ± 4.7 |
| 21 | 229.2 ± 3.3 | 228.4 ± 3.7 |
| 28 | 249.1 ± 3.3 | 252.1 ± 4.0 |
| 35 | 297.7 ± 7.4 | 292.0 ± 4.9 |
| 42 | 310.7 ± 7.7 | 304.7 ± 4.9 |
| 49 | 334.7 ± 9.2 | 331.0 ± 4.8 |
| 56 | 343.3 ± 10.3 | 348.0 ± 5.3 |
| 63 | 367.9 ± 10.1 | 369.7 ± 6.0 |
| 70 | 378.5 ± 11.9 | 380.5 ± 6.4 |
| 77 | 396.5 ± 13.9 | 398.8 ± 6.9 |
| 84 | 405.9 ± 13.2 | 405.3 ± 6.9 |
| 91 | 414.5 ± 13.4 | 413.7 ± 7.6 |
| 98 | 420.5 ± 13.0 | 419.7 ± 8.1 |
| 105 | 432.8 ± 13.5 | 436.6 ± 8.4 |
| 112 | 442.7 ± 14.9 | 441.1 ± 8.5 |
| 119 | 454.3 ± 15.8 | 451.0 ± 9.1 |
| 126 | 458.5 ± 16.2 | 456.7 ± 9.5 |
| 133 | 464.7 ± 16.4 | 460.9 ± 9.7 |
| 161 | 470.4 ± 16.5 | 467.7 ± 9.3 |
| 168 | 473.7 ± 16.8 | 467.3 ± 8.2 |
| 177 | 472.4 ± 12.4 | 467.4 ± 6.1 |

*Note: The differences are not significant ($p > 0.05$).*

There were no differences in the parameters characterizing the antioxidant status of the rats (Table 5.259). Elevation of superoxide dismutase activity in erythrocytes by 6% and the content of hepatic MDA by 11% revealed on Day 30 in the test group in comparison to the control rats were within the physiological ranges characteristic of the growing animals [11]. On Day 180, no differences were observed.

On the whole, the parameters characterizing protective and adaptive processes remained within the physiological norm (Tables 5.260 and 5.261). Moderate elevation of activity of some enzymes involved in metabolism of xenobiotics was established in test rats in comparison with the control animals. On Day 30, the contents of aminopyrine N-demethylase and CDNB-glutathione transferase were elevated by 23% and 39%, respectively. On Day

**Table 5.251** Absolute and Relative Weight of Internal Organs of Control and Test Rats ($M \pm m$, $n = 7$)

| Organ | | 30 Days | | 180 Days | |
|---|---|---|---|---|---|
| | | Control | Test | Control | Test |
| Kidneys | Abs.[a], g | 1.90 ± 0.05 | 1.97 ± 0.09 | 2.51 ± 0.17 | 2.67 ± 0.17 |
| | Rel.[b], g/100g | 0.723 ± 0.012 | 0.708 ± 0.027 | 0.547 ± 0.022 | 0.572 ± 0.031 |
| Liver | Abs., g | 8.55 ± 0.18 | 9.18 ± 0.37 | 11.32 ± 0.30 | 11.86 ± 0.64 |
| | Rel., g/100g | 3.248 ± 0.048 | 3.313 ± 0.096 | 2.487 ± 0.051 | 2.544 ± 0.120 |
| Spleen | Abs., g | 1.86 ± 0.12 | 2.03 ± 0.25 | 1.38 ± 0.05 | 1.52 ± 0.14 |
| | Rel., g/100g | 0.707 ± 0.036 | 0.732 ± 0.084 | 0.304 ± 0.011 | 0.323 ± 0.025 |
| Heart | Abs., g | 0.96 ± 0.02 | 1.03 ± 0.03 | 1.23 ± 0.06 | 1.31 ± 0.04 |
| | Rel., g/100g | 0.366 ± 0.011 | 0.375 ± 0.006 | 0.269 ± 0.009 | 0.280 ± 0.007 |
| Testicles | Abs., g | 2.64 ± 0.23 | 2.79 ± 0.09 | 3.26 ± 0.09 | 3.63 ± 0.20 |
| | Rel., g/100g | 0.998 ± 0.076 | 1.010 ± 0.033 | 0.719 ± 0.029 | 0.778 ± 0.037 |
| Hypophysis | Abs., mg | 8.20 ± 1.02 | 8.17 ± 0.48 | 9.17 ± 1.14 | 7.83 ± 0.31 |
| | Rel., mg/100g | 3.142 ± 0.379 | 2.955 ± 0.168 | 2.033 ± 0.265 | 1.686 ± 0.084 |
| Adrenal glands | Abs., mg | 27.50 ± 2.57 | 30.00 ± 3.10 | 30.67 ± 3.34 | 22.50 ± 1.89 |
| | Rel., mg/100g | 10.39 ± 0.86 | 10.81 ± 1.03 | 6.788 ± 0.817 | 4.847 ± 0.436 |

Note: The differences are insignificant ($p > 0.05$).
[a]Absolute weight of internal organs.
[b]Relative weight of internal organs (per 100g body weight).

**Table 5.252** Microscopic Examination of Internal Organs in Rats Fed Diet with Conventional Maize (Control) or Transgenic Maize Line MON 88017 (Test) (Combined Data Obtained on Experimental Days 30 and 180)

| Organ | Control | Test |
|---|---|---|
| Liver | Clear trabecular structure; no alterations in hepatocytes or portal ducts | No differences from control |
| Kidneys | Usual appearance of cortical and medullar substance; no alterations in glomeruli or pelvis epithelium | No differences from control |
| Lung | Alveolar space is air-filled; no alterations in bronchi and blood vessels | No differences from control |
| Spleen | Large folliculi with wide clear marginal zones and reactive centers; splenic pulp is moderately plethoric | No differences from control |
| Small intestine | Preserved villous epithelium; usual infiltration in villous stroma | No differences from control |
| Testicle | Clearly definable spermatogenesis and age-related spermiogenesis | No differences from control |

**Table 5.253** Area of Structural Components (%) in Ileum Wall of Rats Fed Diet with Conventional Maize (Control) or Transgenic Maize Line MON 88017 (Test) ($M \pm m$, $n = 25$)

| Structural Component | 30 Day Control | 30 Day Test | 180 Day Control | 180 Day Test |
|---|---|---|---|---|
| Mucous membrane:[a] | 82.9 ± 1.7 | 82.6 ± 1.4 | 82.5 ± 1.5 | 81.6 ± 1.3 |
| intestinal crypts[b] | 84.3 ± 1.3 | 83.9 ± 1.5 | 84.1 ± 1.3 | 83.8 ± 1.4 |
| lamina propria[b] | 12.3 ± 1.6 | 12.7 ± 1.2 | 12.5 ± 1.5 | 12.9 ± 1.3 |
| Smooth muscle layer[b] | 0.9 ± 0.1 | 1.2 ± 0.2 | 0.9 ± 0.1 | 1.0 ± 0.2 |
| Submucous layer[a] | 3.4 ± 0.2 | 3.6 ± 0.2 | 3.6 ± 0.2 | 3.5 ± 0.1 |
| Muscular tunic[a] | 13.1 ± 1.2 | 12.9 ± 1.1 | 13.2 ± 1.1 | 12.9 ± 1.1 |
| Serous membrane[a] | 1.1 ± 0.1 | 1.1 ± 0.1 | 1.2 ± 0.1 | 1.1 ± 0.1 |

*Note:* The differences are not significant ($p > 0.05$).
[a]Percentage of total area of the small intestine on histological sections.
[b]Percentage of area of mucous membrane.

**Table 5.254** Absolute Number of Various Epithelial Cells in Intestinal Villi and Ileum Crypts in Rats Fed Diet with Conventional Maize (Control) or Transgenic Maize Line MON 88017 (Test) ($M \pm m$, $n = 25$)

| Structural Component | 30 Day Control | 30 Day Test | 180 Day Control | 180 Day Test |
|---|---|---|---|---|
| *Villous epithelial cells* | | | | |
| Border cells | 55.6 ± 2.7 | 54.8 ± 2.3 | 55.1 ± 2.3 | 54.9 ± 2.2 |
| Goblet cells | 31.8 ± 1.9 | 32.1 ± 1.7 | 31.9 ± 1.9 | 32.0 ± 1.8 |
| Undifferentiated cells | 2.8 ± 0.5 | 2.7 ± 0.7 | 2.8 ± 0.4 | 2.7 ± 0.3 |
| Total | 89.2 ± 2.9 | 89.6 ± 2.3 | 89.7 ± 2.7 | 89.9 ± 2.1 |
| *Epithelial cells in half crypt* | | | | |
| Border cells | 27.3 ± 1.9 | 26.4 ± 2.1 | 27.1 ± 1.9 | 26.8 ± 2.0 |
| Goblet cells | 17.0 ± 1.5 | 16.9 ± 1.7 | 17.3 ± 1.5 | 16.9 ± 1.3 |
| Undifferentiated cells | 5.4 ± 1.0 | 5.7 ± 1.0 | 5.4 ± 1.0 | 5.5 ± 0.9 |
| Total | 48.3 ± 3.5 | 46.9 ± 3.7 | 47.3 ± 3.4 | 46.9 ± 3.3 |

*Note:* The differences are not significant ($p > 0.05$).

180, the content of epoxide hydrolase was elevated by 18%. Such small elevation in activity of these enzymes not accompanied by complex manifestations of stress of the protective and adaptive systems in test rats attests to the absence of adverse effects of chronic intake of transgenic maize on the protective and adaptive potencies of the organism.

**Table 5.255** Morphometric Parameters of Liver, Kidney, and Spleen of Rats Fed Diet with Conventional Maize (Control) or Transgenic Maize Line MON 88017 (Test) ($M \pm m$, $n = 25$)

| Structural Component | 30 Day Control | 30 Day Test | 180 Day Control | 180 Day Test |
|---|---|---|---|---|
| **Liver** | | | | |
| Area of capillary bed, %[a] | 33.2 ± 1.6 | 34.1 ± 1.4 | 35.2 ± 1.5 | 35.1 ± 1.3 |
| Number of binucleate hepatocytes per 100 hepatocytes | 3.5 ± 0.4 | 3.2 ± 0.5 | 3.5 ± 0.3 | 3.2 ± 0.3 |
| **Kidneys** | | | | |
| Number of glomeruli in visual field at 400× | 11.2 ± 0.9 | 11.0 ± 0.8 | 11.7 ± 0.9 | 11.9 ± 0.8 |
| **Spleen, number of lymphoid type cells in the area of 880 μm²** | | | | |
| Lymphoid nodule reproduction site | 27.5 ± 1.7 | 27.0 ± 1.6 | 28.5 ± 1.4 | 28.0 ± 1.3 |
| Lymphoid nodule mantle | 30.1 ± 1.6 | 29.9 ± 1.7 | 30.9 ± 1.8 | 31.1 ± 1.9 |
| Periarterial lymphoid sheath | 34.3 ± 1.8 | 33.5 ± 1.5 | 34.5 ± 1.8 | 34.4 ± 1.6 |

Note: The differences are insignificant ($p > 0.05$).
[a] The total area of liver on histological section is taken as 100%.

**Table 5.256** Hematological Parameters of Blood Serum of Rats Fed Diet with Conventional Maize (Control) or Transgenic Maize Line MON 88017 (Test) ($M \pm m$, $n = 28$)

| Parameter | 30 Day Control | 30 Day Test | 180 Day Control | 180 Day Test | Standard Values [28,45,48] |
|---|---|---|---|---|---|
| Total erythrocyte count, ×10⁶/μL | 8.386 ± 0.184 | 7.647 ± 0.131* | 8.764 ± 0.209 | 8.756 ± 0.169 | 4.4–8.9 |
| Hemoglobin concentration, g/L | 159.6 ± 3.0 | 153.5 ± 1.7 | 144.9 ± 2.8 | 146.4 ± 2.8 | 86–173 |
| Hematocrit, vol.% | 49.27 ± 0.98 | 44.27 ± 0.72* | 43.70 ± 0.85 | 45.13 ± 0.88 | 31.4–51.9 |
| MCV, μm³ | 58.86 ± 0.51 | 58.17 ± 0.81 | 49.71 ± 0.42 | 51.43 ± 0.53 | 50.6–93.8 |
| MCH, pg | 19.47 ± 0.21 | 19.80 ± 0.31 | 16.54 ± 0.13 | 16.76 ± 0.14 | 16–21 |
| MCHC, g/dL | 33.14 ± 0.11 | 34.07 ± 0.23* | 33.13 ± 0.22 | 32.49 ± 0.19 | 24.7–36.8 |
| ESR, mm/h | 2.000 ± 0.308 | 0.714 ± 0.184* | 1.857 ± 0.261 | 1.571 ± 0.202 | 0–5 |
| Leukocytes, 10³/μL | 9.158 ± 1.237 | 7.267 ± 0.780 | 6.086 ± 0.666 | 6.686 ± 0.630 | 1.4–34.3 |
| Basophils, % | 0 | 0 | 0 | 0 | 0–1 |
| Eosinophils, % | 1.286 ± 0.286 | 0.857 ± 0.459 | 5.429 ± 0.896 | 1.714 ± 0.565* | 0.0–5.5 |
| Stab neutrophils, % | 1.571 ± 0.369 | 4.429 ± 1.429 | 4.000 ± 1.662 | 2.286 ± 0.522 | 18–36 |
| Segmentonuclear neutrophils, % | 20.29 ± 1.74 | 17.29 ± 3.68 | 33.00 ± 1.56 | 29.29 ± 2.16 | |
| Metamyelocyte, % | 0 | 0 | 0 | 0 | 0 |
| Lymphocytes, % | 74.29 ± 1.44 | 75.14 ± 4.41 | 57.00 ± 2.81 | 63.00 ± 3.39 | 42.3–98.0 |
| Monocytes, % | 2.333 ± 0.760 | 1.714 ± 0.522 | 1.429 ± 0.481 | 4.000 ± 0.976* | 0.0–7.9 |
| Platelets, ×10³/μL | 749.9 ± 25.4 | 704.8 ± 29.5 | 834.0 ± 40.0 | 749.7 ± 24.7 | 409–1250 |

ESR, erythrocyte sedimentation rate; MCH, mean cell hemoglobin; MCHC, mean cell hemoglobin concentration; MCV, mean cell volume.
*Difference significant at $p < 0.05$.

Table 5.257 Blood Serum Biochemical Parameters of Rats Fed Diet with Conventional Maize (Control) or Transgenic Maize Line MON 88017 (Test) ($M \pm m$, $n = 28$)

| Parameter | 30 Day Control | 30 Day Test | 180 Day Control | 180 Day Test | Standard Values [28,45,48] |
|---|---|---|---|---|---|
| Total protein, g/L | 72.42 ± 3.25 | 65.93 ± 3.09 | 74.01 ± 3.40 | 75.97 ± 1.80 | 56–82 |
| Albumin, g/L | 42.85 ± 1.62 | 36.55 ± 3.00 | 43.99 ± 1.21 | 49.67 ± 2.59 | 25–48 |
| Triglycerides, mM/L | 1.759 ± 0.141 | 1.533 ± 0.140 | 1.161 ± 0.068 | 1.277 ± 0.109 | 0.3–1.6 |
| Total bilirubin, µM/L | 4.429 ± 0.751 | 3.857 ± 0.595 | 3.571 ± 0.481 | 5.000 ± 0.617 | 1–4 |
| Conjugated bilirubin, µM/L | 0 | 0 | 0 | 0 | 0 |
| Urea, mM/L | 7.586 ± 0.674 | 6.114 ± 0.566 | 8.357 ± 0.958 | 8.600 ± 0.487 | 4.0–10.0 |
| Creatinine, µM/L | 59.71 ± 6.27 | 63.33 ± 8.67 | 92.00 ± 5.44 | 84.00 ± 7.75 | 13–92 |
| Glucose, mM/L | 2.477 ± 0.251 | 2.013 ± 0.435 | 5.330 ± 0.318 | 6.356 ± 0.563 | 4.5–10.0 |
| Cholesterol, mM/L | 2.673 ± 0.313 | 2.413 ± 0.271 | 2.222 ± 0.081 | 2.781 ± 0.224* | 0.6–4.3 |
| γ-glutamyl transferase, U/L | 7.680 ± 2.516 | 9.300 ± 2.930 | 8.643 ± 0.878 | 6.340 ± 0.929 | 0–3 |
| Lactate dehydrogenase, U/L | 4006 ± 206 | 3924 ± 331 | 3485 ± 339 | 2387 ± 460 | <5800 |
| Alpha-amylase, U/L | 565.7 ± 69.0 | 565.1 ± 22.6 | 621.3 ± 55.5 | 780.0 ± 86.6 | <3207 |
| Creatine phosphokinase, U/L | 9665 ± 1033 | 8645 ± 626 | 5253 ± 946 | 4670 ± 1031 | ± 200% |
| Alkaline phosphatase, U/L | 436.4 ± 9.0 | 471.0 ± 15.2 | 171.2 ± 7.7 | 200.5 ± 21.4 | 112–814 |
| Alanine aminotransferase, U/L | 76.00 ± 7.90 | 62.40 ± 12.31 | 69.29 ± 10.78 | 76.00 ± 10.76 | 33–120 |
| Aspartate aminotransferase, U/L | 209.1 ± 19.1 | 175.4 ± 59.0 | 214.3 ± 14.6 | 213.3 ± 20.8 | 60–236 |
| Lipase, U/L | 17.83 ± 1.52 | 26.83 ± 1.58* | 11.00 ± 0.41 | 13.50 ± 1.26 | <30 |

*Difference from control significant at $p < 0.05$.

Thus, the chronic 180-day toxicological experiment on rats fed diet with transgenic maize line MON 88017 did not reveal any toxic effects. The values of all examined parameters were within the range of physiological variations characteristic for rats.

## Assessment of Potential Genotoxicity of Transgenic Maize MON 88017

The study of potential genotoxicity of transgenic maize line MON 88017 was carried out on male C57Bl/6 mice fed diet with transgenic (test group) or conventional (control group) maize for 30 days. Initial body weight of mice was 19–22 g. The mice were fed the standard semi-synthetic casein diet (Table 5.162) supplemented with conventional or transgenic milled maize grain (2.4 g/day/mouse). Assessment of potential genotoxicity is based on revealing DNA damage by alkaline gel electrophoresis of the cells isolated from bone marrow, liver, and rectum (DNA-comet assay) [8], as well as on detection of mutagenic activity by counting the chromosomal aberrations in metaphasic cells of mouse bone marrow [3]. No fewer than 100 cells of each micropreparation were analyzed. The toxicological studies were conducted

**Table 5.258** Urinary Biochemical Parameters of Rats Fed Diet with Conventional Maize (Control) or Transgenic Maize Line MON 88017 (Test) ($M \pm m$, $n = 25$)

| Parameter | 30 Day | | 180 Day | |
|---|---|---|---|---|
| | Control | Test | Control | Test |
| pH | 5.8 ± 0.1 | 6.0 ± 0.1 | 5.8 ± 0.1 | 5.8 ± 0.1 |
| Daily diuresis, mL | 2.4 ± 0.3 | 3.0 ± 0.7 | 6.2 ± 0.4 | 6.4 ± 1.0 |
| Relative density, g/mL | 1.25 ± 0.03 | 1.20 ± 0.03 | 1.15 ± 0.01 | 1.15 ± 0.01 |
| Creatinine, mg/mL | 1.68 ± 0.20 | 1.60 ± 0.30 | 1.29 ± 0.12 | 1.12 ± 0.16 |
| Creatinine, mg/day | 3.96 ± 0.54 | 4.07 ± 0.58 | 7.72 ± 0.82 | 6.39 ± 0.81 |

*Note: The differences are not significant ($p > 0.05$).*

**Table 5.259** Antioxidant Status of Rats Fed Diet with Conventional Maize (Control) or Transgenic Maize Line MON 88017 (Test) ($M \pm m$, $n = 28$)

| Parameter | 30 Day | | 180 Day | |
|---|---|---|---|---|
| | Control | Test | Control | Test |
| *Enzymatic activity of antioxidant protection system* | | | | |
| Glutathione reductase, μM/min/g Hb | 35.39 ± 1.42 | 37.93 ± 1.49 | 39.49 ± 2.35 | 38.81 ± 1.01 |
| Glutathione peroxidase, μM/min/g Hb | 62.64 ± 2.10 | 65.09 ± 1.75 | 59.79 ± 2.04 | 61.59 ± 2.97 |
| Catalase, mM/min/g Hb | 469.5 ± 18.7 | 497.9 ± 16.4 | 530.2 ± 12.4 | 500.8 ± 13.8 |
| Superoxide dismutase, Unit/min/g Hb | 1869 ± 17 | 1977 ± 4* | 1936 ± 41 | 1953 ± 53 |
| *Content of LPO products* | | | | |
| erythrocytic MDA, nM/ml | 5.366 ± 0.453 | 5.491 ± 0.313 | 4.880 ± 0.245 | 4.866 ± 0.512 |
| Blood serum MDA, nM/ml | 6.854 ± 0.233 | 6.941 ± 0.434 | 6.222 ± 0.226 | 6.699 ± 0.162 |
| Hepatic MDA, nM/g | 476.4 ± 20.9 | 528.9 ± 9.2* | 420.9 ± 20.1 | 409.2 ± 38.3 |

*Difference from control significant at $p < 0.05$.

in accordance with Methodical Guidelines of the Ministry of Health of the Russian Federation MUK 2.3.2.970-00 and MU 2.3.2.2306-07 [8,9].

On Day 30 of the dietary experiment, the body weight of mice was in the range 24–26 g. The data of cytogenetic studies of mouse bone marrow and the mean parameters of chromosomal aberrations in the test group did not significantly differ from the corresponding values for control mice (Table 5.262) and did not surpass the level of spontaneous mutagenesis characteristic of C57Bl/6 mice [6]. There were no differences in the degree of structural DNA damage in bone marrow, liver, and rectum (Table 5.263). The study of DNA integrity and the level of chromosomal aberrations in test mice fed diet

**Table 5.260** Activity of Enzymes Involved in Metabolism of Xenobiotics and Content of Hepatic Protein in Rats Fed Diet with Conventional Maize (Control) or Transgenic Maize Line MON 88017 (Test) ($M \pm m$, $n = 25$)

| Parameter | 30 Day | | 180 Day | |
|---|---|---|---|---|
| | Control | Test | Control | Test |
| Cytochrome P450, nM/mg protein | 0.68 ± 0.05 | 0.74 ± 0.05 | 1.20 ± 0.05 | 1.05 ± 0.05 |
| Cytochrome $b_5$, nM/mg protein | 0.78 ± 0.03 | 0.82 ± 0.07 | 0.98 ± 0.04 | 0.93 ± 0.04 |
| Aminopyrine N-demethylation, nM/min/mg protein | 5.03 ± 0.24 | 6.20 ± 0.28* | 9.37 ± 0.32 | 9.85 ± 0.35 |
| Benzpyrene hydroxylation, Fl-units/min/mg protein | 7.08 ± 0.78 | 7.00 ± 1.19 | 8.87 ± 0.48 | 9.42 ± 0.57 |
| Acetylesterase, µM/min/mg protein | 5.28 ± 0.12 | 5.20 ± 0.43 | 5.45 ± 0.07 | 5.55 ± 0.07 |
| Epoxide hydrolase, nM/min/mg protein | 4.17 ± 0.26 | 5.66 ± 1.05 | 5.40 ± 0.30 | 6.36 ± 0.24* |
| UDP-Glucuronosil transferase, nM/min/mg protein | 16.0 ± 0.9 | 20.2 ± 2.4 | 27.9 ± 0.8 | 30.2 ± 2.7 |
| CDNB-Glutathione transferase, µM/min/mg protein | 0.94 ± 0.05 | 1.31 ± 0.07* | 1.29 ± 0.02 | 1.32 ± 0.07 |
| Microsomal protein, mg/g | 15.1 ± 0.5 | 14.2 ± 0.5 | 15.5 ± 0.3 | 14.5 ± 0.5 |
| Cytosolic protein, mg/g | 95.0 ± 1.9 | 94.4 ± 2.4 | 82.8 ± 2.0 | 78.4 ± 1.0 |

*Difference from control significant at $p < 0.05$.

**Table 5.261** Activity of Hepatic Lysosomal Enzymes in Rats Fed Diet with Conventional Maize (Control) or Transgenic Maize Line MON 88017 (Test) ($M \pm m$, $n = 25$)

| Parameter | 30 Day | | 180 Day | |
|---|---|---|---|---|
| | Control | Test | Control | Test |
| *Total activity, µM/min/g tissue* | | | | |
| Arylsulfatase A and B | 2.16 ± 0.05 | 2.37 ± 0.04* | 2.40 ± 0.08 | 2.34 ± 0.03 |
| β-Galactosidase | 2.57 ± 0.07 | 2.62 ± 0.11 | 2.22 ± 0.08 | 2.14 ± 0.04 |
| β-Glucuronidase | 2.29 ± 0.06 | 2.45 ± 0.07 | 2.49 ± 0.07 | 2.32 ± 0.05 |
| *Non-sedimentable activity, % total activity* | | | | |
| Arylsulfatase A and B | 3.17 ± 0.14 | 3.35 ± 0.06 | 2.85 ± 0.09 | 2.93 ± 0.08 |
| β-Galactosidase | 5.91 ± 0.16 | 6.17 ± 0.33 | 7.30 ± 0.52 | 7.21 ± 0.21 |
| β-Glucuronidase | 7.44 ± 0.30 | 7.66 ± 0.25 | 5.61 ± 0.22 | 5.44 ± 0.29 |

*Difference from control significant at $p < 0.05$.

with transgenic maize line MON 88017 revealed no genotoxic effects of the transgenic diet in comparison with conventional maize.

## Assessment of Potential Impact of Transgenic Maize Line MON 88017 on Immune System in Studies on Mice

The immunomodulating effects of transgenic maize line MON 88017 were examined during 45 days on CBA and C57Bl/6 mice with an initial body

**Table 5.262** Chromosomal Damage in Bone Marrow Cells in Mice Fed Diet with Conventional Maize (Control) or Transgenic Maize Line MON 88017 (Test) ($M \pm m$)

| Group | Number of Cells | By 100 Cells | | | | | Total Number of Damaged Metaphases, % |
|---|---|---|---|---|---|---|---|
| | | Gaps | Single Fragments | Paired Fragments | Crossovers | Cell with Multiple Damages[a] | |
| Control | 500 | 0.6 | 1.4 | – | 0.2 | – | $2.2 \pm 0.7$ |
| Test | 500 | 0.8 | 1.6 | – | – | – | $2.4 \pm 0.7$ |

*Note: The differences are not significant ($p > 0.05$).*
[a]More than 5 chromosomal aberrations in a cell.

**Table 5.263** Degree of Structural DNA Damage in Internal Organs of Mice Fed Diet with Conventional Maize (Control) or Transgenic Maize Line MON 88017 (Test) ($M \pm m$)

| Group | Bone Marrow | | Liver | | Rectum | |
|---|---|---|---|---|---|---|
| | Number of Cells | Damaged DNA, % | Number of Cells | Damaged DNA, % | Number of Cells | Damaged DNA, % |
| Control | 507 | $6.6 \pm 1.2$ | 504 | $6.8 \pm 1.1$ | 503 | $8.6 \pm 1.2$ |
| Test | 501 | $6.7 \pm 1.1$ | 513 | $7.4 \pm 1.2$ | 508 | $8.7 \pm 1.3$ |

*Note: The differences are not significant ($p > 0.05$).*

weight of 18–20 g. During the entire period of the experiment, mice were maintained on standard vivarium diet (Table 5.114). Milled maize grain was included in the feed (3 g/mouse/day) replacing the oatmeal and providing equal amount of nutrients and calories. Assessment of immune-modulating and sensitization effects of transgenic maize were carried out in four tests:

- Effect on humoral component of immune system—by determining the level of hemagglutination to sheep erythrocytes (SE);
- Effect on cellular component of immune system— by reaction of delayed hypersensitivity to SE;
- Sensitization effect— by the test of sensitivity to histamine;
- Effect on natural resistance of mice to infection—by injection with *Salmonella typhimurium*.

The studies were conducted in accordance with Methodical Guidelines of the Ministry of Health of the Russian Federation MUK 2.3.2.970-00 and MU 2.3.2.2306-07 [8,9].

The test on the humoral component of the immune system showed that dynamics of production of the antibodies against SE in control mice was similar to that in the test group for CBA and C57Bl/6 mice, which attests to the absence of any extra immunomodulating effect of transgenic maize MON 88017 in comparison with conventional maize (Table 5.264). Some insignificant differences in the

**Table 5.264** Level of Antibodies Raised Against SE in Mice Fed Diet with Conventional Maize (Control) or Transgenic Maize Line MON 88017 (Test) ($M \pm m$, $n \geq 10$)

| SE Post-Injection Day | Antibody Titer | | | |
| --- | --- | --- | --- | --- |
| | CBA Mice | | C57Bl/6 Mice | |
| | Control | Test | Control | Test |
| 7 | 107 ± 21 | 171 ± 43 | 128 ± 74 | 85 ± 21 |
| 14 | 107 ± 21 | 171 ± 43 | 128 ± 74 | 85 ± 21 |
| 21 | 36 ± 28 | 96 ± 32 | 56 ± 37 | 29 ± 17 |

Note: The differences are not significant ($p > 0.05$).

**Table 5.265** Delayed Hypersensitivity Reaction Index in Mice Fed Diet with Conventional Maize (Control) or Transgenic Maize Line MON 88017 (Test) ($M \pm m$, $n \geq 10$)

| CBA Mice | | C57Bl/6 Mice | |
| --- | --- | --- | --- |
| Control | Test | Control | Test |
| 35.5 ± 2.7 | 32.1 ± 3.4 | 37.6 ± 5.6 | 36.1 ± 6.1 |

Note: The differences are not significant ($p > 0.05$).

**Table 5.266** Susceptibility to *Salmonella Typhimurium* in Mice Fed Diet with Conventional Maize (Control) or Transgenic Maize Line MON 88017 (Test) ($M \pm m$, $n \geq 10$)

| Parameter | CBA Mice | | C57Bl/6 Mice | |
| --- | --- | --- | --- | --- |
| | Control | Test | Control | Test |
| $LD_{50}$ | $220 \times 10^2$ | $850 \times 10^{2*}$ | $0.87 \times 10^2$ | $3.2 \times 10^{2*}$ |
| Mortality of mice | Start: post-injection week 1 | | Start: post-injection week 1 | |
| | Finish: post-injection Day 18 | | Finish: post-injection Day 18 | |
| Mean lifetime | 4.1 to 5.0 day | | 7.2 to 8.0 day | |

*Difference from control group significant at $p < 0.05$.

antibody levels observed on Day 21 after injection of antigen remained within the physiological range, reflecting individual sensitivity of the animals. Similarly, the delayed hypersensitivity reaction did not reveal a modulating effect of transgenic maize line MON 88017 on the cellular component of the immune system (Table 5.265). The tests on sensitization effect and the action on natural resistance of mice to *Salmonella typhimurium* revealed no adverse effects of the transgenic maize.

Table 5.266 shows that test mice of both subgroups (CBA and C57Bl/6) demonstrated significantly higher resistance to infection with *Salmonella typhimurium* than did the control animals. The revealed differences in $LD_{50}$ values remained within the physiological ranges for this parameter, which were $100-2000 \times 10^2$ and $0.1-10 \times 10^2$ for CBA and C57Bl/6 mice, respectively, according to the data of the I. I. Mechnikov State Research Institute for Vaccines and Sera. These data showed that *Salmonella typhimurium* produced a typical infection of murine typhus both in test and control mice. The CBA mice (insensitive to *Salmonella typhimurium*) were more resistant to infection than were the C57Bl/6 mice (susceptible to murine typhus).

Therefore, transgenic maize line MON 88017 demonstrated no extra sensitizing or immunomodulating potencies in comparison with conventional maize.

### Assessment of Potential Allergenicity of Transgenic Maize line MON 88017

A 30-day subchronic experiment was carried out on male Wistar rats with an initial body weight of $180 \pm 10$ g. Over the entire duration of the experiment, the rats were maintained on standard vivarium diet (Table 5.114). Milled maize grain was included into the feed (3.5 g/rat/day) partially replacing the oatmeal and providing an equal amount of nutrients and calories. Potential allergenicity was assessed by severity of generalized anaphylaxis and by concentration of circulating specific immunoglobulin antibodies (the sum of $IgG_1$ and $IgG_4$ fractions) after intraperitoneal sensitization of mature rats with a nutrient antigen (ovalbumin) followed by intravenous injection of an anaphylaxis-provoking dose of this protein to sensitized animals. The studies were conducted in accordance to Methodical Guidelines of the Ministry of Health of the Russian Federation MUK 2.3.2.970-00 and MU 2.3.2.2306-07 [8,9].

On Day 29 of the dietary experiment, the body weights of control and test rats were $317 \pm 8$ and $305 \pm 5$ g, respectively ($p > 0.05$). There were no significant differences in severity of anaphylactic shock (Table 5.267) or intensity

**Table 5.267** Severity of Active Anaphylactic Shock in Rats Fed Diet with Conventional Maize (Control) or Transgenic Maize Line MON 88017 (Test) ($M \pm m$)

| Group[a] | Anaphylactic Index | Severe Reactions, % | Mortality, % |
|---|---|---|---|
| Control | 2.95 | 54.5 | 54.5 |
| Test | 2.79 | 50.0 | 50.0 |
| p | >0.05[b] | >0.05[c] | >0.05[c] |

[a]n = 22 (control group) and n = 24 (test group).
[b]Mann-Whitney non-parametric rank test.
[c]Two-sided Fisher angular conversion U-test.

**Table 5.268** Humoral Response Intensity in Rats Fed Diet with Conventional Maize (Control) or Transgenic Maize Line MON 88017 (Test) ($M \pm m$)

| Group[a] | $D_{492}$ | Concentration of Antibodies, mg/mL | Logarithm of Antibody Concentration |
|---|---|---|---|
| Control | 0.764 ± 0.030 | 6.2 ± 0.7 | 0.727 ± 0.052 |
| Test | 0.806 ± 0.026 | 7.1 ± 0.8 | 0.797 ± 0.047 |
| *Statistical analysis* | | | |
| Student t-test | >0.05 | >0.05 | >0.05 |
| ANOVA variance homogeneity test, P | >0.05 | >0.05 | >0.05 |
| Mann-Whitney non-parametrical rank test | | >0.05 | |

[a] $n = 22$ (control group) and $n = 24$ (test group).

of humoral immune response (Table 5.268). Analysis of data distributions in both groups of the rats performed with ANOVA test attested to variance homogeneity of the examined parameters ($p > 0.05$). Assessment of severity of active anaphylactic shock and intensity of humoral immune response in rats fed diet with transgenic maize line MON 88017 attested to the absence of allergenic effect of this maize variety.

### Assessment of Technological Parameters of Maize Line MON 88017

The studies of functional and technological properties of transgenic maize line MON 88017 were conducted in accordance with Methodical Guidelines of the Ministry of Health of the Russian Federation "Medical and Biological Assessment of Food Products Derived from Genetically Modified Sources": MUK 2.3.2.970-00. The results demonstrated the following.

- Quality of the examined samples of maize grain met the requirements of Russian National Standards GOST 13634-90 "Maize. Procurement and Supply Requirements".
- The control and test samples of maize grain were processed identically. No difficulties or differences were observed in the technological process when producing starch from transgenic maize MON 88017 or conventional maize. According to Russian National Standards GOST P 51985-2002 "Maize Starch. General Technological Requirements" the quality of derived maize starch was superior.
- The amylograms of dry control and test starch obtained on Brabender amylograph did not differ.

- Gelatinization temperature of transgenic starch was somewhat lower than that of control starch. After exposure at 95°C for 15 min and cooling to 60°C, viscosity values of starch gelatins (7%) of both samples were identical.
- The data on gelatinization properties revealed insignificant differences in the properties of the control and test starch samples. These differences can result from varying degrees of grain ripening.
- The structural and mechanical properties of the extrusive products derived from transgenic maize line MON 88017 were practically identical to those obtained from conventional maize.

Thus, the study of functional and technological properties revealed no significant differences between the examined grain samples of transgenic maize line MON 88017 and its conventional counterpart.

### Conclusions

Peer review of the data submitted by the applicant and the results of complex medical and biological studies of transgenic maize line MON 88017 tolerant to glyphosate and resistant against corn rootworm *Diabrotica* spp., attest to the absence of any toxic, genotoxic, sensitization, immunomodulating, or allergenic effects of this maize line. By biochemical composition, transgenic maize line MON 88017 was identical to conventional maize.

In accordance with Federal Law No.52-FZ "On Sanitary and Epidemiological Population Welfare" (March 30, 1999), transgenic maize line MON 88017 tolerant to glyphosate and resistant against corn rootworm *Diabrotica* spp. has been passed registered for food use, listed in the State Register, and licensed for use in the territory of the Russian Federation, import into the territory of the Russian Federation, and placing on the market without restrictions (State Registration Certificate No. 77.99.34.11.U.3259.5.07 on May 08, 2007).

## 5.2.8 DIABROTICA-RESISTANT TRANSGENIC MAIZE LINE MIR604

### Molecular Characteristics of Transgenic Maize Line MIR604

#### Recipient Organism

Maize (*Zea mays* L.) is characterized by a long history of safe use as human food and animal feed.

#### Donor Organism

The naturally occurring *cry3A* gene from *Bacillus thuringiensis* subsp. *Tenebrionis* was modified to version *mcry3A* gene. Modification resulted in inclusion of the recognition site of cathepsin-G serine protease into δ-endotoxin Cry3A, which increased the toxicity of mCry3A against *Diabrotica virgifera virgifera* and *Diabrotica longicornis barbery*.

The *pmi* gene of phosphomannose isomerase was isolated from *E.coli* [33,39].

### Method of Genetic Transformation

Transgenic maize line MIR604 was produced by agrobacterial transformation of the maize genome with binary plasmid vector pZM26 incorporating an expression cassette of the δ-endotoxin gene (*mcry3A*) and an expression cassette of the phosphomannose isomerase selective gene (*pmi*). The expression cassette of the *mcry3A* gene contained the following genetic elements:

- MTL—promoter of metallothionein-like maize gene providing expression predominantly in the roots of *Zea mays L.*;
- *mcry3A* gene, which is a modified version of the naturally occurring *cry3A* gene from *Bacillus thuringiensis* subsp. *tenebrionis*. Modification resulted in inclusion of the recognition site of cathepsin-G serine protease into δ-endotoxin Cry3A, which increased toxicity of mCry3A against *Diabrotica virgifera virgifera* and *Diabrotica longicornis barbery*;
- NOS 3′ terminator (the non-translated fragment of the nopaline synthase gene) isolated from the Ti-plasmid of *Agrobacterium tumefaciens*.

The expression cassette of the *pmi* gene contained the following genetic elements:

- ZmUbiInt, the constitutive promoter and the first intron of the maize polyubiquitin gene responsible for expression in monocotyledonous plants;
- *pmi* gene of phosphomannose isomerase, which catalyzes isomerization of mannose-6-phosphate into fructose-6-phosphate;
- NOS 3′ terminator (the non-translated fragment of the nopaline synthase gene) isolated from the Ti-plasmid of *Agrobacterium tumefaciens*;

Resistance of transgenic maize against damage by insect pests results from the presence of the *mcry3A* gene in the plant genome, encoding synthesis of the protein composed of 598 amino acid residues ($M_r = 67\,kDa$) which is toxic for *Diabrotica* spp. The presence of the *pmi* selective gene responsible for synthesis of phosphomannose isomerase ($M_r = 45\,kDa$, 391 amino acid residues) made it possible to select the modified maize cells during cultivation on growth medium.

### Global Registration Status of Maize Line MIR604

Table 5.269 shows the registration status of transgenic maize line MIR604 at the time of registration in Russia [19].

**Table 5.269** Registration Status of Transgenic Maize Line MIR604 in Various Countries

| Country | Date of Approval | Application |
| --- | --- | --- |
| Australia | 2006 | Food |
| Canada | 2007 | Food, feed, environmental release |
| Korea | 2007 | Food, feed |
| Mexico | 2007 | Food, feed |
| USA | 2007 | Food, feed, environmental release |
| Taiwan | 2007 | Food |
| Philippines | 2007 | Food, feed |
| Japan | 2007 | Food, feed, environmental release |

Note: The current registration status of the transgenic crop is on http://www.biotradestatus.com/

## Safety Assessment of Transgenic Maize Line MIR604 Conducted in the Russian Federation

The studies were conducted in accordance with Methodical Guidelines of the Ministry of Health of the Russian Federation MUK 2.3.2.970-00 and MU 2.3.2.2306-07 [8,9]. PCR analysis of the maize test and control samples was performed to confirm the identity of the transformation event and its absence in the conventional control line.

### Biochemical Composition of Transgenic Maize Grain

The content of proteins, fatty acids, lipids, and carbohydrates in the grain of transgenic maize line MIR604 did not significantly differ from the corresponding values for the conventional maize (Table 5.270). The contents of vitamins and minerals in the examined samples (Table 5.271) varied within the limits characteristic of maize *Zea mays* L. [1,10,15].

The content of toxic elements in transgenic and conventional maize samples (Table 5.272) did not surpass the acceptable levels specified by Sanitary Norms SanPiN 2.3.2.1078-01 (section 1.4.1.). Thus, the grain of transgenic maize line MIR604 and isogenic conventional maize did not significantly differ by biochemical composition. The safety parameters of the grain of conventional and transgenic maize line MIR604 complied with the requirements of the regulations valid in the Russian Federation [8].

### Toxicological Assessment of Transgenic Maize Line MIR604

Chronic 170-day experiment was carried out on male Wistar rats with an initial body weight of 80–90 g. During the entire experiment, the animals were fed a standard semi-synthetic casein diet with conventional maize (control group) or transgenic maize line MIR604 (test group). Composition of the diets is given in Table 5.164. Milled maize grain was included in the feed,

**Table 5.270** Comparative Content of Proteins, Fatty Acids, Lipids, and Carbohydrates in Transgenic and Conventional Maize ($M \pm m$; $n = 6$–$8$)

| Parameter | Conventional Maize | Transgenic Maize Line MIR604 |
|---|---|---|
| Total protein, g/100 g | 11.64 ± 0.05 | 11.09 ± 0.16 |
| *Carbohydrates, g/100 g* | | |
| Fructose | 0.185 ± 0.010 | 0.147 ± 0.015 |
| Glucose | 0.189 ± 0.009 | 0.161 ± 0.014 |
| Sucrose | 2.043 ± 0.149 | 1.733 ± 0.120 |
| *Dietary fibers, g/100 g* | | |
| Insoluble | 8.737 ± 0.761 | 9.430 ± 0.380 |
| Soluble | 2.410 ± 0.235 | 3.143 ± 0.127 |
| Total | 11.147 ± 0.892 | 12.567 ± 0.503 |
| *Lipids* | | |
| Whole lipids, g/100 g | 6.667 ± 0.145 | 6.767 ± 0.088 |
| *Fatty acid composition of whole lipids and fractions, rel. %* | | |
| Total PFA | 50.763 ± 1.968 | 51.217 ± 1.487 |
| Myristic 14:0 | 0.073 ± 0.007 | 0.063 ± 0.003 |
| Pentadecanoic Σ 15:0 | 0.047 ± 0.003 | 0.047 ± 0.003 |
| Palmitic 16:0 | 14.800 ± 0.199 | 14.520 ± 0.053 |
| Palmitoleic 16:1 | 0.260 ± 0.042 | 0.220 ± 0.015 |
| Margaric Σ 17:0 | 0.180 ± 0.012 | 0.173 ± 0.012 |
| Stearic 18:0 | 2.617 ± 0.219 | 2.627 ± 0.146 |
| Oleic Σ 18:1 cys | 30.010 ± 1.580 | 29.743 ± 1.452 |
| Linoleic 18:2 | 48.603 ± 2.404 | 49.130 ± 1.824 |
| α-Linolenic 18:3 ω-3 | 2.160 ± 0.440 | 2.087 ± 0.339 |
| Arachidic 20:0 | 0.587 ± 0.037 | 0.610 ± 0.045 |
| Eicosenoic Σ 20:1 | 0.320 ± 0.068 | 0.310 ± 0.032 |
| Docosanoic 22:0 | 0.283 ± 0.020 | 0.307 ± 0.032 |
| Docosenoic Σ 22:1 | 0.063 ± 0.009 | 0.070 ± 0.012 |

replacing diet ingredients on an isocaloric basis with due account for the content of proteins, fat, and carbohydrates.

The toxicological studies were conducted in accordance with Methodical Guidelines of the Ministry of Health of the Russian Federation MUK 2.3.2.970-00 [8]. During the entire experiment, daily feed intake and general condition of the animals were documented. Body weight was measured every week. Tissue specimens were taken on experimental Days 30 and 170.

**Table 5.271** Comparative Content of Vitamins and Minerals in Transgenic and Conventional Maize Grain ($M \pm m$; $n = 6–8$)

| Parameter | Conventional Maize | Transgenic Maize Line MIR604 |
|---|---|---|
| Vitamin $B_1$, mg/100 g | 0.323 ± 0.026 | 0.327 ± 0.035 |
| Vitamin $B_2$, mg/100 g | 0.097 ± 0.007 | 0.140 ± 0.021 |
| Vitamin $B_6$, mg/100 g | 0.383 ± 0.009 | 0.367 ± 0.003 |
| Carotenoids (total), mg/100 g | 1.417 ± 0.062 | 1.497 ± 0.077 |
| Vitamin E ($\beta+\gamma$), mg/100 g | 6.137 ± 0.254 | 6.383 ± 0.313 |
| Sodium, mg/kg | 22.82 ± 4.68 | 18.50 ± 3.31 |
| Calcium, mg/kg | 6.450 ± 0.517 | 6.247 ± 1.867 |
| Magnesium, mg/kg | 491.7 ± 9.5 | 509.0 ± 10.6 |
| Iron, mg/kg | 14.063 ± 0.528 | 12.200 ± 1.334 |
| Potassium, mg/kg | 4707 ± 141 | 4565 ± 93 |
| Zinc, mg/kg | 8.230 ± 0.221 | 8.657 ± 0.292 |
| Copper, mg/kg | 0.663 ± 0.059 | 0.632 ± 0.090 |
| Selenium, µg/kg | 195.3 ± 4.3 | 182.3 ± 5.5 |
| Phosphorus, % $P_2O_2$ | 0.730 ± 0.006 | 0.750 ± 0.006 |

**Table 5.272** Analysis of Toxic Elements in Maize Grain

| Parameter | Conventional Maize | Transgenic Maize Line MIR604 | Acceptable Levels |
|---|---|---|---|
| *Toxic elements, mg/kg* | | | |
| Lead | ≤0.001 ± 0.000 | ≤0.001 ± 0.000 | 0.5 |
| Arsenic | ≤0.10 ± 0.00 | ≤0.10 ± 0.00 | 0.2 |
| Cadmium | 0.004 ± 0.001 | ≤0.001 ± 0.000 | 0.1 |
| Mercury | not detected | not detected | 0.03 |
| *Pesticides, mg/kg* | | | |
| Hexachlorocyclohexane | not detected | not detected | 0.5 |
| DDT and its metabolites | not detected | not detected | 0.02 |
| Benzpyrene | not detected | not detected | 0.001 |
| *Mycotoxins, mg/kg* | | | |
| Aflatoxin $B_1$ | not detected | not detected | 0.005 |
| Deoxynivalenol | not detected | not detected | 0.7 |
| T2-toxin | not detected | not detected | 1.0 |
| Zearalenone | not detected | not detected | 1.0 |
| Fumonisin $B_1 + B_2$ | 0.007 ± 0.003 | not detected | – |

### 5.2.8 Diabrotica-Resistant Transgenic Maize Line MIR604

**Table 5.273** Body Weight (g) of Rats Fed Diet with Conventional Maize (Control) or Transgenic Maize Line MIR604 (Test) ($M \pm m$, $n = 30$)

| Duration of Experiment, Day | Control | Test |
|---|---|---|
| 0 | 80.5 ± 0.7 | 80.5 ± 0.7 |
| 7 | 116.7 ± 2.4 | 127.6 ± 4.6 |
| 14 | 156.2 ± 3.2 | 157.4 ± 6.2 |
| 21 | 193.3 ± 7.1 | 202.8 ± 5.8 |
| 28 | 221.5 ± 5.5 | 233.6 ± 9.7 |
| 35 | 256.3 ± 6.2 | 260.3 ± 10.7 |
| 42 | 293.8 ± 6.3 | 296.0 ± 9.8 |
| 49 | 309.0 ± 6.6 | 315.4 ± 10.2 |
| 56 | 329.1 ± 7.7 | 332.0 ± 10.7 |
| 63 | 343.3 ± 7.9 | 348.8 ± 12.6 |
| 70 | 365.1 ± 9.1 | 368.3 ± 13.3 |
| 77 | 375.0 ± 8.4 | 386.1 ± 13.9 |
| 84 | 384.2 ± 9.5 | 391.4 ± 14.4 |
| 91 | 397.7 ± 9.0 | 408.0 ± 15.6 |
| 98 | 414.6 ± 9.0 | 418.3 ± 14.3 |
| 105 | 427.7 ± 9.8 | 424.0 ± 14.2 |
| 112 | 432.4 ± 10.7 | 434.8 ± 15.1 |
| 119 | 438.1 ± 8.9 | 440.6 ± 15.1 |
| 126 | 443.5 ± 9.0 | 445.5 ± 16.1 |
| 133 | 450.9 ± 9.0 | 446.2 ± 14.6 |
| 140 | 451.4 ± 8.6 | 449.8 ± 16.7 |
| 147 | 453.8 ± 10.9 | 460.8 ± 14.7 |
| 154 | 459.3 ± 12.0 | 463.0 ± 17.9 |
| 161 | 459.5 ± 11.8 | 467.3 ± 13.8 |
| 170 | 465.5 ± 14.8 | 472.0 ± 24.9 |

*Note: The differences are not significant ($p > 0.05$).*

During the dietary experiment, the general condition of the rats was similar and satisfactory in control and test groups. No animal mortality was observed in either group. By outward appearance, the condition of hair, behavior, and growth rate of the test rats did not differ from that of the control animals. Daily intake of feed was 20–22 g/day. Body weight of the rats did not significantly differ between control and test groups (Table 5.273). The absolute and relative weights of internal organs in the test rats did not differ from the corresponding values for the control animals (Figure 5.13; Table 5.274). Morphological examination of liver, kidney, lungs, spleen, small intestine, testicles, and prostate as well as structural morphometric analysis of small intestine, liver, kidneys, and spleen revealed no differences at any time in the experiment (Tables 5.275 to 5.278).

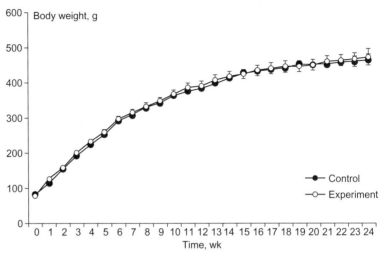

**FIGURE 5.13**
Comparative dynamics of body weight of rats fed diet with transgenic maize line MIR604 (test) or its conventional counterpart (control).

**Table 5.274** Absolute and Relative Weights of Internal Organs of Control and Test Rats ($M \pm m$, $n = 25$)

| Parameter | | 30 Days Control | 30 Days Test | 170 Days Control | 170 Days Test |
|---|---|---|---|---|---|
| Liver, g | abs.[a] | 10.87 ± 0.54 | 11.35 ± 0.47 | 12.24 ± 0.72 | 13.73 ± 0.46 |
| | rel.[b] | 3.733 ± 0.165 | 3.960 ± 0.155 | 2.585 ± 0.105 | 2.819 ± 0.089 |
| Kidneys, g | abs. | 2.083 ± 0.101 | 1.942 ± 0.062 | 2.550 ± 0.114 | 2.798 ± 0.085 |
| | rel. | 0.710 ± 0.026 | 0.677 ± 0.020 | 0.539 ± 0.013 | 0.575 ± 0.017 |
| Spleen, g | abs. | 2.010 ± 0.140 | 1.915 ± 0.151 | 1.538 ± 0.105 | 1.638 ± 0.053 |
| | rel. | 0.683 ± 0.036 | 0.667 ± 0.047 | 0.324 ± 0.016 | 0.339 ± 0.021 |
| Heart, g | abs. | 0.975 ± 0.023 | 1.010 ± 0.021 | 1.245 ± 0.045 | 1.277 ± 0.025 |
| | rel. | 0.333 ± 0.008 | 0.353 ± 0.011 | 0.265 ± 0.013 | 0.262 ± 0.007 |
| Testicles, g | abs. | 2.813 ± 0.153 | 2.860 ± 0.073 | 3.185 ± 0.094 | 3.517 ± 0.149 |
| | rel. | 0.963 ± 0.043 | 0.998 ± 0.026 | 0.676 ± 0.025 | 0.720 ± 0.016 |
| Hypophysis, mg | abs. | 8.500 ± 0.671 | 7.667 ± 1.174 | 12.33 ± 0.96 | 10.67 ± 1.28 |
| | rel. | 2.902 ± 0.166 | 2.688 ± 0.439 | 2.604 ± 0.172 | 2.168 ± 0.209 |
| Adrenal glands, mg | abs. | 20.33 ± 3.63 | 18.20 ± 3.35 | 21.33 ± 2.40 | 25.50 ± 2.59 |
| | rel. | 6.968 ± 1.232 | 6.360 ± 1.154 | 4.529 ± 0.517 | 5.184 ± 0.419 |
| Seminal vesicles, mg | abs. | 332.5 ± 46.5 | 365.0 ± 14.8 | 730.0 ± 31.5 | 693.3 ± 36.8 |
| | rel. | 114.8 ± 16.4 | 127.6 ± 6.1 | 155.6 ± 10.1 | 141.8 ± 4.0 |
| Prostate, mg | abs. | 144.2 ± 26.1 | 164.7 ± 23.45 | 385.0 ± 34.5 | 468.3 ± 57.5 |
| | rel. | 48.93 ± 7.77 | 57.71 ± 8.23 | 82.66 ± 9.05 | 94.69 ± 9.41 |

Note: The differences are insignificant ($p > 0.05$).
[a]Absolute weight of internal organs.
[b]Relative weight of internal organs (per 100 g body weight).

### 5.2.8 Diabrotica-Resistant Transgenic Maize Line MIR604

**Table 5.275** Microscopic Examination of Internal Organs in Rats Fed Diet with Conventional Maize (Control) or Transgenic Maize Line MIR604 (Test) (Combined Data Obtained on Experimental Days 30 and 170)

| Organ | Control | Test |
| --- | --- | --- |
| Liver | Clear trabecular structure; no alterations in hepatocytes or portal ducts | No differences from control |
| Kidneys | Usual appearance of cortical and medullary substance; no alterations in glomeruli or pelvis epithelium | No differences from control |
| Lung | Alveolar space is air-filled; no alterations in bronchi or blood vessels | No differences from control |
| Spleen | Large folliculi with wide clear marginal zones and reactive centers; splenic pulp is moderately plethoric | No differences from control |
| Small intestine | Preserved villous epithelium; usual infiltration in villous stroma | No differences from control |
| Testicle | Clearly definable spermatogenesis and age-related spermiogenesis | No differences from control |

**Table 5.276** Area of Structural Components (%) in Ileum Wall of Rats Fed Diet with Conventional Maize (Control) or Transgenic Maize Line MIR604 (Test) ($M \pm m$, $n = 25$)

| Structural Component | 30 Days | | 170 Days | |
| --- | --- | --- | --- | --- |
| | Control | Test | Control | Test |
| Mucous membrane:[a] | 82.9 ± 1.5 | 82.7 ± 1.4 | 84.1 ± 1.2 | 83.7 ± 1.1 |
| intestinal crypts[b] | 84.3 ± 1.2 | 83.9 ± 1.1 | 84.0 ± 1.1 | 84.1 ± 1.3 |
| lamina propria[b] | 12.5 ± 1.4 | 12.3 ± 1.3 | 12.3 ± 1.2 | 12.0 ± 1.1 |
| smooth muscle layer[b] | 1.3 ± 0.2 | 1.2 ± 0.2 | 1.2 ± 0.1 | 1.2 ± 0.1 |
| Submucous layer[a] | 3.3 ± 0.1 | 3.5 ± 0.2 | 3.4 ± 0.1 | 3.5 ± 0.2 |
| Muscular layer[a] | 13.3 ± 1.2 | 13.2 ± 1.0 | 13.3 ± 1.0 | 13.2 ± 1.1 |
| Serous membrane[a] | 1.1 ± 0.1 | 1.1 ± 0.1 | 1.0 ± 0.1 | 1.0 ± 0.1 |

Note: The differences are not significant ($p > 0.05$).
[a]Percentage of total area of ileum wall on histological sections.
[b]Percentage of area of mucous membrane.

Post-mortem examination revealed no alterations in the internal organs in either group of rats. Similarly, histological examinations of internal organs performed on Days 30 and 170 did not reveal any statistically significant differences (Table 5.275).

Morphological examination of liver, kidney, lungs, spleen, small intestine, testicles, and prostate as well as structural morphometric analysis of small intestine, liver, kidneys, and spleen revealed no differences at any time of the experiment (Tables 5.276 to 5.278).

**Table 5.277** Absolute Number of Various Epithelial Cells in Intestinal Villi and Ileum Crypts in Rats Fed Diet with Conventional Maize (Control) or Transgenic Maize Line MIR604 (Test) ($M \pm m$, $n = 25$)

| Structural Component, abs. Units | 30 Days | | 170 Days | |
|---|---|---|---|---|
| | Control | Test | Control | Test |
| Number of villous epithelial cells: | $89.5 \pm 2.1$ | $89.9 \pm 2.0$ | $90.5 \pm 2.1$ | $89.8 \pm 1.9$ |
| border cells | $55.6 \pm 2.2$ | $55.2 \pm 2.3$ | $53.0 \pm 2.3$ | $53.1 \pm 2.2$ |
| goblet cells | $31.5 \pm 1.5$ | $31.1 \pm 1.4$ | $30.0 \pm 1.3$ | $30.1 \pm 1.4$ |
| undifferentiated cells | $2.6 \pm 0.3$ | $2.7 \pm 0.3$ | $2.6 \pm 0.1$ | $2.5 \pm 0.1$ |
| Number of epithelial cells in ½ crypt: | $45.3 \pm 3.0$ | $46.1 \pm 3.3$ | $47.1 \pm 2.5$ | $46.9 \pm 2.7$ |
| border cells | $26.3 \pm 1.9$ | $26.1 \pm 2.0$ | $26.0 \pm 2.0$ | $26.2 \pm 2.1$ |
| goblet cells | $17.0 \pm 1.6$ | $16.9 \pm 1.5$ | $17.0 \pm 1.3$ | $16.9 \pm 1.2$ |
| undifferentiated cells | $5.3 \pm 1.1$ | $5.5 \pm 1.2$ | $5.0 \pm 1.1$ | $5.1 \pm 1.2$ |

Note: The differences are not significant ($p > 0.05$).

**Table 5.278** Morphometric Parameters of Liver, Kidney, and Spleen of Rats Fed Diet with Conventional Maize (Control) or Transgenic Maize Line MIR604 (Test) ($M \pm m$, $n = 25$)

| Structural Component | 30 Days | | 170 Days | |
|---|---|---|---|---|
| | Control | Test | Control | Test |
| *Liver* | | | | |
| Area of capillary bed, %[a] | $34.9 \pm 1.2$ | $34.7 \pm 1.1$ | $37.1 \pm 1.2$ | $36.7 \pm 1.3$ |
| Number of binucleate hepatocytes per 100 hepatocytes | $3.1 \pm 0.2$ | $3.2 \pm 0.3$ | $3.4 \pm 0.4$ | $3.5 \pm 0.3$ |
| *Kidneys* | | | | |
| Number of glomeruli in visual field at 400× | $11.3 \pm 0.9$ | $11.5 \pm 0.9$ | $11.7 \pm 0.8$ | $11.7 \pm 0.6$ |
| *Spleen, number of lymphoid type cells on the area of 880 µm²* | | | | |
| Lymphoid nodule reproduction site | $29.1 \pm 1.9$ | $29.0 \pm 1.9$ | $29.3 \pm 2.3$ | $29.5 \pm 2.1$ |
| Lymphoid nodule mantle | $31.9 \pm 1.7$ | $32.7 \pm 1.8$ | $33.6 \pm 2.2$ | $33.9 \pm 2.0$ |
| Periarterial lymphoid sheath | $34.0 \pm 1.9$ | $34.1 \pm 1.7$ | $33.1 \pm 1.9$ | $33.5 \pm 1.8$ |

Note: The differences are not significant ($p > 0.05$).
[a]The total area of liver on histological section is taken as 100%.

During the entire experiment, the hematological parameters of peripheral blood of control and test rats remained within the physiological range (Table 5.279). The biochemical parameters of blood serum (Table 5.280) and urine (Table 5.281) in the test group did not significantly differ from the analogous parameters for the control rats. Elevation of the content of triglycerides and moderation of activity of lactate dehydrogenase, creatine phosphokinase,

**Table 5.279** Hematological Parameters of Blood Serum of Rats Fed Diet with Conventional Maize (Control) or Transgenic Maize Line MIR604 (Test) ($M \pm m$, $n = 28$)

| Parameter | 30 Days | | 180 Days | | Standard Values [28, 45, 48] |
|---|---|---|---|---|---|
| | Control | Test | Control | Test | |
| Total erythrocyte count, ×10$^6$/μL | 7.410 ± 0.174 | 8.084 ± 0.189* | 9.667 ± 0.081 | 10.650 ± 1.120 | 4.4–8.9 |
| Hemoglobin concentration, g/L | 145.1 ± 2.4 | 151.9 ± 3.3 | 131.0 ± 2.9 | 129.1 ± 1.1 | 86–173 |
| Hematocrit, vol.% | 42.50 ± 1.02 | 44.66 ± 0.98 | 42.80 ± 1.10 | 42.24 ± 0.71 | 31.4–51.9 |
| MCV (erythrocyte), μm$^3$ | 55.43 ± 0.75 | 55.14 ± 0.46 | 43.57 ± 1.11 | 43.29 ± 0.81 | 50.6–93.8 |
| MCH, pg | 19.20 ± 0.21 | 18.79 ± 0.22 | 13.34 ± 0.33 | 13.01 ± 0.20 | 16–21 |
| MCHC, g/dL | 34.62 ± 0.29 | 33.97 ± 0.18 | 30.63 ± 0.23 | 30.26 ± 0.29 | 24.7–36.8 |
| ESR, mm/h | 2.429 ± 0.481 | 2.143 ± 0.340 | 2.000 ± 0.436 | 2.571 ± 0.369 | 0–5 |
| Leukocytes, ×10$^3$/μL | 6.267 ± 0.596 | 7.250 ± 0.437 | 7.761 ± 0.411 | 7.229 ± 0.438 | 1.4–34.3 |
| Basophils, % | 0 | 0 | 0 | 0 | 0–1 |
| Eosinophils, % | 0.714 ± 0.360 | 1.000 ± 0.436 | 2.571 ± 0.429 | 2.857 ± 0.800 | 0.0–5.5 |
| Stab neutrophils, % | 3.714 ± 1.107 | 2.857 ± 0.404 | 5.167 ± 1.302 | 5.833 ± 1.108 | 18–36 |
| Segmentonuclear neutrophils, % | 21.86 ± 2.61 | 17.83 ± 2.65 | 34.50 ± 1.66 | 25.57 ± 2.42* | |
| Metamyelocytes, % | 0 | 0 | 0 | 0 | 0 |
| Lymphocytes, % | 72.71 ± 2.72 | 74.25 ± 3.75 | 63.67 ± 3.76 | 62.43 ± 2.82 | 42.3–98.0 |
| Monocytes, % | 0.571 ± 0.297 | 0.714 ± 0.286 | 0.857 ± 0.459 | 0.571 ± 0.297 | 0.0–7.9 |
| Platelets, ×10$^3$/μL | 756.3 ± 29.4 | 822.9 ± 11.9 | 764.0 ± 30.5 | 754.7 ± 53.5 | 409–1250 |

ESR, erythrocyte sedimentation rate; MCH, mean cell hemoglobin; MCHC, mean cell hemoglobin concentration; MCV, mean cell volume.
*Differences from control significant at $p < 0.05$.

aspartate aminotransferase, and lipase in blood serum of test rats revealed on Day 30, and decrease in the content of creatinine observed on Day 170, remained within the physiological range of these parameters [28,45,48]. Such differences have no diagnostic value, because they result from peculiarities of metabolic status and statistical variations.

Examination of antioxidant status and activity of the enzymes involved in protective and adaptive processes, which are the system biomarkers reflecting adaptation level of the entire organism to the environment, revealed no adverse effects of the transgenic maize on rats (Tables 5.282 to 5.285).

There were no differences in the parameters characterizing the antioxidant status of the rats (Table 5.282). Elevation of blood serum MDA revealed on Day 30 in the test group in comparison with the control rats remained within the physiological range characteristic of the growing animals. By Day 170, no differences were observed.

**Table 5.280** Biochemical Parameters of Blood Serum of Rats Fed Diet with Conventional Maize (Control) or Transgenic Maize Line MIR604 (Test) ($M \pm m$, $n = 28$)

| Parameter | 30 Days Control | 30 Days Test | 170 Days Control | 170 Days Test | Standard Values [28, 45, 48] |
|---|---|---|---|---|---|
| Total protein, g/L | 73.16 ± 3.63 | 69.36 ± 3.46 | 76.88 ± 0.59 | 75.69 ± 1.83 | 56–82 |
| Albumin, g/L | 39.59 ± 2.67 | 43.86 ± 1.14 | 41.30 ± 1.68 | 47.39 ± 2.52 | 25–48 |
| Triglycerides, mM/L | 0.900 ± 0.097 | 1.576 ± 0.080* | 0.913 ± 0.048 | 0.723 ± 0.127 | 0.3–1.6 |
| Total bilirubin, µM/L | 4.571 ± 0.812 | 4.857 ± 0.459 | 4.667 ± 0.882 | 3.143 ± 0.404 | 1–4 |
| Conjugated bilirubin, µM/L | 0 | 0 | 0 | 0 | 0 |
| Urea, mM/L | 7.957 ± 0.524 | 8.614 ± 0.599 | 9.414 ± 0.633 | 11.300 ± 2.082 | 4.0–10.0 |
| Creatinine, µM/L | 58.00 ± 5.22 | 64.33 ± 11.20 | 72.25 ± 4.44 | 55.75 ± 3.33* | 13–92 |
| Glucose, mM/L | 6.979 ± 0.540 | 6.873 ± 0.192 | 4.486 ± 0.389 | 4.243 ± 0.103 | 4.5–10.0 |
| Cholesterol, mM/L | 2.696 ± 0.325 | 2.457 ± 0.164 | 3.069 ± 0.207 | 2.711 ± 0.391 | 0.6–4.3 |
| γ-glutamyl transferase, U/L | 9.214 ± 1.150 | 9.500 ± 1.841 | 4.050 ± 0.670 | 2.200 ± 1.041 | 0–3 |
| Lactate dehydrogenase, U/L | 2467 ± 188 | 1944 ± 77* | 4359 ± 705 | 3894 ± 405 | <5800 |
| Alpha-amylase, U/L | 1073 ± 67 | 1214 ± 69 | 861.4 ± 36.7 | 868.9 ± 111.0 | <3207 |
| Creatine phosphokinase, U/L | 5680 ± 685 | 3634 ± 332* | 5359 ± 397 | 5081 ± 1181 | ± 200% |
| Alkaline phosphatase, U/L | 515.9 ± 54.9 | 564.7 ± 46.8 | 248.9 ± 32.5 | 222.0 ± 41.1 | 112–814 |
| Alanine aminotransferase, U/L | 44.29 ± 7.30 | 50.14 ± 9.32 | 66.17 ± 5.76 | 74.14 ± 4.17 | 33–120 |
| Aspartate aminotransferase, U/L | 215.6 ± 21.1 | 154.2 ± 17.6* | 275.5 ± 29.7 | 280.6 ± 19.3 | 60–236 |
| Lipase, U/L | 29.00 ± 1.79 | 19.14 ± 0.46* | 16.14 ± 1.65 | 15.29 ± 1.17 | <30 |

*Differences from control significant at $p < 0.05$.

**Table 5.281** Urinary Biochemical Parameters of Rats Fed Diet with Conventional Maize (Control) or Transgenic Maize Line MIR604 (Test) ($M \pm m$, $n = 25$)

| Parameter | 30 Days Control | 30 Days Test | 170 Days Control | 170 Days Test |
|---|---|---|---|---|
| pH | 5.91 ± 0.17 | 5.86 ± 0.14 | 6.00 ± 0.20 | 6.03 ± 0.08 |
| Daily diuresis, mL | 3.3 ± 0.3 | 3.3 ± 0.4 | 3.6 ± 0.3 | 3.4 ± 0.1 |
| Relative density, g/mL | 1.16 ± 0.10 | 1.18 ± 0.12 | 1.19 ± 0.07 | 1.20 ± 0.08 |
| Creatinine, mg/mL | 1.26 ± 0.17 | 1.37 ± 0.20 | 1.62 ± 0.16 | 1.61 ± 0.14 |
| Creatinine, mg/day | 4.69 ± 0.22 | 4.92 ± 0.29 | 5.63 ± 0.32 | 5.38 ± 0.31 |

The differences are not significant ($p > 0.05$).

On the whole, the parameters characterizing protective and adaptive processes remained within the normal physiological range (Tables 5.283 and 5.284). On Day 30, the test rats demonstrated a small decrease in activity of CDNB-glutathione transferase and elevation of activity of β-galactosidase

### 5.2.8 Diabrotica-Resistant Transgenic Maize Line MIR604

**Table 5.282** Antioxidant Status of Rats Fed Diet with Conventional Maize (Control) or Transgenic Maize Line MIR604 (Test) ($M \pm m$, $n = 28$)

| Parameter | 30 Days | | 170 Days | |
|---|---|---|---|---|
| | Control | Test | Control | Test |
| *Enzymatic activity of antioxidant protection system* | | | | |
| Glutathione reductase, µM/min/g Hb | 40.94 ± 1.46 | 39.10 ± 0.87 | 43.23 ± 2.29 | 44.83 ± 1.64 |
| Glutathione peroxidase, µM/min/g Hb | 64.69 ± 1.59 | 65.20 ± 2.16 | 75.66 ± 2.45 | 75.18 ± 2.91 |
| Catalase, mM/min/g Hb | 512.3 ± 20.9 | 499.3 ± 14.9 | 550.7 ± 17.6 | 560.0 ± 18.8 |
| Superoxide dismutase, Unit/min/g Hb | 2020 ± 42 | 1912 ± 38 | 2135 ± 58 | 2147 ± 38 |
| *Content of LPO products* | | | | |
| erythrocytic MDA, nM/mL | 4.740 ± 0.184 | 4.748 ± 0.140 | 4.599 ± 0.228 | 4.693 ± 0.415 |
| Blood serum MDA, nM/mL | 5.824 ± 0.345 | 6.909 ± 0.284* | 6.208 ± 0.519 | 6.909 ± 0.234 |
| Hepatic MDA, nM/g | 282.5 ± 9.9 | 277.4 ± 11.4 | 279.1 ± 18.9 | 266.9 ± 11.1 |

*Difference from control significant at $p < 0.05$.

**Table 5.283** Activity of Enzymes Involved in Metabolism of Xenobiotics and Content of Hepatic Proteins in Rats Fed Diet with Conventional Maize (Control) or Transgenic Maize Line MIR604 (Test) ($M \pm m$, $n = 25$)

| Parameter | 30 Days | | 170 Days | |
|---|---|---|---|---|
| | Control | Test | Control | Test |
| Cytochrome P450, nM/mg protein | 0.65 ± 0.07 | 0.70 ± 0.05 | 0.99 ± 0.04 | 0.97 ± 0.06 |
| Cytochrome $b_5$, nM/mg protein | 0.56 ± 0.06 | 0.58 ± 0.04 | 0.87 ± 0.02 | 0.86 ± 0.04 |
| Aminopyrine N-demethylation, nM/min/mg protein | 7.70 ± 0.74 | 7.53 ± 0.58 | 8.00 ± 0.45 | 7.22 ± 0.56 |
| Benzpyrene hydroxylation, Fl-units/min/mg protein | 7.47 ± 0.85 | 7.47 ± 0.56 | 6.18 ± 0.60 | 5.63 ± 0.58 |
| Acetylesterase, µM/min/mg protein | 4.83 ± 0.06 | 4.91 ± 0.10 | 5.71 ± 0.10 | 5.25 ± 0.18 |
| Epoxide hydrolase, nM/min/mg protein | 8.17 ± 0.56 | 7.64 ± 0.78 | 9.88 ± 1.19 | 8.39 ± 1.02 |
| UDP-Glucuronosil transferase, nM/min/mg protein | 24.0 ± 1.5 | 25.7 ± 1.2 | 22.6 ± 2.1 | 19.8 ± 0.5 |
| CDNB-Glutathione transferase, µM/min/mg protein | 0.81 ± 0.03 | 0.70 ± 0.02* | 1.09 ± 0.04 | 1.20 ± 0.06 |
| Microsomal protein, mg/g | 14.9 ± 0.4 | 15.2 ± 0.4 | 14.6 ± 0.5 | 13.7 ± 0.5 |
| Cytosolic protein, mg/g | 100.8 ± 1.8 | 101.7 ± 0.6 | 87.7 ± 2.9 | 88.9 ± 2.8 |

*Difference from control significant at $p < 0.05$.

and β-glucuronidase. No differences in these parameters were observed on Day 170. Such small variations in activity of these enzymes not accompanied by the complex manifestations of stress in the protective and adaptive systems of experimental rats attest to the absence of adverse effects of chronic intake of transgenic maize on the protective and adaptive systems of the organism.

Thus, the chronic 170-day toxicological experiment on rats fed diet with transgenic maize line MIR604 attested to the absence of any toxic effects. The

**Table 5.284** Activity of Hepatic Lysosomal Enzymes in Rats Fed Diet with Conventional Maize (Control) or Transgenic Maize Line MIR604 (Test) ($M \pm m$, $n = 25$)

|  | 30 Days | | 170 Days | |
| --- | --- | --- | --- | --- |
| Parameter | Control | Test | Control | Test |
| *Total activity, µM/min/g tissue* | | | | |
| Arylsulfatase A and B | 2.15 ± 0.06 | 2.23 ± 0.06 | 2.22 ± 0.05 | 2.14 ± 0.03 |
| β-Galactosidase | 2.19 ± 0.05 | 2.38 ± 0.05* | 1.96 ± 0.03 | 1.97 ± 0.05 |
| β-Glucuronidase | 2.19 ± 0.02 | 2.36 ± 0.07* | 2.39 ± 0.01 | 2.29 ± 0.04 |
| *Non-sedimentable activity, % total activity* | | | | |
| Arylsulfatase A and B | 3.14 ± 0.12 | 3.10 ± 0.09 | 3.35 ± 0.08 | 3.27 ± 0.07 |
| β-Galactosidase | 5.57 ± 0.20 | 5.42 ± 0.29 | 5.79 ± 0.18 | 5.85 ± 0.24 |
| β-Glucuronidase | 6.50 ± 0.17 | 6.25 ± 0.26 | 5.71 ± 0.28 | 5.72 ± 0.30 |

*Difference from control significant at $p < 0.05$.

**Table 5.285** Chromosomal Damage to Bone Marrow Cells in Mice Fed Diet with Conventional Maize (Control) or Transgenic Maize Line MIR604 (Test) ($M \pm m$)

| Group | Number of Cells | By 100 Cells | | | | | Total Number of Damaged Metaphases, % |
| --- | --- | --- | --- | --- | --- | --- | --- |
| | | Gaps | Single Fragments | Paired Fragments | Crossovers | Cell with Multiple Damages[a] | |
| Control | 500 | 1.0 | 0.6 | 0.2 | 0.2 | – | 2.0 ± 0.6 |
| Test | 500 | 0.8 | 1.4 | – | – | – | 2.2 ± 0.7 |

Note: The differences are not significant ($p > 0.05$).
[a]More than 5 chromosomal aberrations in a cell.

values of all examined parameters were within the range of physiological variations characteristic of the rats.

### Assessment of Potential Genotoxicity of Transgenic Maize MIR604

The study of potential genotoxicity of transgenic maize line MIR604 was carried out on male C57Bl/6 mice fed diet with transgenic (test group) or conventional (control group) maize for 30 days. The initial body weight of mice was 20–22 g. The mice were fed a standard semi-synthetic casein diet (3.5 g/day/mouse) with conventional or transgenic milled maize grain (Table 5.164). Assessment of potential genotoxicity is based on revealing DNA damage by alkaline gel electrophoresis of cells isolated from bone marrow, liver, and rectum (DNA-comet assay) [2], as well as detection of mutagenic

**Table 5.286** Degree of Structural DNA Damage in Internal Organs of Mice Fed Diet with Conventional Maize (Control) or Transgenic Maize Line MIR604 (Test) ($M \pm m$)

| Group | Bone Marrow | | Liver | | Rectum | |
|---|---|---|---|---|---|---|
| | Number of cells | Damaged DNA, % | Number of cells | Damaged DNA, % | Number of cells | Damaged DNA, % |
| Control | 500 | 7.9 ± 0.7 | 500 | 4.5 ± 1.4 | 500 | 9.5 ± 0.8 |
| Test | 500 | 8.3 ± 1.6 | 500 | 5.7 ± 1.1 | 500 | 11.0 ± 0.5 |

Note: The differences are not significant ($p > 0.05$).

activity by counting the chromosomal aberrations in metaphasic cells of mouse bone marrow [6]. No fewer than 100 cells of each micropreparation were analyzed. The toxicological studies were conducted in accordance with Methodical Guidelines of the Ministry of Health of the Russian Federation MUK 2.3.2.970-00 and MU 2.3.2.2306-07 [8,9].

On Day 30, the body weight of mice was in the range 23–26 g. The data of cytogenetic studies of mouse bone marrow, the mean values of chromosomal aberrations in the test group, did not significantly differ from the corresponding values for control mice (Table 5.285) and did not surpass the level of spontaneous mutagenesis characteristic for C57Bl/6 mice. There were no differences in the degree of structural DNA damage in bone marrow, liver, and rectum (Table 5.286). Examination of DNA integrity and the level of chromosomal aberrations in test mice fed diet with transgenic maize line MIR604 revealed no extra genotoxic effects in comparison with control mice.

## Assessment of Potential Impact of Transgenic Maize Line MIR604 on Immune System in Studies on Mice

The immunomodulating effects of transgenic maize line MIR604 were examined for 45 days on CBA and C57Bl/6 mice with an initial body weight of 18–20 g. During the entire length of the experiment, the mice were maintained on standard vivarium diet (Table 5.114). Milled maize grain was included into the feed (3 g/mouse/day) replacing the oatmeal and providing an equal amount of nutrients and calories. The details of the experiment are described in section 5.2.7.

The test on the humoral component of the immune system showed that dynamics of production of the antibodies raised against SE in control mice was similar to that in the test group for CBA and C57Bl/6 mice, which attests to the absence of a significant difference in the immunomodulating effect (Table 5.287). Some insignificant differences in the antibody levels remained within the physiological range, reflecting the individual sensitivity of the animals. The delayed hypersensitivity reaction did not reveal a modulating effect

**Table 5.287** Level of Antibodies Raised Against SE in Mice Fed Diet with Conventional Maize (Control) or Transgenic Maize Line MIR604 (Test) ($M \pm m$, $n \geq 10$)

| SE Post-Injection Day | CBA Mice | | C57Bl/6 Mice | |
|---|---|---|---|---|
| | Control | Test | Control | Test |
| 7 | $128 \pm 45$ | $107 \pm 21$ | $256 \pm 81$ | $269 \pm 100$ |
| 21 | $80 \pm 59$ | $97 \pm 31$ | $85 \pm 21$ | $85 \pm 43$ |

Note: The differences are not significant ($p > 0.05$).

**Table 5.288** Delayed hypersensitivity Reaction in Mice Fed Diet with Conventional Maize (Control) or Transgenic Maize Line MIR604 (Test) ($M \pm m$, $n \geq 10$)

| Parameter | CBA Mice | | C57Bl/6 Mice | |
|---|---|---|---|---|
| | Control | Test | Control | Test |
| Delayed hypersensitivity reaction index | $7.5 \pm 5.1$ | $6.8 \pm 2.2$ | $4.2 \pm 1.7$ | $6.3 \pm 4.6$ |

Note: The differences are not significant ($p > 0.05$).

**Table 5.289** Natural Resistance against *Salmonella typhimurium* in Mice Fed Diet with Conventional Maize (Control) or Transgenic Maize Line MIR604 (Test) ($M \pm m$, $n \geq 10$)

| Parameter | CBA Mice | | C57Bl/6 Mice | |
|---|---|---|---|---|
| | Control | Test | Control | Test |
| $LD_{50}$ Mortality of mice | $79 \times 10^3$ Beginning: post-injection week 1 Termination: post-injection Day 18 | $87 \times 10^3$ | $0.54 \times 10^3$ Beginning: post-injection week 1 Termination: post-injection Day 18 | $0.34 \times 10^3$ |
| Mean lifetime | 5.2 to 7.4 day | | 4.5 to 8.0 day | |

of transgenic maize line MIR604 on the cellular component of the immune system (Table 5.288).

The tests on sensitization and on the natural resistance of mice to *Salmonella typhimurium* revealed no adverse effects of the transgenic maize (Table 5.289). Infection with *S. typhimurium* induced typical manifestations of the disease in both groups of mice, the CBA group (insensitive to *S. typhimurium*) being more resistant against infection than the C57Bl/6 mice (sensitive to *S. typhimurium*).

Therefore, these studies revealed no sensitizing or immunomodulating potencies of transgenic maize line MIR604 in comparison with conventional maize.

Table 5.290 Severity of Active Anaphylactic Shock in Rats Fed Diet with Conventional Maize (Control) or Transgenic Maize Line MIR604 (Test) ($M \pm m$)

| Group[a] | Anaphylactic Index | Severe Reactions, % | Mortality, % |
|---|---|---|---|
| Control | 2.52 | 48.0 | 48.0 |
| Test | 2.32 | 48.0 | 48.0 |
| p | >0.05[b] | >0.05[c] | >0.05[c] |

[a] n = 22 (control group) and n = 24 (test group).
[b] Mann-Whitney non-parametrical rank test.
[c] Two-sided Fisher angular conversion U-test.

## Assessment of Potential Allergenicity of Transgenic Maize Line MIR604

A 30-day subchronic experiment was carried out on male Wistar rats with initial body weight of $180 \pm 10$ g. During the entire duration of the experiment, the rat were maintained on standard vivarium diet (Table 5.114). Milled maize grain was included into the feed (3.5 g/rat/day), partially replacing the oatmeal and providing equal amount of nutrients and calories. Potential allergenicity was assessed by severity of generalized anaphylaxis and by concentration of circulating specific immunoglobulin antibodies (the sum of $IgG_1$ and $IgG_4$ fractions) after intraperitoneal sensitization of mature rats with a nutrient antigen (ovalbumin) followed by intravenous injection of an anaphylaxis-provoking dose of this protein to sensitized animals. The studies were conducted in accordance with Methodical Guidelines of the Ministry of Health of the Russian Federation MUK 2.3.2.970-00 and MU 2.3.2.2306-07 [8,9].

On Day 29, the body weights of control and test rats were $292 \pm 6$ and $302 \pm 6$ g, respectively ($p > 0.05$). There were no significant differences in severity of anaphylactic shock (Table 5.290) or intensity of humoral immune response (Table 5.291). Analysis of distributions of the examined parameters in both groups of rats performed with an ANOVA test attested to their variance homogeneity ($p > 0.05$).

Assessment of severity of active anaphylactic shock and intensity of humoral immune response in rats fed diet with transgenic maize line MIR604 attest to the absence of allergenic effect of this maize variety.

## Assessment of Technological Parameters of Maize Line MIR604

The studies of functional and technological properties of transgenic maize line MIR604 were conducted in accordance with Methodical Guidelines of the Ministry of Health of the Russian Federation "Medical and Biological

**Table 5.291** Humoral Response Intensity in Rats Fed Diet with Conventional Maize (Control) or Transgenic Maize Line MIR604 (Test) ($M \pm m$)

| Group[a] | $D_{492}$ | Concentration of Antibodies, mg/mL | Logarithm of Antibody Concentration |
|---|---|---|---|
| Control | $0.808 \pm 0.041$ | $6.1 \pm 0.7$ | $0.708 \pm 0.056$ |
| Test | $0.814 \pm 0.031$ | $5.8 \pm 0.5$ | $0.720 \pm 0.043$ |
| *Statistical analysis* | | | |
| Student t-test | >0.05 | >0.05 | >0.05 |
| ANOVA variance homogeneity test, p | >0.05 | >0.05 | >0.05 |
| Mann-Whitney non-parametrical rank test | | >0.05 | |

[a] n=25 (control group) and n= 25 (test group).

Assessment of Food Products Derived from Genetically Modified Sources": MUK 2.3.2.970-00. The study resulted in the following conclusions.

- The quality of the examined samples of maize grain met the requirements of Russian National Standards GOST 13634-90 "Maize. Procurement and Supply Requirements".
- The control and test samples of maize grain were processed identically. No differences or difficulties were observed in the technological process when producing starch from transgenic maize MIR604 or conventional maize. According to Russian National Standards GOST P 51985-2002 "Maize Starch. General Technological Requirements" the quality of derived maize starch was superior.
- The amylograms of 7% water suspension of test maize starch obtained on Brabender amylograph did not significantly differ from the control amylograms of maize starches.
- Gelatinization temperature and viscosity of 7% starch paste derived from transgenic maize were somewhat lower than those of control maize. These differences can be caused by the different values of amylase/amylopectin ratio, which can be determined by the climatic environment of the growing plant and the degree of grain ripening and is not a key parameter.
- The starches derived from control and transgenic maize are characterized by identical gel-producing properties.
- The structural and mechanical properties of the extrusive products derived from transgenic maize line MIR604 were practically identical to those obtained from the grains of conventional maize.

Thus, the study of functional and technological properties revealed no significant differences between the examined grain samples of transgenic maize line MIR604 and its conventional counterpart.

## Conclusions
Peer review of the data submitted by the applicant and the results of complex medical and biological studies of transgenic maize line MIR604 resistant against damage by rootworm *Diabrotica* spp., attest to the absence of any toxic, genotoxic, sensitization, immunomodulating, or allergenic effects of this maize line. By biochemical composition, transgenic maize line MIR604 was identical to conventional maize.

In accordance with Federal Law No. 52-FZ "On Sanitary and Epidemiological Population Welfare" (March 30, 1999), transgenic maize line MIR604 resistant against rootworm *Diabrotica* spp. has been registered for food use, listed in the State Register, and licensed for use in the territory of the Russian Federation, import into the territory of the Russian Federation, and placing on the market without restrictions (State Registration Certificate No. 77.99.26.11.U.5763.7.07 on July 20, 2007).

## 5.2.9 MAIZE LINE 3272 PRODUCING ALPHA-AMYLASE ENZYME

### Molecular Characteristics of Maize Line 3272
Maize line 3272, producing alpha-amylase enzyme (thermotolerant alpha-amylase, KF 3.2.1.1.), was developed as a source of native alpha-amylase for use in the manufacture of ethanol fuel without the use of bacteriogenous amylase.

### Recipient Organism
Maize (*Zea mays L.*) has a long-term history of safe use as human food and animal feed.

### Donor Organism
The chimeric gene *amy797E* produces AMY797E protein ($M = 50.2\,kD$), constructed from DNA from three hyperthermophilic organisms of *Thermococcales* order (genes BD5031, BD5064, and BD5063). The chimeric protein AMY797E is optimized for process specifications to manufacture ethanol fuel from dry corn meal, and it reaches maximum activity at temperature >80°C, pH 4.5 and low content of $Ca^{2+}$ ions [20].

The *pmi* gene of phosphomannose isomerase was isolated from *E. coli* [33,39,40].

### Method of Genetic Transformation

The maize line 3272 was developed by *Agrobacterium*-mediated transformation of corn genome by the plasmid vector pNOV7013, containing an expression cassette of a gene of alpha-amylase (*amy797E*) and an expression cassette of a gene of phosphomannose isomerase (*pmi*). The expression cassette of gene *amy797E* includes the following genetic elements: GZein, the promoter of the zein gene of corn, that provides an endosperm-specific *amy797E* gene expression; chimeric gene *amy797E*; PEPC9—an intron of the gene of phosphoenolpyruvate carboxylase of corn, 35S—the terminator of cauliflower mosaic virus (CaMV). The structure of the expression cassette of gene *pmi* includes the following genetic elements: ZmUbiInt, constitutive promoter and first intron of a gene of poly-ubiquitin of corn, that provides expression in monocotyledons; *pmi*, a gene of selective protein phosphomannose isomerase, catalyzing reaction of an isomerization mannose-6-phosphate into fructose-6-phosphate; NOS, terminator (untranslated gene region nopaline synthase), isolated from the Ti-plasmid of *Agrobacterium tumefaciens* [19,58].

The presence of the selective gene *pmi*, determining PMI protein synthesis, phosphomannose isomerase enzyme (EC 5.3.1.8, $M = 45$ kDa, 391 amino-acid residues), allows cells to grow on a nutrient solution containing mannose as the prevailing or sole source of carbon [20,23,28], and provides means for selection of the modified corn cells in the phase of cultivation on nutrient solution. The use of PMI protein as a selective marker, and the use of mannose as a selective agent, is an effective alternative to traditional marker systems in which selective agents confer tolerance to antibiotics or herbicides.

### Global Registration Status of Maize Line 3272

Table 5.292 shows the countries that had issued permission to use maize line 3272 at the time of registration in Russia [19].

**Table 5.292** Global Regulatory Status of GM maize line 3272

| Country | Registration Date | Scope |
| --- | --- | --- |
| Australia and New Zealand | 2008 | Food, feed |
| Canada | 2008 | Food, feed Environment |
| Mexico | 2008 | Food, feed |
| Philippines | 2008 | Food, feed |
| USA | 2007 | Food, feed |

*Note: The current registration status of the transgenic crop is on http://www.biotradestatus.com/*

## 5.2.9 Maize Line 3272 Producing Alpha-Amylase Enzyme

### Safety Assessment of Maize Line 3272 Conducted in the Russian Federation

Research was conducted in accordance with the requirements outlined in methodological instructive regulations MUK 2.3.2.970-00 "Medico-biological assessment of food products, derived from genetically modified sources" [8] and MU 2.3.2.2306-07 "Medico-biological safety assessment of genetically-engineered and modified organisms of plant origin" [9]. PCR analysis of the maize test and control samples was performed to confirm the identity of the transformation event and its absence in the conventional control line.

### *Assessment of Nutritional Quality and Food Safety Parameters of GM Maize Line 3272*

Assessment of nutritional qualities and food safety parameters of maize line 3272 showed no significant difference from its traditional analogue (Tables 5.293 and 5.294). Content of nutrients, mineral substances, and microelements in the studied corn samples fell within the limits of biological variations typical for *Zea mays L.*[1,10,15]

Content of toxic elements (Table 5.295) in samples of GM maize and its traditional analogue did not exceed the acceptable levels established by Sanitary Regulations and norms 2.3.2.1078-01 (it. 1.4.1.) [5]. As shown in Table 5.295, maize samples slightly differed in fumonisin content; therefore their intake with the diet was carefully calculated: in samples of conventional maize, fumonisin content was below the level of detection, hence, in the diet of the control group fumonisin content was ~0.00 mg/kg feed (0.000 mg/kg body mass); the diet including GM maize contained fumonisins in the amount of 0.023–0.037 mg/kg feed (0.0015–0.0037 mg/kg body mass). Comparison of the obtained results with fumonisins NOAEL for rats— <15 mg/kg feed or 0.25 mg/kg body mass (NOAEL for kidneys, which are most sensitive to fumonisin activity) [41,54,55]—shows absence of the possibility of any toxic biological effect of fumonisins, as their level in the diet of the experimental group animals was much lower than NOAEL. Some differences found in mycotoxins content have no biological significance.

Thus, assessment of nutritional quality and food safety parameters demonstrated composition equivalence of GM maize line 3272 to its traditional analogue, as well as conformity to safety requirements adopted in the Russian Federation.

### *Assessment of Potential Toxicity of GM Maize Line 3272 in Experiment in Rats*

An experiment of duration 180 days was conducted on male Wistar rats with an initial weight of 70–90 g. Rats received a semi-synthetic casein diet (diet composition described in Table 5.164). Ground maize grain was included into the feed, replacing diet ingredients with account taken of proteins, fats, and carbohydrates

**Table 5.293** Protein and Fat Content, Carbohydrate and Lipid Composition of Maize Grain

| Parameter | Conventional Maize | Maize Line 3272 |
|---|---|---|
| Crude protein, g/100 g | 9.175 ± 0.090 | 9.570 ± 0.051 |
| *Carbohydrate composition, g/100 g* | | |
| Fructose | 0.122 ± 0.005 | 0.142 ± 0.002 |
| Glucose | 0.212 ± 0.003 | 0.226 ± 0.003 |
| Sucrose | 1.517 ± 0.007 | 1.550 ± 0.025 |
| *Dietary fibers g/100 g* | | |
| Insoluble | 8.09 ± 0.45 | 8.92 ± 0.20 |
| Soluble | 2.04 ± 0.28 | 3.09 ± 0.15 |
| Total | 10.14 ± 0.17 | 12.01 ± 0.09 |
| *Lipids* | | |
| Crude lipids, g/100 g | 5.233 ± 0.120 | 5.633 ± 0.186 |
| *Fatty acid composition of crude lipids and fractions, % rat* | | |
| Sum of PUFA | 52.26 ± 0.74 | 50.58 ± 0.50 |
| Myristic 14:0 | 0.077 ± 0.003 | 0.070 ± 0.006 |
| Pentadecanoic Σ 15:0 | 0.040 ± 0.006 | 0.047 ± 0.003 |
| Palmitic 16:0 | 15.87 ± 0.30 | 16.30 ± 0.13 |
| Palmitic-oleic 16:1 | 0.220 ± 0.021 | 0.223 ± 0.020 |
| Heptadecoic Σ 17:0 | 0.133 ± 0.009 | 0.153 ± 0.003 |
| Stearic 18:0 | 2.233 ± 0.043 | 2.520 ± 0.040 |
| Oleic Σ 18:1 cis-9 | 27.69 ± 0.21 | 28.60 ± 0.20 |
| Linoleic 18:2 | 50.20 ± 0.83 | 48.10 ± 0.22 |
| α-Linoleic 18:3 ω-3 | 2.053 ± 0.090 | 2.480 ± 0.147 |
| Arachic 20:0 | 0.503 ± 0.034 | 0.633 ± 0.024 |
| Eicosenoic Σ 20:1 | 0.473 ± 0.064 | 0.433 ± 0.026 |
| Docosanoic 22:0 | 0.227 ± 0.022 | 0.333 ± 0.043 |
| Docosenoic Σ 22:1 | 0.063 ± 0.009 | 0.080 ± 0.010 |

Average data shown, $M \pm m$; $n = 6–8$.

content in the introduced product, observing isocaloric principle. The rats were divided into two groups: the test group, which received GM maize line 3272 with their diet, and the control group, which received conventional corn.

Sample collection was carried out after 30 and 180 days. During the experiment the palatability of the feed, body mass, and overall condition of the animals were assessed.

No mortality was observed during the experiment in the test or control groups, and the overall condition of the animals was satisfactory. Feed

## 5.2.9 Maize Line 3272 Producing Alpha-Amylase Enzyme

**Table 5.294** Vitamins and Minerals Content in Maize Grain

| Parameter | Conventional Maize | Maize Line 3272 |
|---|---|---|
| Vitamin $B_1$, mg/100 g | 0.119 ± 0.002 | 0.117 ± 0.001 |
| Vitamin $B_2$, mg/100 g | 0.354 ± 0.016 | 0.360 ± 0.006 |
| Carotenoids (sum), mg/100 g | 0.770 ± 0.035 | 0.790 ± 0.012 |
| Vitamin E (β+γ), mg/100 g | 6.513 ± 0.319 | 6.537 ± 0.317 |
| Sodium, mg/kg | 82.47 ± 1.88 | 70.87 ± 6.39 |
| Calcium, mg/kg | 20.60 ± 3.01 | 17.83 ± 1.65 |
| Magnesium, mg/kg | 727.3 ± 26.1 | 760.0 ± 26.1 |
| Iron, mg/kg | 10.73 ± 0.50 | 10.40 ± 1.02 |
| Potassium, mg/kg | 4266 ± 99 | 4246 ± 145 |
| Zinc, mg/kg | 10.57 ± 0.32 | 11.07 ± 0.18 |
| Copper, mg/kg | 0.873 ± 0.061 | 0.733 ± 0.068 |
| Selenium, mg/kg | 199.7 ± 0.9 | 194.3 ± 0.7 |
| Phosphorus, % $P_2O_5$ | 0.743 ± 0.003 | 0.770 ± 0.032 |

*Average data shown, $M \pm m$; $n = 6–8$.*

**Table 5.295** Harmful Substances in Maize Grain Samples

| Parameter | Conventional Maize | Maize Line 3272 | Acceptable Levels (SanRN 2.3.2.1078-01) |
|---|---|---|---|
| *Toxic elements, mg/kg* | | | |
| Lead | ≤0.001 ± 0.000 | ≤0.001 ± 0.000 | 0.5 |
| Arsenic | n/d | n/d | 0.2 |
| Cadmium | ≤0.001 ± 0.000 | ≤0.001 ± 0.000 | 0.1 |
| Mercury | n/d | n/d | 0.03 |
| *Pesticides, mg/kg* | | | |
| Hexachlorocyclohexane | n/d | n/d | 0.5 |
| DDT and its metabolites | n/d | n/d | 0.02 |
| Hexachlorane | n/d | n/d | – |
| Aldrin | n/d | n/d | – |
| Heptachlor | n/d | n/d | – |
| Kelthane | n/d | n/d | – |
| Benzapyrene | n/d | n/d | 0.001 |
| *Mycotoxins, mg/kg* | | | |
| Aflatoxin $B_1$ | n/d | n/d | 0.005 |
| Deoxynyvalenol | n/d | n/d | 0.7 |
| T2 toxin | n/d | n/d | 1.0 |
| Zearalenone | n/d | n/d | 1.0 |
| Fumonisin $B_1$ | n/d | 0.083 ± 0.019 | – |
| Fumonisin $B_2$ | n/d | n/d | – |

*Average data shown, $M \pm m$; $n = 6–8$.*

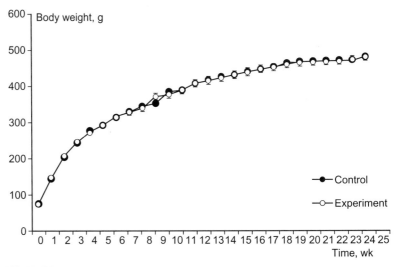

**FIGURE 5.14**
Comparative dynamics of body weight of rats fed diet with transgenic maize line 3272 (test) or its conventional counterpart (control).

consumption was 20–22 g diet per rat per 24 hours. No significant differences in appearance, coat condition, behavior, growth rate, or body weight of the animals were detected between the test and control groups (Figure 5.14 and Table 5.296). Weekly weight gain of both groups of rats corresponded to levels characteristic of animals of the breed and age [20,23,38].

The absolute and relative weight of internal organs in control and test rats were determined at 30 and 180 days after the onset of the experiment (Table 5.297).

Absolute and relative weight of internal organs of the rats in the test group did not differ from the control group on Day 30 (Table 5.297). Some decrease in absolute weights of liver and heart observed in the test group rats on Day 180 remained within the physiological variations characteristic of rats of the corresponding age, i.e. 8.00–15.00 g (liver) and 0.75–1.50 g (heart). The relative weights of internal organs of test rats on Day 180 did not significantly differ from the corresponding values for the control rats fed conventional maize.

Post-mortem examination revealed no alterations in the internal organs in both groups of rats. Similarly, histological examinations of internal organs performed on Days 30 and 180 did not reveal any statistically significant differences (Table 5.298). Morphometric analysis of the structure of small intestine, liver, kidneys, and spleen did not detect any differences between the groups throughout the whole experiment (Tables 5.299 to 5.301).

According to the data described in Table 5.302, blood parameters were within norm in general. Calculated values characterizing conditions of erythrocytes

## 5.2.9 Maize Line 3272 Producing Alpha-Amylase Enzyme

**Table 5.296** Body Weight (g) of Rats Fed Diet with Conventional Maize (Control) or Transgenic Maize Line 3272 (Test) ($M \pm m$, $n = 30$)

| Duration of Experiment, Day | Control | Test |
|---|---|---|
| 0 | 79.5 ± 0.8 | 79.5 ± 0.7 |
| 7 | 153.7 ± 2.7 | 156.9 ± 2.6 |
| 14 | 216.6 ± 5.0 | 218.4 ± 3.3 |
| 21 | 256.7 ± 5.4 | 259.6 ± 4.4 |
| 28 | 292.1 ± 5.7 | 286.9 ± 5.1 |
| 35 | 307.8 ± 8.2 | 308.0 ± 5.2 |
| 42 | 330.0 ± 5.4 | 329.0 ± 5.1 |
| 49 | 347.9 ± 5.4 | 345.1 ± 5.6 |
| 56 | 362.6 ± 5.4 | 357.9 ± 5.7 |
| 63 | 370.1 ± 8.0 | 389.7 ± 7.6 |
| 70 | 403.9 ± 6.6 | 396.8 ± 6.8 |
| 77 | 409.3 ± 8.7 | 409.4 ± 7.3 |
| 84 | 428.8 ± 7.0 | 429.1 ± 8.1 |
| 91 | 439.2 ± 6.4 | 436.6 ± 7.3 |
| 98 | 448.1 ± 8.0 | 444.0 ± 7.2 |
| 105 | 453.7 ± 5.4 | 453.3 ± 8.0 |
| 112 | 463.4 ± 8.5 | 460.2 ± 7.4 |
| 119 | 469.6 ± 7.6 | 470.4 ± 7.5 |
| 126 | 477.6 ± 7.5 | 475.2 ± 7.6 |
| 133 | 488.0 ± 7.1 | 483.4 ± 8.0 |
| 140 | 492.6 ± 7.4 | 488.4 ± 7.5 |
| 147 | 492.6 ± 8.0 | 490.9 ± 7.4 |
| 154 | 493.9 ± 7.5 | 492.7 ± 6.9 |
| 161 | 495.1 ± 8.8 | 493.0 ± 6.3 |
| 168 | 496.5 ± 7.1 | 497.9 ± 7.3 |
| 175 | 506.1 ± 3.0 | 504.2 ± 9.6 |

were an exception: after 30 days of the experiment, mean corpuscular volume, mean corpuscular hemoglobin, and mean corpuscular hemoglobin concentration in rats of the test group were, by 5%, 7% and 2% ($p < 0.05$) respectively, higher than in rats of the control group. However, after 180 days of the experiment there was no difference between the groups. The results of the hematological assessment of the rats RBC parameters (mean corpuscular volume, mean corpuscular hemoglobin, and mean corpuscular hemoglobin concentration) normally vary within 85%, 95%, and 49%, respectively [28,45,48]); this, together with the absence of a tendency to maintain the observed differences

**Table 5.297** Absolute and Relative Weight of Internal Organs of Rats Fed Diet with Conventional Maize or Transgenic Maize Line 3272 ($M \pm m$, $n = 6–8$)

| Organ | | 30 Days | | 180 Days | |
|---|---|---|---|---|---|
| | | Control | Test | Control | Test |
| Liver | Abs.[a], g | 10.25 ± 0.44 | 10.64 ± 0.48 | 12.94 ± 0.43 | 11.61 ± 0.36* |
| | Rel.[b], g/100 g | 3.434 ± 0.126 | 3.485 ± 0.104 | 2.624 ± 0.088 | 2.430 ± 0.071 |
| Kidney | Abs., g | 2.022 ± 0.096 | 2.140 ± 0.089 | 2.988 ± 0.139 | 2.707 ± 0.083 |
| | Rel., g/100 g | 0.677 ± 0.028 | 0.700 ± 0.015 | 0.605 ± 0.025 | 0.566 ± 0.016 |
| Spleen | Abs., g | 2.018 ± 0.221 | 2.170 ± 0.227 | 1.728 ± 0.134 | 1.588 ± 0.099 |
| | Rel., g/100 g | 0.675 ± 0.072 | 0.719 ± 0.087 | 0.350 ± 0.027 | 0.333 ± 0.022 |
| Heart | Abs., g | 1.002 ± 0.015 | 1.043 ± 0.041 | 1.392 ± 0.034 | 1.248 ± 0.036* |
| | Rel., g/100 g | 0.336 ± 0.008 | 0.342 ± 0.006 | 0.282 ± 0.007 | 0.261 ± 0.007 |
| Testicles | Abs., g | 2.862 ± 0.066 | 2.998 ± 0.106 | 3.612 ± 0.167 | 3.412 ± 0.129 |
| | Rel., g/100 g | 0.961 ± 0.030 | 0.984 ± 0.031 | 0.732 ± 0.031 | 0.714 ± 0.025 |
| Hypophysis | Abs., mg | 9.400 ± 1.030 | 10.00 ± 1.29 | 9.500 ± 0.619 | 9.667 ± 0.715 |
| | Rel., mg/100 g | 3.107 ± 0.328 | 3.242 ± 0.325 | 1.927 ± 0.129 | 2.028 ± 0.161 |
| Adrenal glands | Abs., mg | 25.00 ± 2.45 | 30.00 ± 3.15 | 28.33 ± 2.81 | 25.50 ± 1.78 |
| | Rel., mg/100 g | 8.397 ± 0.901 | 9.680 ± 1.017 | 5.729 ± 0.544 | 5.331 ± 0.345 |
| Seminal vesicles | Abs., mg | 379.5 ± 29.8 | 349.0 ± 56.1 | 780.0 ± 56.3 | 648.3 ± 34.0 |
| | Rel., mg/100 g | 126.7 ± 10.2 | 114.1 ± 17.5 | 157.9 ± 10.8 | 136.0 ± 8.2 |
| Prostate | Abs., mg | 195.0 ± 36.9 | 156.7 ± 21.0 | 515.0 ± 67.7 | 390.0 ± 50.8 |
| | Rel., mg/100 g | 64.16 ± 11.09 | 50.91 ± 5.97 | 104.3 ± 13.4 | 81.8 ± 10.8 |

[a]Absolute weight of internal organs.
[b]Relative weight of internal organs (per 100 g body weight).
*$p < 0.05$ in comparison with control.

**Table 5.298** Microscopic Examination of Internal Organs in Rats Fed Diet with Conventional Maize (Control) or Transgenic Maize Line 3272 (Test) (Combined Data Obtained on Experimental Days 30 and 180)

| Organ | Control | Test |
|---|---|---|
| Liver | Clear trabecular structure; no alterations in hepatocytes or portal ducts | No differences from control |
| Kidneys | Usual appearance of cortical and medullar substance; no alterations in glomeruli or pelvis epithelium | No differences from control |
| Lung | Alveolar space is air-filled; no alterations in bronchi or blood vessels | No differences from control |
| Spleen | Large folliculi with wide clear marginal zones and reactive centers; splenic pulp is moderately plethoric | No differences from control |
| Small intestine | Preserved villous epithelium; usual infiltration in villous stroma | No differences from control |
| Testicle | Clearly definable spermatogenesis and age-related spermiogenesis | No differences from control |

### 5.2.9 Maize Line 3272 Producing Alpha-Amylase Enzyme

**Table 5.299** Area of Structural Components (%) of the Surface of Ileum Wall in Rats Fed Diet with Conventional Maize (Control) or Transgenic Maize Line 3272 (Test) ($M \pm m$, $n = 25$)

| Structural Component | 30 Days Control | 30 Days Test | 180 Days Control | 180 Days Test |
|---|---|---|---|---|
| Mucous membrane[a] | 83.0 ± 1.1 | 82.8 ± 1.2 | 81.9 ± 1.3 | 82.1 ± 1.4 |
|   intestinal crypts[b] | 84.3 ± 1.2 | 83.9 ± 1.1 | 84.1 ± 1.2 | 84.3 ± 1.1 |
|   lamina propria[b] | 12.3 ± 1.0 | 12.2 ± 1.1 | 12.3 ± 1.3 | 12.4 ± 1.2 |
|   myomere[b] | 1.1 ± 0.2 | 1.1 ± 0.1 | 1.2 ± 0.2 | 1.2 ± 0.2 |
| Submucous layer[a] | 3.4 ± 0.1 | 3.3 ± 0.1 | 3.4 ± 0.1 | 3.3 ± 0.2 |
| Muscular layer[a] | 13.2 ± 1.1 | 13.3 ± 1.1 | 13.2 ± 1.2 | 13.3 ± 1.1 |
| Serosa[a] | 1.1 ± 0.1 | 1.1 ± 0.1 | 1.1 ± 0.1 | 1.1 ± 0.1 |

[a]Percentage of total area of ileum wall on histological sections.
[b]Percentage of area of mucous membrane.

**Table 5.300** Absolute Quantity of Different Types of Epithelial Cells in Intestinal Villi and Ileum Crypt in Rats Fed Diet with Conventional Maize (Control) or Transgenic Maize Line 3272 (Test) ($M \pm m$, $n = 25$)

| Structural Component, abs. Units | 30 Days Control | 30 Days Test | 180 Days Control | 180 Days Test |
|---|---|---|---|---|
| Quantity of villous epithelial cells: | 94.1 ± 1.9 | 93.9 ± 2.0 | 95.1 ± 1.9 | 94.9 ± 2.0 |
|   limbic cells | 55.0 ± 2.1 | 55.1 ± 2.0 | 54.6 ± 2.0 | 54.8 ± 2.1 |
|   goblet cells | 31.1 ± 1.1 | 31.0 ± 1.3 | 30.5 ± 1.3 | 30.1 ± 1.4 |
|   non-differentiated cells | 2.4 ± 0.3 | 2.4 ± 0.2 | 2.5 ± 0.1 | 2.5 ± 0.2 |
| Epithelial cells quantity per ½ of crypt: | 47.0 ± 3.2 | 47.1 ± 3.3 | 48.3 ± 3.1 | 48.1 ± 3.3 |
|   limbic cells | 26.1 ± 1.5 | 25.8 ± 1.3 | 27.1 ± 1.7 | 26.8 ± 1.9 |
|   goblet cells | 16.1 ± 1.1 | 16.1 ± 1.2 | 16.0 ± 1.1 | 16.1 ± 1.2 |
|   non-differentiated cells | 5.2 ± 1.1 | 5.3 ± 1.1 | 5.1 ± 1.0 | 5.0 ± 1.1 |

between the groups, supports the conclusion that the use of GM maize line 3272 in feed has no effect on peripheral blood structure of rats.

Biochemical assessment of blood serum (Table 5.303) and urine (Table 5.304) in rats of the control and test groups showed no significant difference. After 180 days of the experiment, a decrease of total protein content and increase of triglyceride content in blood serum in rats of the test group was observed, but the values fell within the limits of physiological fluctuations characteristic for these parameters [28,45,48]. It should be noted that,

**Table 5.301** Morphometric Parameters of Liver, Kidneys, and Spleen in Rats Fed Diet with Conventional Maize (Control) or Transgenic Maize Line 3272 (Test) ($M \pm m$, $n = 25$)

| Parameter | 30 Days | | 180 Days | |
|---|---|---|---|---|
| | Control | Test | Control | Test |
| *Liver* | | | | |
| Capillary surface, %[a] | 36.6 ± 1.2 | 36.5 ± 1.0 | 37.1 ± 1.1 | 36.9 ± 1.0 |
| Quantity of binuclear hepatocytes (per 100 hepatocytes) | 3.2 ± 0.2 | 3.1 ± 0.1 | 3.2 ± 0.1 | 3.1 ± 0.1 |
| *Kidneys* | | | | |
| Quantity of glomerules in sight (magnification 400×) | 11.3 ± 0.5 | 11.2 ± 0.6 | 11.3 ± 0.4 | 11.1 ± 0.6 |
| *Spleen, number of lymphoid series cells over the area of 880 μm²* | | | | |
| Germinal center of lymphoid tubercle | 30.4 ± 2.0 | 30.5 ± 1.9 | 29.5 ± 2.1 | 30.1 ± 1.9 |
| Pallial zone of lymphoid tubercle | 31.2 ± 1.8 | 31.3 ± 1.9 | 31.5 ± 1.9 | 31.3 ± 1.8 |
| Periarterial lymphoid sheath | 32.3 ± 1.9 | 32.2 ± 1.8 | 33.0 ± 1.8 | 33.1 ± 1.7 |

[a] Overall area of liver on section determined as 100%.

**Table 5.302** Hematological Factors of Blood Serum in Rats Fed Diet with Conventional Maize (Control) or Transgenic Maize Line 3272 (Test) ($M \pm m$, $n = 28$)

| Parameter | 30 Days | | 180 Days | | Normal Values According to [28,45,48] |
|---|---|---|---|---|---|
| | Control | Test | Control | Test | |
| Total erythrocyte count, ×10⁶/μL | 7.989 ± 0.205 | 7.777 ± 0.228 | 7.009 ± 0.504 | 7.860 ± 0.192 | 4.4 – 8.9 |
| Hemoglobin concentration, g/L | 121.9 ± 3.3 | 126.4 ± 2.8 | 117.6 ± 1.2 | 120.6 ± 2.2 | 86 – 173 |
| Hematocrit, vol.% | 42.49 ± 1.16 | 43.34 ± 0.87 | 35.00 ± 2.63 | 40.09 ± 0.73 | 31.4 – 51.9 |
| MCV (erythrocyte), μm³ | 47.27 ± 0.71 | 49.43 ± 0.57* | 49.94 ± 0.47 | 51.14 ± 0.89 | 50.6 – 93.8 |
| MCH, pg | 13.53 ± 0.22 | 14.44 ± 0.16* | 15.01 ± 0.24 | 15.31 ± 0.36 | 16 – 21 |
| MCHC, g/dL | 28.64 ± 0.06 | 29.20 ± 0.23* | 30.14 ± 0.40 | 30.03 ± 0.28 | 24.7 – 36.8 |
| ESR, mm/h | 1.714 ± 0.421 | 3.143 ± 0.800 | 3.000 ± 0.548 | 2.857 ± 0.769 | 0 – 5 |
| Leukocytes, 10³/μL | 11.34 ± 2.00 | 9.41 ± 1.18 | 7.550 ± 1.299 | 7.043 ± 0.752 | 1.4 – 34.3 |
| Basophils, % | 0 | 0 | 0 | 0 | 0 – 1 |
| Eosinophils, % | 1.286 ± 0.522 | 1.286 ± 0.644 | 3.857 ± 0.962 | 3.571 ± 0.922 | 0.0 – 5.5 |
| Stab neutrophils, % | 3.600 ± 0.927 | 3.857 ± 0.595 | 4.500 ± 1.443 | 4.800 ± 0.490 | 18 – 36 |
| Segmentonuclear neutrophils, % | 21.86 ± 4.68 | 21.29 ± 2.08 | 32.50 ± 3.25 | 27.57 ± 2.69 | 0.4 – 50.8 |
| Metamyelocytes, % | 0 | 0 | 0 | 0 | 0 |
| Lymphocytes, % | 71.57 ± 5.11 | 70.71 ± 2.77 | 50.14 ± 4.75 | 61.29 ± 3.94 | 42.3 – 98.0 |
| Monocytes, % | 0.429 ± 0.202 | 2.714 ± 1.085 | 1.429 ± 0.685 | 1.714 ± 0.778 | 0.0 – 7.9 |
| Platelets, 10³/μL | 730.7 ± 30.8 | 688.0 ± 24.7 | 729.3 ± 86.1 | 696.6 ± 73.1 | 409 – 1250 |

*Variation from control significant at $p < 0.05$.

**Table 5.303** Biochemical Parameters of Blood Serum in Rats Fed Diet with Conventional Maize (Control) or Transgenic Maize Line 3272 (Test) ($M \pm m$, $n = 28$)

| Parameter | 30 Days Control | 30 Days Tests | 180 Days Control | 180 Days Test | Normal Values According to [28,45,48] |
|---|---|---|---|---|---|
| Total protein, g/L | 74.71 ± 4.51 | 74.53 ± 2.62 | 88.23 ± 2.83 | 77.40 ± 2.62* | 56-82 |
| Triglycerides, mmol/L | 1.011 ± 0.259 | 0.973 ± 0.137 | 0.605 ± 0.067 | 0.797 ± 0.037* | 0.3-1.6 |
| Total bilirubin, µM/L | 3.429 ± 0.612 | 3.000 ± 0.436 | 3.857 ± 0.508 | 3.429 ± 0.369 | 1-4 |
| Direct bilirubin, µM/L | 0 | 0 | 0 | 0 | 0 |
| Glucose, mmol/L | 5.553 ± 0.340 | 5.941 ± 0.353 | 4.130 ± 0.465 | 3.333 ± 0.189 | 4.5-10.0 |
| Alkaline phosphatase, U/L | 536.1 ± 27.3 | 752.4 ± 106.2 | 479.4 ± 47.4 | 399.1 ± 41.4 | 112-814 |
| Alanine aminotransferase, U/L | 101.3 ± 29.7 | 95.1 ± 6.2 | 66.83 ± 9.01 | 56.00 ± 4.93 | 33-120 |
| Lipase, U/L | 21.43 ± 3.43 | 28.83 ± 2.50 | 17.71 ± 2.23 | 17.86 ± 1.71 | <30 |

*Variation from control significant at $p < 0.05$.

**Table 5.304** Biochemical Parameters of Urine in Rats Fed Diet with Conventional Maize (Control) or Transgenic Maize Line 3272 (Test) ($M \pm m$, $n = 25$)

| Parameter | 30 Days Control | 30 Days Test | 180 Days Control | 180 Days Test |
|---|---|---|---|---|
| pH | 5.91 ± 0.10 | 6.02 ± 0.17 | 6.00 ± 0.18 | 6.01 ± 0.12 |
| Daily urine, mL | 4.23 ± 0.32 | 4.34 ± 0.39 | 4.18 ± 0.30 | 4.26 ± 0.30 |
| Relative density, g/mL | 1.15 ± 0.08 | 1.14 ± 0.07 | 1.14 ± 0.07 | 1.15 ± 0.07 |
| Creatinine, mg/mL | 1.24 ± 0.08 | 1.25 ± 0.12 | 1.64 ± 0.08 | 1.65 ± 0.11 |
| Creatinine, mg/24 h | 5.35 ± 0.41 | 5.53 ± 0.50 | 6.89 ± 0.42 | 7.11 ± 0.44 |

in studies *in vivo* in certain cases, statistically non-uniform distribution of values of some parameters in groups, influencing size of the average ($M \pm m$), takes place; differences in the range of physiological standard are conditional to individual peculiarities of living organism metabolic status, as well as to random selection factor.

Assessment of the antioxidant status and activity of enzymes participating in protective-adaptive processes, as well as system biomarkers that reflect the level of an organism's adaptation to the environment, revealed no significant difference between the test and control groups (Tables 5.305 to 5.307).

As shown in Table 5.305, there were no significant differences in the factors characterizing antioxidant status. An increase in MDA content in blood serum

**Table 5.305** Antioxidant Status in Rats Fed Diet with Conventional Maize (Control) or Transgenic Maize Line 3272 (Test) ($M \pm m$, $n = 28$)

| Parameter | 30 Days | | 180 Days | |
|---|---|---|---|---|
| | Control | Test | Control | Test |
| *Enzyme strength of antioxidant protection system* | | | | |
| Glutathione reductase, µM/min/g Hb | 50.81 ± 3.31 | 47.55 ± 2.08 | 47.70 ± 2.82 | 43.29 ± 2.30 |
| Glutathione peroxidase, µM/min/g Hb | 81.33 ± 2.60 | 80.05 ± 3.35 | 79.42 ± 4.55 | 80.97 ± 1.37 |
| Catalase, mmol/min/g Hb | 673.3 ± 29.4 | 669.6 ± 27.5 | 659.2 ± 32.7 | 680.3 ± 26.0 |
| Superoxide dismutase, AU/min/g Hb | 2207 ± 72 | 2181 ± 63 | 2391 ± 66 | 2468 ± 29 |
| *Content of lipid peroxidation products* | | | | |
| Erythrocyte MDA, nmol/mL | 4.260 ± 0.440 | 4.097 ± 0.260 | 4.311 ± 0.222 | 4.690 ± 0.283 |
| Serum MDA, nmol/mL | 7.280 ± 0.175 | 8.324 ± 0.113* | 6.181 ± 0.264 | 6.415 ± 0.250 |
| Liver MDA, nmol/g | 296.9 ± 10.6 | 290.3 ± 11.1 | 294.9 ± 9.9 | 274.6 ± 8.8 |

*Variation from control significant at $p < 0.05$.

of rats of the test group detected on the 30th day fell within the limits of physiological fluctuations characteristic of growing animals. By the 180th day, no differences between the groups were observed.

Activity of enzymes of xenobiotic metabolism, as well as general and nonsedimenting activity of enzymes of liver lysosomes had no statistically significant differences between rats of control and test groups throughout the duration of the experiment (Tables 5.306 and 5.307).

Thus, results of 180-day toxicological experiments in rats fed increased quantities of GM maize line 3272 with their diet (46% caloric value) demonstrate the absence of any toxic effect. Values of all evaluated factors fell within the limits of physiological fluctuations characteristic of rats.

### Genotoxicity Assessment of GM Maize Line 3272 in an Experiment in Mice

Assessment of potential genotoxicity of GM maize line 3272 was conducted in an experiment in male mice of C57Bl/6 line sensitive to genotoxic influence, receiving GM maize line 3272 and its conventional counterpart with diet for 30 days. Initial body mass of mice was 20.1 ± 0.3 g. The mice received a semi-synthetic casein diet (Table 5.162). Ground maize grain was added to the diet at the rate of 4 g/mouse/24 h. Evaluation of potential genotoxicity included identification of DNA damage by the method of alkaline gel-electrophoresis of isolated cells of bone marrow, liver, and rectum (DNA-comet assay) [2], and detection of mutagenic activity by the assessment of

### 5.2.9 Maize Line 3272 Producing Alpha-Amylase Enzyme

**Table 5.306** Activity of Enzymes of Xenobiotic Metabolism and Protein Content in Liver of Rats Fed Diet with Conventional Maize (Control) or Transgenic Maize Line 3272 (Test) ($M \pm m$, $n = 24$)

| Parameter | 30 Days | | 180 Days | |
|---|---|---|---|---|
| | Control | Test | Control | Test |
| Cytochrome P450, nmol/mg protein | 0.76 ± 0.06 | 0.78 ± 0.06 | 0.90 ± 0.06 | 0.92 ± 0.03 |
| Cytochrome $b_5$, nmol/mg protein | 0.69 ± 0.04 | 0.72 ± 0.05 | 0.89 ± 0.04 | 0.94 ± 0.05 |
| N-Demethylation of amidopyrine, nmol/min/mg protein | 8.53 ± 0.11 | 8.23 ± 0.34 | 9.42 ± 0.17 | 9.27 ± 0.27 |
| Benzpyrene hydroxylation, Fl-units/min/mg protein | 6.58 ± 0.33 | 6.83 ± 0.44 | 8.12 ± 0.32 | 7.78 ± 0.09 |
| Acetyl esterase, µM/min/mg protein | 5.10 ± 0.04 | 5.28 ± 0.07 | 5.35 ± 0.27 | 5.58 ± 0.09 |
| Epoxide hydrolase, nmol/min/mg protein | 11.6 ± 0.9 | 12.2 ± 0.7 | 10.2 ± 0.3 | 10.2 ± 0.3 |
| UDP-Glucuronosyl transferase, nmol/min/mg protein | 22.2 ± 1.9 | 25.1 ± 2.8 | 22.5 ± 1.2 | 20.8 ± 0.6 |
| HDNB-glutathione S-transferase, µM/min/mg protein | 0.82 ± 0.04 | 0.81 ± 0.04 | 1.22 ± 0.05 | 1.22 ± 0.05 |
| Microsomal protein, mg/g | 14.9 ± 0.3 | 15.3 ± 0.4 | 15.3 ± 0.5 | 16.0 ± 0.3 |
| Cytosolic protein, mg/g | 96.8 ± 2.7 | 99.4 ± 2.4 | 85.6 ± 2.1 | 85.6 ± 2.1 |

**Table 5.307** Activity of Enzymes of Liver Lysosomes in Rats Fed Diet with Conventional Maize (Control) or Transgenic Maize Line 3272 (Test) ($M \pm m$, $n = 24$)

| Parameter | 30 Days | | 180 Days | |
|---|---|---|---|---|
| | Control | Test | Control | Test |
| *Total activity, µM/min × g tissue* | | | | |
| Arylsulfatases A and B | 2.34 ± 0.04 | 2.38 ± 0.02 | 2.29 ± 0.06 | 2.36 ± 0.03 |
| β-Galactosidase | 2.33 ± 0.08 | 2.33 ± 0.09 | 2.44 ± 0.09 | 2.35 ± 0.09 |
| β-Glucuronidase | 2.35 ± 0.08 | 2.30 ± 0.03 | 2.19 ± 0.05 | 2.24 ± 0.03 |
| *Non-sedimenting activity, % of total* | | | | |
| Arylsulfatases A and B | 3.25 ± 0.04 | 3.20 ± 0.06 | 2.98 ± 0.08 | 2.97 ± 0.13 |
| β-Galactosidase | 5.70 ± 0.37 | 5.59 ± 0.36 | 6.58 ± 0.39 | 6.50 ± 0.40 |
| β-Glucuronidase | 5.23 ± 0.29 | 5.18 ± 0.17 | 6.25 ± 0.15 | 6.40 ± 0.19 |

chromosomal aberrations in metaphase cells of mice bone marrow [5]. No fewer than 100 cells from each microslide were analyzed.

Mouse body mass was 22–26 g after 30 days from the start of the experiment. As shown in Tables 5.308 and 5.309, average parameters of chromosomal aberrations in mice bone marrow cells of control and test groups were not significantly different and did not exceed the spontaneous mutagenesis level characteristic of mice of C57Bl/6 line. No differences between the control and test groups in the levels of DNA structure damage in bone marrow, liver, or rectum were detected.

**Table 5.308** Chromosomal Damage in Bone Marrow Cells in Mice Fed Diet with Conventional Maize (Control) or Transgenic Maize Line 3272 (Test) ($M \pm m$)

| Group | Number of Cells | Per 100 Cells | | | | | Total of Damaged Metaphases, % |
|---|---|---|---|---|---|---|---|
| | | Genes | Individual Fragments | Paired Fragments | Exchanges | Cells with Multiple Injuries[a] | |
| Control | 500 | 0.8 | 1.2 | – | – | – | 2.00 ± 0.62 |
| Test | 500 | 1.0 | 0.8 | – | – | – | 1.80 ± 0.59 |

[a]More than five chromosomal aberrations in a cell.

**Table 5.309** Level of DNA Structure Damage in Organs in Mice Fed Diet with Conventional Maize (Control) or Transgenic Maize Line 3272 (Test) ($M \pm m$).

| Group | Bone Marrow | | Liver | | Rectum | |
|---|---|---|---|---|---|---|
| | Number of Cells | DNA Damaged, % | Number of Cells | DNA Damaged, % | Number of Cells | DNA Damaged, % |
| Control | 500 | 7.3 ± 0.2 | 500 | 5.8 ± 0.2 | 500 | 7.5 ± 0.3 |
| Test | 500 | 7.5 ± 0.2 | 500 | 5.7 ± 0.2 | 500 | 7.4 ± 0.2 |

Note: Differences not significant ($p > 0.05$).

Thus, the results of DNA integrity and level of chromosomal aberrations in the experiment in mice demonstrate the absence of genotoxic effect of maize line 3272 compared with its conventional counterpart.

### Immunological Assessment of GM Maize Line 3272 in Mice

The immunomodulating effects of transgenic maize line 3272 were examined during 45 days on CBA and C57Bl/6 mice with an initial body weight of 18–20 g. Throughout the entire duration of the experiment, the mice were maintained on standard vivarium diet (Table 5.114). Milled maize grain was included into the feed (3 g/mouse/day), replacing the oatmeal and providing an equal amount of nutrients and calories. The experimental details are described in section 5.2.7.

The assessment of the immune system humoral component state showed that the dynamics of formation of sheep erythrocytes antibodies in mice of the control group was similar to that of mice of the test group (both in mice of CBA and C57Bl/6 lines): increase of antibody titer at the 7th day after administration of ram erythrocytes was maintained at high level up to the 21st day (Table 5.310). Results of the analysis demonstrate the absence of a

### 5.2.9 Maize Line 3272 Producing Alpha-Amylase Enzyme

**Table 5.310** Level of Antibodies Raised Against SE in Mice Fed Diet with Conventional Maize (Control) or Transgenic Maize Line 3272 (Test)

| Number of Days After Administration of Sheep Erythrocytes | CBA Line | | C57Bl/6 Line | |
|---|---|---|---|---|
| | Control | Test | Control | Test |
| 7  | 102 ± 44 | 106 ± 30 | 12 ± 5 | 21 ± 10 |
| 14 | 48 ± 20  | 30 ± 19  | 3 ± 2  | 13 ± 10 |
| 21 | 147 ± 74 | 140 ± 62 | 9 ± 7  | 21 ± 13 |

*Average data shown, $M \pm m$ from $n = 10$; $p > 0.05$.*

**Table 5.311** Delayed Hypersensitivity Test Index in Mice Fed Diet with Conventional Maize (Control) or Transgenic Maize Line 3272 (Test)

| CBA Line | | C57Bl/6 Line | |
|---|---|---|---|
| Control | Test | Control | Test |
| 24.4 ± 6.6 | 16.6 ± 9.7 | 10.2 ± 8.9 | 3.4 ± 3.1 |

*Average data shown, $M \pm m$ from $n = 10$; $p > 0.05$.*

modulating effect of GM maize on the humoral component of the immune system. Certain fluctuations in antibody levels during the observation period in mice of both groups can be explained by differences in individual sensitivity to the sensitizing factor.

Assessment of the condition of the cellular component of the immune system by a delayed hypersensitivity test demonstrated the absence of a modulating effect of GM maize line 3272 (Table 5.311).

Assessment of sensitizing effect and effect on mouse response to *Salmonella typhimurium* demonstrated no negative influence of GM maize. When infected with *Salmonella typhimurium*, mice of control and test groups had a typical infection; mice of the CBA line (insensitive to *Salmonella typhimurium*) were more resistant to the infection than were mice of the C57Bl/6 line (sensitive to *Salmonella typhimurium*).

Thus, results of immunological assessment of GM maize line 3272 in an experiment in mice of oppositely reacting lines demonstrated absence of any immunomodulating and sensitizing effect of the maize line 3272 compared with its conventional counterpart.

#### Allergological Assessments of Maize Line 3272 in Experiments in Rats

An experiment of 30 days duration was conducted in male Wistar rats with an initial body mass of $170 \pm 10$ g. Throughout the experiment the rats received the general diet specified in Table 5.114.

**Table 5.312** Reaction of Active Anaphylactic Shock in Rats Fed Diet with Conventional Maize (Control) or Transgenic Maize Line 3272 (Test)

| Group[a] | Anaphylactic Index | Heavy Reactions, % | Mortality, % |
|---|---|---|---|
| Control | 3.13 | 69.6 | 69.6 |
| Test | 3.13 | 66.7 | 66.7 |
| P | >0.05[b] | >0.05[c] | >0.05[c] |

[a]23 rats in the control group, 24 rats in the test group.
[b]Mann-Whitney nonparametric rank test.
[c]Bi-directional test U with Fisher angular transformation.

**Table 5.313** Intensity of Humoral Immune Response in Rats Fed Diet with Conventionl Maize (Control) or Transgenic Maize Line 3272 (Test)

| Group[a] | $D_{492}$ | AT Level, mg/mL | Log of AT Level |
|---|---|---|---|
| Control | 0.913 ± 0.032 | 19.4 ± 1.7 | 1.244 ± 0.048 |
| Test | 0.807 ± 0.042 | 14.9 ± 1.7 | 1.078 ± 0.070 |
| *Statistical analysis* | | | |
| t-Student test | >0.05 | >0.05 | >0.05 |
| Test for homogeneity of distribution, ANOVA | >0.05 | >0.05 | >0.05 |
| Mann-Whitney nonparametric rank test | >0.05 | | |

Average data shown, M ± m.
[a]21 rats in the control group, 23 rats in the test group (samples from 2 rats in the control group and 1 rat in the test group were not obtained).

Ground maize grain was added to the feed at the rate of 3.5 g/rat/24 h, excluding oatmeal quantity, equivalent in caloric value and nutrient materials content. Estimation of potential allergenic capacity included determination of the severity of system anaphylaxis development and level of circulating sensitizing antibodies (subclasses $IgG_1 + IgG_4$) at intra-abdominal sensitization of adult rats by a food antigen—chicken egg ovalbumin—with subsequent intravenous administration of an anaphylaxis–provoking dose of the same protein to sensitized animals.

Body mass of rats of the control and test groups at the 29th day of the experiment was 251 ± 8 and 242 ± 5 g, respectively; differences between groups were not significant ($p > 0.05$ according to Mann-Whitney criterion), and value distribution in the groups was homogeneous ($p > 0.05$ according to ANOVA test). Severity of active anaphylactic shock reaction (Table 5.312), as well as intensity of humoral immune response (Table 5.313), in rats of the test group had no statistically significant differences from similar factors for the rats of the control group. ANOVA analysis of factors distribution in the groups showed their homogeneity ($p > 0.05$).

Thus, allergological studies of maize line 3272 in rats demonstrate the absence of an allergenic effect of the GM maize line compared with its conventional counterpart.

### *Assessment of Technological Parameters of Maize Line 3272*
The study of functional-technological properties GM maize line 3272 was conducted according to the requirements outlined in Methodological Regulations of the Ministry of Health of the Russian Federation "Medico-biologic assessment of food products, derived from genetically modified sources": MUK 2.3.2.970-00.

### Results
- The quality of the samples of maize grain corresponds to GOST 3634-81 "Maize. Technical specifications".
- Samples of maize grain were processed in equal modes. When producing starch from samples of maize 3272 and its conventional counterpart, no differences or difficulties in the technological process were detected.
- Samples of maize starch conform to GOST P51985-2002 requirements "Starch maize. General specifications".
- Amylograms of 6% water suspensions of maize starch samples from GM maize line 3272 recorded by Brabender amylograph are characteristic of corn starch from conventional raw materials.
- The temperature of gelatinization of starch obtained from GM maize line 3272 is slightly higher than that of starch obtained from reference maize. Viscosity of 6% starch paste from maize 3272 is higher than viscosity of starch paste from reference maize, at all temperatures of viscosity determination. Observed differences may be conditional on difference in amylose/amylopectin ratio, which depends on climatic conditions of growing and grain maturity and is of no biological importance.
- Starch obtained from GM maize line 3272 and starch obtained from the reference sample have identical gelling properties.
- Structural and mechanical properties of extrusion-type products produced from grain of GM maize line 3272 practically coincide with similar parameters of extrusion-type products produced from grain of the corn reference sample.

Thus, the results of the assessment of the technological properties demonstrate the absence of significant difference between GM maize line 3272 and its conventional counterpart.

### *Conclusions*
Expert assessment of the data provided by the applicant, and results of complex biomedical research of GM maize line 3272, synthesizing alpha-amylase enzyme, by the Russian authorities demonstrate the absence of any toxic,

genotoxic, sensitizing, immunomodulating, or allergenic effect in this maize line, as well as its composition equivalence to its traditional analogue.

In accordance with the Federal law of March 30, 1999, No.52-FZ "On Sanitary and Epidemiological Safety of the Population", GM maize line 3272, synthesizing alpha-amylase enzyme, has registered for food use, and can be safely used on the territory of the Russian Federation, imported to the territory of the Russian Federation, and placed on the market without restrictions (State Registration Certificate No. 77.99.26.11.Y.2009.4.10 of April 5, 2010).

# Subchapter 5.3

# Rice

## 5.3.1 RICE LINE LLRICE62 TOLERANT TO GLUFOSINATE AMMONIUM

### Molecular Characteristics of Rice Line LLRICE62

*Recipient Organism*

Rice (*Oryza sativa L.*) is characterized by a long history of safe use as human food. At present, it is cultivated in more than 100 countries. Its global production volume is the second highest of any crop.

*Donor Organism*

The donor of the *bar* gene, *Streptomyces hygroscopicus* strain HP632, is an aerobic soil bacterium nonpathogenic to humans, animals, or plants. The gene *bar* encodes phosphinothricin acetyltransferase (PAT), which acetylates glufosinate ammonium and thereby neutralizes the effect of this herbicide on plants.

*Method of Genetic Transformation*

Plasmid DNA was incorporated into the genome of rice line Bengal by the method of direct DNA transformation. The nucleotide sequence of the *bar* gene isolated from *Streptomyces hygroscopicus* was modified to enhance expression in the plant. The activity of this gene is controlled by the 35S promoter

### 5.3.1 Rice Line LLRICE62 Tolerant to Glufosinate Ammonium

**Table 5.314** Registration Status of Transgenic Rice Line LLRICE62 in Various Countries

| Country | Date of Approval | Scope |
|---|---|---|
| USA | 1999 | Environmental release |
|  | 2000 | Food, feed |

*Note: The current registration status of the transgenic crop is on http://www.biotradestatus.com/*

and 35S terminator from cauliflower mosaic virus. Molecular analysis of rice line LLRICE62 showed that genomic transformation resulted only in insertion of the *bar* gene and the DNA sequences controlling its expression [19].

## Global Registration Status of Rice Line LLRICE62

Table 5.314 shows the countries that had granted registration to use the transgenic rice line LLRICE62 at the time of registration in Russia.

## Safety Assessment of Transgenic Rice Line LLRICE62 Conducted in the Russian Federation

Studies were conducted in accordance with the requirements of the Ministry of Health of the Russian Federation authorized for risk and safety assessment of food derived from GM sources [8]. PCR analysis of the rice test and control samples was performed to confirm the identity of the transformation event and its absence in the conventional control line.

### *Biochemical Composition of Transgenic Rice*

The content of proteins and the amino acid composition in the grain of transgenic rice line LLRICE62 did not significantly differ from the corresponding values for the conventional rice (Table 5.315). The content of starch in the grain of the two rice cultivars did not significantly differ (Table 5.316). There were no significant differences in the contents of lipids and fatty acids (Table 5.317). The insignificant variations in the content of pentadecanoic and Heptadecenoic acids remained within the range characteristic of rice: respectively, 0.01–0.05 and 0.02–0.06 g/100 g (data of the State Research Institute of Nutrition, RAMS). Similarly, there were no significant differences in the content of vitamins of group B or group E (Table 5.318).

The content of minerals in conventional rice and transgenic rice line LLRICE62 was also similar (Table 5.319). The revealed insignificant changes remained within the range of physiological variations characteristic of

**Table 5.315** Protein Content (%) and Amino Acid Composition (g/100 g protein) in Rice Grain

| Ingredient | Conventional Rice | Transgenic Rice Line LLRICE62 |
|---|---|---|
| Protein, % | 9.09 | 8.13 |
| *Amino acids* | | |
| Lysine | 4.76 | 5.36 |
| Histidine | 3.89 | 3.69 |
| Arginine | 6.89 | 7.12 |
| Aspartic acid | 9.49 | 10.27 |
| Threonine | 3.62 | 3.81 |
| Serine | 4.28 | 4.03 |
| Glutamic acid | 18.58 | 18.42 |
| Proline | 6.44 | 6.93 |
| Glycine | 4.53 | 4.02 |
| Alanine | 5.05 | 4.85 |
| Cysteine | 1.97 | 1.92 |
| Valine | 5.21 | 5.03 |
| Methionine | 2.25 | 2.01 |
| Isoleucine | 4.69 | 4.34 |
| Leucine | 7.73 | 7.49 |
| Tyrosine | 3.21 | 2.87 |
| Phenylalanine | 4.55 | 4.22 |

**Table 5.316** Carbohydrate Content (g/100 g) in Rice Grain

| Carbohydrate | Conventional Rice | Transgenic Rice Line LLRICE62 |
|---|---|---|
| Starch | 53.2 | 50.5 |

rice (potassium, 2000–4000 mg/kg; iron, 10–25 mg/kg; and magnesium, 1500–2500 mg/kg).

The content of heavy metals in the grain of conventional rice and transgenic rice line LLRICE62 did not surpass the limits acceptable according to the regulations valid in the Russian Federation [5] (Table 5.320).

Thus, the grain of transgenic rice line LLRICE62 and isogenic conventional rice did not significantly differ by chemical composition [1,16]. The safety parameters of both rice cultivars did not surpass the limits acceptable according to the regulations valid in the Russian Federation (SanPin).

### 5.3.1 Rice Line LLRICE62 Tolerant to Glufosinate Ammonium

**Table 5.317** Lipids and Fatty Acids Content (Rel. %) in Rice Grain

| Ingredient | Conventional Rice | Transgenic Rice Line LLRICE62 |
|---|---|---|
| Lipids, % | 2.10 | 2.68 |
| *Fatty acids* | | |
| Lauric 12:0 | 0.08 | 0.07 |
| Myristic 14:0 | 0.38 | 0.44 |
| Pentadecanoic 15:0 | 0.02 | 0.04 |
| Palmitic 16:0 | 15.57 | 16.57 |
| Palmitoleic 16:1 | 0.22 | 0.21 |
| Margaric (heptadecanoic) 17:0 | 0.04 | 0.04 |
| Heptadecenoic 17:1 | 0.02 | 0.04 |
| Stearic 18:0 | 1.71 | 1.77 |
| cis-9-Oleic 18:1 | 37.75 | 38.53 |
| trans-11-Vaccenic 18:1 | 2.02 | 1.91 |
| Linoleic 18:2 | 38.88 | 37.26 |
| Linolenic 18:3 | 1.04 | 1.05 |
| Arachidic 20:0 | 0.66 | 0.60 |
| Gondoic 20:1 | 0.57 | 0.48 |
| Behenic 22:0 | 0.05 | 0.03 |
| Erucic 22:1 | 0.34 | 0.33 |
| Lignocerinic 24:0 | 0.61 | 0.52 |
| Tetracosenic (24:1) | 0.07 | 0.08 |

**Table 5.318** Vitamins Content (mg/100 g product) in Rice Grain

| Vitamin | Conventional Rice | Transgenic Rice Line LLRICE62 |
|---|---|---|
| Vitamin $B_1$ | 0.46 | 0.42 |
| Vitamin $B_2$ | 0.02 | 0.02 |
| Vitamin E ($\alpha$-tocopherol) | 0.68 | 0.76 |

**Table 5.319** Mineral Composition (mg/kg) in Rice Grain

| Ingredient | Conventional Rice | Transgenic Rice Line LLRICE62 |
|---|---|---|
| Sodium | 46.8 | 56.9 |
| Calcium | 15.4 | 15.7 |
| Magnesium | 2108 | 1704 |
| Iron | 20.8 | 13.9 |
| Potassium | 3023 | 2430 |
| Phosphorus | 270 | 270 |

**Table 5.320** Analysis of Toxic Elements in Rice Grain

| Ingredient | Conventional Rice | Transgenic Rice Line LLRICE62 |
|---|---|---|
| Arsenic, mg/kg | <0.1 | <0.1 |
| Lead, mg/kg | 0.051 | 0.023 |
| Cadmium, mg/kg | 0.036 | 0.008 |
| Deoxynivalenol, mg/kg | Not detected | Not detected |
| Zearalenone, mg/kg | Not detected | Not detected |
| Aflatoxin $B_1$, mg/kg | Not detected | Not detected |

**Table 5.321** Rat Diet with Milled Rice (g/100 g feed)

| Ingredient | Weight, g |
|---|---|
| Milled rice | 35.5 |
| Casein | 20.7 |
| Corn starch | 24.7 |
| Sunflower-seed oil | 5.0 |
| Lard | 5.0 |
| Salt mix[a] | 4.0 |
| Liposoluble vitamin mix[a] | 0.10 |
| Water-soluble vitamin mix[a] | 1.00 |
| Bran | 4.0 |
| Total | 100 |

[a]The contents of mixes are given in Table 5.39 to 5.41.

### Toxicological Assessment of Transgenic Rice Line LLRICE62

The experiment was carried out on male Wistar rats with an initial body weight of 70–80 g. After admission to the vivarium of the State Research Institute of Nutrition, the rats were placed in quarantine for 10 days. At the onset of feeding experimental diets, the body weight of the rats was 85–90 g. During the entire experiment, the rats were fed a standard semi-synthetic diet (Table 5.321) with conventional (control) or transgenic (test) rice.

The milled rice was added to the rat diet instead of the corresponding amount of casein, starch, and bran to preserve the balance of basic nutrients. After a 10-day adaptation period to the control diet, the rats were randomized into control and test groups and fed the examined diets. The test rats were fed diet with transgenic rice line LLRICE62 (36 g/100 g ration), while the control rats were fed diet with the same amount of conventional rice.

### 5.3.1 Rice Line LLRICE62 Tolerant to Glufosinate Ammonium

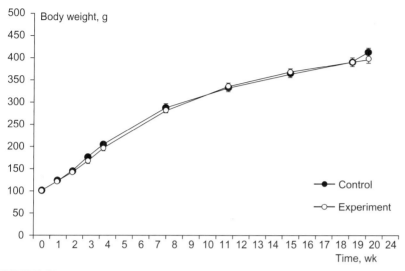

**FIGURE 5.15**
Comparative dynamics of body weight of rats fed diet with transgenic rice line LLRICE62 (test) or conventional rice (control).

Throughout the entire duration of the experiment, the general condition of the rats was similar in the control and test groups. No animal mortality was observed during this study. Daily rice intake (by one rat) was similar in both groups: 5.7 g (month 1), 7.2 g (months 2 and 3), and 7.9 g (months 4, 5, and 6). There were no differences in absolute body weight of the rats (Figure 5.15; Table 5.322).

The absolute and relative weights of internal organs did not significantly differ between control and test rats (Table 5.323).

### Assessment of Biochemical Parameters

During the 180-day chronic experiment, the content of total protein and glucose as well as activity of alanine aminotransferase, aspartate aminotransferase, and alkaline phosphatase in sera of the rats fed diet with transgenic rice line LLRICE62 (test group) did not significantly differ from the corresponding parameters of the control rats fed diet with conventional rice (Table 5.324).

Throughout the entire duration of the 180-day experiment, there were no differences in the urinary biochemical parameters between the test and control groups of animals (Table 5.325).

The study examined the potential effect of transgenic rice line LLRICE62 on the content of LPO products in rats of both groups. On Day 30, the blood

**Table 5.322** Comparative Body Weight of Rats (g) Fed Diet with Conventional Rice (Control) or Transgenic Rice Line LLRICE62 (Test) ($M \pm m$; $n = 25$)

| Duration of Experiment, Weeks | Control | Test |
|---|---|---|
| 0 | 107.5 ± 3.6 | 107.5 ± 2.3 |
| 1 | 132.0 ± 4.2 | 131.5 ± 4.3 |
| 2 | 156.5 ± 3.9 | 153.5 ± 4.6 |
| 3 | 189.5 ± 6.0 | 180.0 ± 5.8 |
| 4 | 220.5 ± 6.6 | 211.5 ± 7.4 |
| 8 | 309.0 ± 8.1 | 302.5 ± 6.7 |
| 12 | 356.5 ± 10.4 | 360.5 ± 8.0 |
| 16 | 390.5 ± 9.5 | 393.5 ± 9.3 |
| 20 | 419.5 ± 9.7 | 418.0 ± 10.6 |
| 24 | 443.0 ± 9.2 | 427.0 ± 10.6 |

*Note: Here and in Tables 5.323 to 5.328 the differences are not significant ($p > 0.05$).*

**Table 5.323** Absolute Weight of Internal Organs of Rats Fed Diet with Conventional Rice (Control) or Transgenic Rice Line LLRICE62 (Test) ($M \pm m$, $n = 6$–8)

| Organ | | 30 Days Control | 30 Days Test | 180 Days Control | 180 Days Test |
|---|---|---|---|---|---|
| Kidneys | Abs.[a], g | 1.65 ± 0.04 | 1.78 ± 0.05 | 2.29 ± 0.06 | 2.33 ± 0.03 |
| | Rel.[b], g/100 g | 0.62 ± 0.02 | 0.65 ± 0.023 | 0.53 ± 0.01 | 0.55 ± 0.01 |
| Liver | Abs., g | 10.11 ± 0.33 | 10.41 ± 2.9 | 12.75 ± 0.50 | 12.39 ± 0.36 |
| | Rel., g/100 g | 3.81 ± 0.04 | 3.80 ± 0.09 | 2.98 ± 0.07 | 2.92 ± 0.08 |
| Spleen | Abs., g | 1.53 ± 0.09 | 1.64 ± 0.11 | 1.23 ± 0.08 | 1.41 ± 0.09 |
| | Rel., g/100 g | 0.58 ± 0.07 | 0.59 ± 0.03 | 0.29 ± 0.04 | 0.33 ± 0.02 |
| Heart | Abs., g | 0.89 ± 0.03 | 0.93 ± 0.023 | 1.19 ± 0.04 | 1.16 ± 0.02 |
| | Rel., g/100 g | 0.34 ± 0.01 | 0.34 ± 0.09 | 0.28 ± 0.01 | 0.28 ± 0.004 |
| Testicles | Abs., g | 2.72 ± 0.16 | 2.32 ± 0.33 | 2.86 ± 0.35 | 3.18 ± 0.11 |
| | Rel., g/100 g | 1.02 ± 0.05 | 0.95 ± 0.12 | 0.66 ± 0.08 | 0.67 ± 0.05 |
| Hypophysis | Abs., mg | 7.20 ± 0.70 | 7.30 ± 0.49 | 10.8 ± 0.73 | 10.4 ± 1.32 |
| | Rel., mg/100 g | 2.70 ± 0.27 | 2.65 ± 0.15 | 2.51 ± 0.30 | 2.47 ± 0.30 |
| Adrenal glands | Abs., mg | 20.6 ± 2.41 | 20.6 ± 1.53 | 21.8 ± 0.17 | 22.0 ± 2.5 |
| | Rel., mg/100 g | 7.70 ± 0.94 | 7.56 ± 0.61 | 5.08 ± 0.17 | 5.28 ± 0.65 |
| Seminal vesicles | Abs., mg | 367.7 ± 70.6 | 394.5 ± 34.6 | 631.7 ± 33.0 | 648.3 ± 43.3 |
| | Rel., mg/100 g | 138.07 ± 26.07 | 143.67 ± 10.10 | 146.9 ± 7.80 | 152.2 ± 8.91 |
| Prostate | Abs., mg | 217.5 ± 44.1 | 211.8 ± 43.6 | 428.3 ± 55.3 | 439.0 ± 68.5 |
| | Rel., mg/100 g | 81.08 ± 15.46 | 76.82 ± 14.91 | 99.8 ± 13.1 | 91.9 ± 16.4 |

[a]Absolute weight of internal organs.
[b]Relative weight of internal organs (per 100 g body weight).

### 5.3.1 Rice Line LLRICE62 Tolerant to Glufosinate Ammonium

**Table 5.324** Biochemical Parameters of Blood Serum of Rats Fed Diet with Conventional Rice (Control) or Transgenic Rice Line LLRICE62 (Test) ($M \pm m$, $n = 6–8$)

| Parameter | 30 Days | | 180 Days | |
|---|---|---|---|---|
| | Control | Test | Control | Test |
| Total protein, g/L | 65.8 ± 2.0 | 65.5 ± 2.4 | 71.8 ± 1.5 | 76.6 ± 3.34 |
| Glucose, mM/L | 5.00 ± 0.61 | 5.15 ± 0.85 | 6.00 ± 0.78 | 6.20 ± 0.66 |
| Alanine aminotransferase, μcat/min/L | 1.84 ± 0.03 | 1.86 ± 0.01 | 2.30 ± 0.04 | 2.25 ± 0.03 |
| Aspartate aminotransferase, μcat/min/L | 0.39 ± 0.04 | 0.38 ± 0.02 | 0.45 ± 0.02 | 0.44 ± 0.18 |
| Alkaline phosphatase, μcat/L | 7.94 ± 0.83 | 7.21 ± 0.71 | 3.37 ± 0.27 | 3.38 ± 0.24 |

**Table 5.325** Urinary Biochemical Parameters of Rats Fed Diet with Conventional Rice (Control) or Transgenic Rice Line LLRICE62 (Test) ($M \pm m$, $n = 6–8$)

| Parameter | 30 Days | | 180 Days | |
|---|---|---|---|---|
| | Control | Test | Control | Test |
| pH | 7.0 | 7.0 | 7.0 | 7.0 |
| Daily diuresis, mL | 2.5 ± 0.3 | 2.7 ± 0.2 | 4.1 ± 0.3 | 4.2 ± 0.5 |
| Relative density, g/mL | 1.08 ± 0.03 | 1.11 ± 0.01 | 1.12 ± 0.02 | 1.14 ± 0.01 |
| Creatinine, mg/mL | 3.90 ± 0.32 | 3.62 ± 0.14 | 2.81 ± 0.29 | 2.57 ± 0.48 |
| Creatinine, mg/day | 9.54 ± 1.01 | 9.50 ± 0.61 | 10.97 ± 0.74 | 9.99 ± 1.67 |

levels of DC and MDA did not significantly differ between the control and test rats. In test rats, hepatic MDA was lower than the control value by 14% (Table 5.326). On Day 180, the content of LPO products in erythrocytes and liver did not significantly differ between groups except for a 19% decrease in serum DC in the control rats.

On experimental Day 30, enzymatic activity of the erythrocytic antioxidant protection system did not significantly differ between groups, except for significant (5%) elevation of SOD activity in test rats in comparison with the control value. However, in 180 days, there was no significant difference in this parameter (Table 5.327).

The data obtained concluded that the antioxidant status of the rats in both groups was in the state of dynamic equilibrium. The revealed differences in a few parameters did not depart from the physiological range; they had no stable trend, and they were not accompanied by similar variations in other parameters characterizing the antioxidant status. Thus, addition of transgenic rice to the diet produced no effect on the antioxidant status of the laboratory animals.

**Table 5.326** Content of LPO Products in Rats Fed Diet with Conventional Rice (Control) or Transgenic Rice Line LLRICE62 (Test) ($M \pm m$, $n = 6–8$)

| Parameter | 30 Days | | 180 Days | |
|---|---|---|---|---|
| | Control | Test | Control | Test |
| *Erythrocytes* | | | | |
| DC, nM/mL | 4.199 ± 0.299 | 3.858 ± 0.269 | 4.228 ± 0.279 | 4.355 ± 0.191 |
| MDA, nM/mL | 4.078 ± 0.249 | 4.289 ± 0.354 | 4.847 ± 0.179 | 4.212 ± 0.249 |
| *Blood serum* | | | | |
| DC, nM/mL | 3.523 ± 0.382 | 3.601 ± 0.252 | 3.816 ± 0.156 | 3.127 ± 0.155* |
| MDA, nM/mL | 5.343 ± 0.234 | 5.755 ± 0.242 | 5.536 ± 0.117 | 5.316 ± 0.162 |
| *Liver* | | | | |
| DC, Unit | 1.061 ± 0.019 | 1.077 ± 0.010 | 1.081 ± 0.018 | 1.070 ± 0.011 |
| MDA, nM/g | 369.2 ± 12.2 | 319.2 ± 10.4* | 329.7 ± 11.1 | 331.8 ± 16.4 |

Note: Here and in Tables 5.327 to 5.329, *$p < 0.05$ in comparison with control.

**Table 5.327** Activity of Enzymes of Erythrocytic Antioxidant Protection System in Rats Fed Diet with Conventional Rice (Control) or Transgenic Rice Line LLRICE62 (Test) ($M \pm m$, $n = 6–8$)

| Parameter | 30 Days | | 180 Days | |
|---|---|---|---|---|
| | Control | Test | Control | Test |
| Glutathione reductase, μmol/min/g Hb | 37.67 ± 1.53 | 38.58 ± 1.19 | 37.03 ± 1.79 | 38.93 ± 1.22 |
| Glutathione peroxidase, μmol/min/g Hb | 55.47 ± 1.10 | 56.25 ± 2.21 | 57.25 ± 0.74 | 56.18 ± 2.28 |
| Catalase, mmol/min/g Hb | 429.8 ± 16.3 | 458.0 ± 13.2 | 456.5 ± 14.3 | 465.0 ± 18.1 |
| Superoxide dismutase, U/min/g Hb | 1815 ± 25 | 1916 ± 37* | 1687 ± 41 | 1679 ± 31 |

The study examined potential effect of transgenic rice line LLRICE62 on the activity of enzymes involved in xenobiotic degradation (Table 5.328). On Day 30, the control and test group rats did not significantly differ. On Day 180 there were no significant differences in the activity of enzymes involved in xenobiotic degradation, except for the content of cytochrome P450 and the rate of benzpyrene hydroxylation, which was lower in the test rats than in the control group. However, these differences remained within the physiological limits of the examined parameters. Such slight decrease in activity of these enzymes, unaccompanied by the complex stress manifestations of the protective and adaptive systems in test rats, attests to the absence of adverse effects of chronic intake of transgenic rice on the protective and adaptive potencies of the organism.

## 5.3.1 Rice Line LLRICE62 Tolerant to Glufosinate Ammonium

**Table 5.328** Activity of Enzymes Involved in Metabolism of Xenobiotics and Protein Content in Liver Microsomes in Rats Fed Diet with Conventional Rice (Control) or Transgenic Rice Line LLRICE62 (Test) ($M \pm m$, $n = 6\text{--}8$)

| Parameter | 30 Days | | 180 Days | |
|---|---|---|---|---|
| | Control | Test | Control | Test |
| Cytochrome P450, nM/mg protein | 0.72 ± 0.04 | 0.73 ± 0.03 | 0.80 ± 0.04 | 0.64 ± 0.01* |
| Cytochrome $b_5$, nM/mg protein | 0.58 ± 0.02 | 0.61 ± 0.02 | 0.82 ± 0.04 | 0.74 ± 0.03 |
| Aminopyrine N-demethylation, nM/min/mg protein | 8.08 ± 0.15 | 7.93 ± 0.33 | 9.60 ± 0.13 | 9.31 ± 0.19 |
| Benzpyrene hydroxylation, Fl-units/min/mg protein | 7.51 ± 0.42 | 7.81 ± 0.33 | 10.73 ± 0.50 | 8.73 ± 0.37* |
| Acetylesterase, µM/min/mg protein | 4.52 ± 0.14 | 4.70 ± 0.14 | 5.65 ± 0.21 | 5.28 ± 0.11 |
| Epoxide hydrolase, nM/min/mg protein | 5.24 ± 0.31 | 5.15 ± 0.42 | 6.00 ± 0.52 | 5.38 ± 0.38 |
| UDP-Glucuronosil transferase, nM/min/mg protein | 28.5 ± 1.8 | 26.4 ± 1.5 | 17.9 ± 1.3 | 16.5 ± 1.5 |
| CDNB-Glutathione transferase, µM/min/mg protein | 0.85 ± 0.04 | 0.84 ± 0.05 | 1.19 ± 0.05 | 1.10 ± 0.03 |
| Protein, mg/g | | | | |
| microsomal | 18.6 ± 0.3 | 17.7 ± 0.4 | 14.6 ± 0.5 | 15.0 ± 0.4 |
| cytosolic | 91.8 ± 1.9 | 93.1 ± 1.9 | 77.8 ± 1.5 | 79.6 ± 1.4 |

**Table 5.329** Total and Non-sedimentable Activity of Hepatic Lysosomal Enzymes of Rats Fed Diet with Conventional Rice (Control) or Transgenic Rice Line LLRICE62 (Test) ($M \pm m$, $n = 6\text{--}8$)

| Parameters | 30 Days | | 180 Days | |
|---|---|---|---|---|
| | Control | Test | Control | Test |
| *Total activity, µM/min/g tissue* | | | | |
| Arylsulfatase A, B | 2.28 ± 0.02 | 2.31 ± 0.02 | 2.11 ± 0.02 | 2.13 ± 0.02 |
| β-Galactosidase | 2.27 ± 0.07 | 2.24 ± 0.05 | 2.27 ± 0.06 | 2.31 ± 0.06 |
| β-Glucuronidase | 2.19 ± 0.04 | 2.19 ± 0.04 | 2.18 ± 0.04 | 2.24 ± 0.03 |
| *Non-sedimentable activity, % total activity* | | | | |
| Arylsulfatase A, B | 3.17 ± 0.06 | 3.12 ± 0.04 | 3.96 ± 0.09 | 3.41 ± 0.10 |
| β-Galactosidase | 5.69 ± 0.07 | 5.60 ± 0.17 | 5.47 ± 0.33 | 5.57 ± 0.27 |
| β-Glucuronidase | 5.32 ± 0.14 | 5.17 ± 0.17 | 5.96 ± 0.17 | 5.61 ± 0.18 |

There were no significant differences in the total and non-sedimentable activity of hepatic lysosomal enzymes examined on experimental Days 30 and 180 (Table 5.329).

Thus, addition of transgenic rice line LLRICE62 to the rat diet for 180 days produced no pronounced changes in the activity of the enzymes involved in xenobiotic degradation, which remained within the physiological boundaries. The absence of differences in total and non-sedimentable activity of hepatic lysosomal enzymes indicates that transgenic rice has no hepatotoxic properties.

Table 5.330 Hematological Parameters of Peripheral Blood Drawn from Rats Fed Diet with Conventional Rice (Control) or Transgenic Rice Line LLRICE62 (Test) ($M \pm m$, $n = 6$–$8$)

| Parameter | 30 Days | | 180 Days | |
|---|---|---|---|---|
| | Control | Test | Control | Test |
| Hemoglobin concentration, g/L | 154.2 ± 2.08 | 148.36 ± 2.37 | 155.2 ± 1.8 | 154.0 ± 1.6 |
| Total erythrocyte count, ×$10^{12}$/L | 6.37 ± 0.14 | 6.96 ± 0.17 | 7.0 ± 0.25 | 6.64 ± 0.33 |
| Hematocrit, vol.% | 50.57 ± 0.26 | 51.5 ± 0.17 | 50.83 ± 0.53 | 50.5 ± 0.9 |
| MCH, pg | 24.6 ± 0.62 | 22.3 ± 0.97 | 22.19 ± 1.16 | 22.7 ± 1.06 |
| MCHC, g/dl | 30.67 ± 0.44 | 28.81 ± 0.6 | 30.36 ± 0.7 | 30.25 ± 0.46 |
| MCV, μ$m^3$ | 80.38 ± 1.08 | 77.03 ± 1.2 | 72.9 ± 2.16 | 73.23 ± 2.0 |
| Total leukocyte count, ×$10^9$/L | 14.82 ± 1.19 | 14.49 ± 0.83 | 15.0 ± 0.8 | 14.78 ± 0.7 |

Table 5.331 Leukogram Parameters of Rats Fed Diet with Conventional Rice (Control) or Transgenic Rice Line LLRICE62 (Test) ($M \pm m$, $n = 6$–$8$)

| Parameter | 30 Days | | 180 Days | |
|---|---|---|---|---|
| | Control | Test | Control | Test |
| *Segmentonuclear neutrophils* | | | | |
| rel., % | 17.89 ± 1.58 | 18.5 ± 1.94 | 24.8 ± 0.7 | 24.7 ± 1.06 |
| abs., ×$10^9$/L | 2.6 ± 0.2 | 2.69 ± 0.27 | 3.76 ± 0.3 | 3.67 ± 0.3 |
| *Eosinophils* | | | | |
| rel., % | 0.67 ± 0.35 | 0.67 ± 0.35 | 1.0 ± 0.35 | 1.17 ± 0.53 |
| abs., ×$10^9$/L | 0.1 ± 0.06 | 0.08 ± 0.04 | 0.147 ± 0.048 | 0.157 ± 0.068 |
| *Lymphocytes* | | | | |
| rel., % | 81.5 ± 1.23 | 81.5 ± 1.23 | 74.17 ± 0.53 | 74.2 ± 0.5 |
| abs., ×$10^9$/L | 12.2 ± 1.15 | 13.5 ± 0.66 | 11.16 ± 0.51 | 10.95 ± 0.5 |

## Hematological Assessments

Hematological parameters of peripheral blood drawn from the rats fed diet with transgenic rice line LLRICE62 (test group) and the control rats fed diet with conventional rice were examined. Over the entire duration of the experiment, there were no significant differences in all examined hematological parameters (Table 5.330). Similarly, there were no significant differences in all leukogram parameters between the control and test rats during 180 days of the experiment (Table 5.331). All hematological parameters remained within the age-related physiological boundaries characteristic of the rats.

Thus, the chronic 180-day hematological study revealed no significant differences in any of the examined parameters between the control rats fed

Table 5.332 Microscopic Examination of Internal Organs in Rats Fed Diet with Conventional Rice (Control) or Transgenic rice line LLRICE62 (Test) (Combined Data Obtained on Experimental Days 30 and 180)

| Organ | Control | Test |
|---|---|---|
| Liver | Clear trabecular structure; no alterations in hepatocytes or portal ducts | No differences from control |
| Kidneys | Usual appearance of cortical and medullar substance; no alterations in glomeruli or pelvis epithelium | No differences from control |
| Lung | Alveolar space is air-filled; no alterations in bronchi and blood vessels; no focal changes | No differences from control |
| Spleen | Middle-size folliculi with wide marginal zones and clear reactive centers; splenic pulp is moderately plethoric | No differences from control |
| Small intestine | Preserved villous epithelium; usual infiltration in villous stroma | No differences from control |
| Testicles | Usual size and appearance of seminiferous tubules; clearly definable spermiogenesis | No differences from control |

the conventional rice and the test rats maintained on the transgenic rice LLRICE62.

## Morphological Assessments

On experimental Day 30, the post-mortem macroscopic examination of internal organs revealed minor bronchial pneumonia in 1 control rat. A similar inspection performed on Day 180 observed a slight atrophy of the testicles in 1 control rat. The histological examinations performed on experimental Days 30 and 180 detected no significant differences between groups in the internal organs of the rats (Table 5.332).

Therefore, the chronic toxicological experiment carried out during 180 days with biochemical, hematological, and morphological examinations revealed no adverse affects of transgenic rice line LLRICE62 on the animals.

## Assessment of Potential Impact of Transgenic Rice Line LLRICE62 on Immune System in Studies on Mice
### Potential Effect on Humoral Component of Immune System

The immunomodulating effect of transgenic rice line LLRICE62 on the humoral component of the immune system was examined on two oppositely reacting mice lines, C57Bl/6 and CBA, by determining the level of hemagglutination to sheep erythrocytes (SE). During 21 days, the control and test mice were fed diet with 2.4 g/day/mouse milled rice derived respectively from conventional rice or the transgenic rice line LLRICE62. The experimental conditions are described in section 5.1.1.

In control and test groups of CBA mice (high-sensitivity animals), antibodies appeared at the titer of 1:83–1:213 on post-immunization Day 7 and remained at this level to Day 21. In both groups of C57Bl/6 mice, antibodies appeared on post-immunization Day 7 at the titer of 1:170–1:192, which insignificantly decreased to 1:117–1:149 on Day 21.

Therefore, the dynamics of antibody synthesis did not differ between test and control groups of either CBA or C57Bl/6 mice.

### Potential Effect on Cellular Component of Immune System

The immunomodulating effect of transgenic rice line LLRICE62 was assessed with delayed hypersensitivity reaction to SE. The experimental conditions are described in section 5.1.1. The dynamics of antibody synthesis did not differ between test and groups of either CBA or C57Bl/6 mice. There were no significant differences in all examined parameters: in the test CBA mice, RI = 82 ± 4, while in the control CBA mice RI = 95 ± 5; in the test C57Bl/6 mice RI = 58 ± 6, while in the control C57Bl/6 mice RI = 73 ± 10. Thus, hypersensitivity reaction to SE showed that the transgenic rice line LLRICE62 produced no effect on the cellular component of the immune system.

### Assessments of Potential Sensibilization Effect of Transgenic Rice

Examination of possible sensitizing action of transgenic rice line LLRICE62 on the immune response to endogenous metabolic products was carried out on mice by testing sensitivity to histamine. Transgenic or conventional rice was added to the diets of, correspondingly, test or control mice for 21 days. Thereafter the mice of both groups were injected intraperitoneally with 2.5 mg histamine hydrochloride dissolved in 0.5 mL physiological solution. The reaction was assessed at 1 h and 24 h by the level of mortality. In this test, there was no mortality nor any differences in behavior between test and control mice, which attest to the absence of any sensitization agent in transgenic rice line LLRICE62.

### Potential Effect of Transgenic Rice on Susceptibility of Mice to *Salmonella typhimurium*

The effect of transgenic rice on susceptibility of mice to infection by salmonella of murine typhus was examined in experiments on C57Bl/6 mice. Mice were injected intraperitoneally with various doses of Strain 415 *Salmonella typhimurium*. During 21 days, the diets of test and control mice were supplemented with transgenic or conventional rice, respectively. The mice of both groups were infected with Strain 415 *Salmonella typhimurium* in doses ranging from $10^2$ to $10^5$ microbial cells per mouse, which differed on a 10-fold basis.

The post-injection observation period was 21 days. The following data were obtained:

- In both groups, the mortality began on post-injection week 1, and all the mice died by post-injection Day 18;
- The mean lifetimes of test and control mice were 9.7 and 9.5 days, respectively;
- $LD_{50}$ values in control and test groups were 72 and 125 microbial cells, correspondingly.

These data showed that *Salmonella typhimurium* produced a typical infection both in control and test mice, and the development of the disease was similar in both groups. Thus, transgenic rice line LLRICE62 did not modify resistance of mice against salmonella of murine typhus.

Taken together the data confirm that transgenic rice line LLRICE62 tolerant to glufosinate ammonium produced no effect on the development of the humoral or cellular immune system. Moreover, it demonstrated no sensitizing potencies and did not affect the resistance of mice to *S. typhimurium*.

### Assessment of Potential Impact of Transgenic Rice Line LLRICE62 on Immune System in Studies on Rats

The study was carried out on male Wistar rats ($n = 48$) weighing initially $180 \pm 10$ g. After a 7-day adaptation period to the standard vivarium diet, the rats for the following 28 days were fed isocaloric diet supplemented with conventional rice (control group) or transgenic rice line LLRICE62 (test group) in the amount of 3.5 g/rat. Both types of milled rice were dissolved in warm boiled water to the consistency of dense curd and supplemented with sunflower-seed oil to improve intake. The base composition of the diet is shown in Table 5.11.

The model of generalized anaphylaxis was developed according to [8] as described in section 5.1.1.

During the entire experiment, the rats of both groups grew normally, which indicates nutritional adequacy of both diets. On experimental Day 29, the body weights of test and control rats were $298 \pm 7$ and $315 \pm 6$ g, respectively ($p > 0.1$).

Table 5.333 shows data on severity of the anaphylactic reaction in control and test rats. By all examined parameters (mortality, severe forms of anaphylactic reaction, and anaphylactic index), there were no differences in the severity of the anaphylactic reactions ($p > 0.1$) between the test and control groups.

Table 5.334 shows the mean values of $D_{492}$, the concentration of antibodies, and common logarithm of this concentration in the control and test groups.

**Table 5.333** Severity of Active Anaphylactic Shock in Rats Fed Diet with Conventional Rice (Control) or Transgenic Rice Line LLRICE62 (Test)

| Group of Rats | AI | Severe Reactions, % | Mortality, % |
|---|---|---|---|
| Control (n = 23) | 3.04 | 56.0 | 56.5 |
| Test (n = 25) | 2.80 | 52.0 | 48.0 |
| $p_{1/2}$ | >0.1 | >0.1 | >0.1 |

**Table 5.334** Parameters of Humoral Immune Response (Level of Specific IgG Antibodies Raised Against Ovalbumin) in Rats Fed Diet with Conventional Rice (Control) or Transgenic Rice Line LLRICE62 (Test) ($M \pm m$)

| Group of Rats | $D_{492}$ | Concentration of Antibodies, mg/mL | Logarithm of Antibody Concentration |
|---|---|---|---|
| Control (n = 23) | 1.006 ± 0.035 | 6.9 ± 0.8 | 0.779 ± 0.046 |
| Test (n = 25) | 0.936 ± 0.041 | 5.7 ± 0.7 | 0.687 ± 0.053 |
| $p_{1/2}$ (Student's t-test) | >0.1 | >0.1 | >0.1 |
| $p_{1/2}$ (Mann-Whitney U-test) | >0.1 | >0.1 | >0.1 |

These data showed that the degree of ovalbumin-induced sensitization of test rats was not greater than that in the control group rats.

This analysis showed that the transgenic rice produced no changes in allergic reactivity and sensitization to a model allergen in comparison with the control rats fed diet with conventional rice.

### Assessment of Potential Genotoxicity of Rice Line LLRICE62

The study of mutagenic action of transgenic (or control) rice in the chronic dietary experiment examined the chromosomal aberrations in somatic cells (bone marrow cells) and the dominant lethal mutations in gametes. The experimental conditions are described in section 5.1.1. In each animal, approximately 70 cells were analyzed at the metaphasic stage of nuclear division taken from 2-month male C57Bl/6 mice weighing 20–25 g. Genetic alterations in the sex cells were examined by revealing the dominant lethal mutations in C57Bl/6 mice (section 5.1.1).

In the study, the progeny was produced from the mice fed the transgenic rice (males for 45 days and females during the entire gestation period). The

### 5.3.1 Rice Line LLRICE62 Tolerant to Glufosinate Ammonium

**Table 5.335** Bone Marrow Cytogenetic Parameters in Mice Fed Diet with Conventional Rice (Control) or Transgenic Rice Line LLRICE62 (Test)

| Group | Number of Analyzed Metaphases | Share of Cells, % | | |
|---|---|---|---|---|
| | | with Chromosomal Aberrations | with Gaps | with Polyploid Chromosome Set |
| Control, C57Bl/6 (n = 5) | 342 | 0.58 ± 0.41 | 0.87 ± 0.50 | 1.16 ± 0.57 |
| Test, C57Bl/6 (n = 5) | 350 | 0.57 ± 0.40 | 1.14 ± 0.56 | 0.85 ± 0.49 |
| *First-generation mice* | | | | |
| Control (n = 5) | 302 | – | 0.66 ± 0.46 | 0.66 ± 0.43 |
| Test (n = 5) | 321 | – | 0.93 ± 0.53 | 0.62 ± 0.43 |

Note: The differences are insignificant ($p > 0.05$).

transgenic rice was fed to female mice during the lactation period and then to the weanlings for 1 month. In the first generation of mice, the bone marrow was isolated from both femoral bones for cytogenetic examination (section 5.1.1).

In the basic generation of the control and test rats, the study revealed only chromosomal abnormalities (single segments and gaps), which can appear spontaneously. These abnormalities are unstable, and they are usually eliminated in subsequent nuclear divisions. In the first generation of test mice, 3 cells demonstrated gaps, and 2 cells had a polyploid chromosome set. Similarly, 2 cells of the control mice progeny had gaps, and another 2 cells had a polyploid chromosome set.

Thus, long-term feeding of both the male parental mice and their first generation with the transgenic rice produced no chromosomal aberrations (Table 5.335).

To examine the dominant lethal mutations in gametes, the test female mice ($n = 64$) were dissected to count and analyze 414 embryos and 451 corpus lutea. In control female mice ($n = 61$), a similar analysis was performed with 402 embryos and 435 corpus lutea. The pre-implantation mortality in the test group (the share of the dead zygotes, embryos, and unfertilized ovules) did not surpass the control value. Similarly, in the test group examined at the stages of the early and late spermatids or mature spermatozoa, the post-implantation embryonic mortality (the most reliable index of mutagenic activity) was low and did not surpass the control value (Table 5.336).

At these stages, there was no induced mortality, which attests to the absence of adverse effects of the transgenic rice on spermiogenesis in mice. The data obtained support the conclusion that transgenic rice line LLRICE62 possesses no mutagenic properties under the experimental conditions described.

Table 5.336 Dominant Lethal Mutations in Sex Cells of Mice Fed Diet with Conventional Rice (Control) or Transgenic Rice Line LLRICE62 (Test)

| Parameter, % | Week 1, Mature Spermatozoa | | Week 2, Late Spermatids | | Week 3, Early Spermatids | |
|---|---|---|---|---|---|---|
| | Control | Test | Control | Test | Control | Test |
| Pre-implantation mortality | 9.74 | 11.25 | 6.62 | 5.55 | 6.15 | 7.48 |
| Post-implantation mortality | 2.87 | 2.11 | 3.54 | 2.20 | 4.91 | 3.67 |
| Survival rate | 87.66 | 86.87 | 90.00 | 97.79 | 89.23 | 89.11 |
| Induced mortality | – | 0 | – | 0 | – | 0 |

## Assessment of Technological Parameters

The assessment of technological parameters was carried out in Moscow State University of Applied Biotechnology (Ministry of Science and Education of the Russian Federation). To characterize the grain of transgenic rice line LLRICE62 tolerant to glufosinate ammonium in comparison with the non-transgenic conventional rice, the moisture and ash contents were determined. In addition, rice starch was produced under laboratory conditions to determine protein mass fraction in the starch, gelatinization temperature and viscosity of the starch gelatins on amylograph, the parameters of thermoplastic extrusion, and the structural and mechanical properties of the extrudates. This study resulted in the following conclusions.

- No differences or difficulties were observed in the technological process when producing starch from transgenic rice LLRICE62 or conventional rice.
- Both control and test starch samples met the requirements of Russian National Standards GOST 10-04-08- 33-41-89 "Rice Starch" and had similar parameters.
- The gelatinization temperature of the starches derived from transgenic rice line LLRICE62 corresponded to the control value of the conventional rice.
- Viscosity of 5% paste of both samples exposed for 15 min to 95°C followed by cooling to 60°C was identical.
- Insignificant differences were revealed in the gel-forming properties of the starches derived from conventional and transgenic rice samples, caused by different degrees of grain ripening.
- The structural and mechanical properties of the extrusive products derived from transgenic rice line LLRICE62 were practically identical to those obtained from the conventional rice.

## Conclusions

By all examined parameters, the data of the complex safety assessment of transgenic rice line LLRICE62 tolerant to glufosinate ammonium attest to the absence of any toxic, genotoxic, immunomodulating, or allergenic effects of this rice line. By biochemical composition, the transgenic rice line LLRICE62 was identical to conventional rice.

Based on results of these studies, the Ministry of Health of the Russian Federation (Department of State Sanitation and Epidemiological Inspectorate) granted the Registration Certificate, which allows the transgenic rice line LLRICE62 to be used in the food industry and placed on the market without restrictions.

# Subchapter 5.4

# Potato

## 5.4.1 POTATO VARIETY SUPERIOR NEWLEAF RESISTANT TO DAMAGE CAUSED BY COLORADO POTATO BEETLE

### Molecular Characterization of Superior NewLeaf Potato Resistant to Damage Caused by Colorado Potato Beetle

*Recipient Organism*

Potato is characterized by high consumer qualities, as a table variety that can also be processed.

### *Donor Organism*

The donor of the *cryIIIA* gene, *Bacillus thuringiensis* subsp. *Tenebrionis*, is a widely spread gram-positive soil bacterium which, during sporulation, produces the proteins that selectively act on a limited group of insects, Colorado potato beetle included. These proteins bind with specific sites in the cells of the digestive system of the insects, where they form ion-selective channels in the plasmalemma, resulting in swelling and lysis of the cells, pronounced disturbance of digestion, and death of the insects [25].

### *Method of Genetic Transformation*

To incorporate the genetic construct into the plant genome, agrobacterial transformation was performed with plasmid vector PV-STBT02 [19]. The genome of transgenic potato resistant to damage caused by Colorado potato beetle contains the *cryIIIA* gene from *Bacillus thuringiensis* subsp. *Tenebrionis* (*B.t.t.*), which encodes synthesis of insecticidal protein responsible for resistance against Colorado potato beetle, and the *nptII* gene from the Tn5 transposon of *E. coli* encoding neomycin phosphotransferase II.

The genome of potato Superior NewLeaf variety resistant against Colorado potato beetle contains the following genetic elements:

- *cryIIIA* coding sequence responsible for resistance against Colorado potato beetle;
- P-35S promoter of cauliflower mosaic virus (CaMV);
- *npt II* marker gene encoding neomycin phosphotransferase from transposon Tn5 of *E. coli*, responsible for tolerance to antibiotics;
- NOS 3' terminator of *nopaline* synthase gene from *Agrobacterium tumefaciens* plasmid;
- *aad* gene of streptomycin adenyltransferase (selective marker).

## Global Registration Status of Transgenic Potato Superior NewLeaf

Table 5.337 shows the countries that had granted registration to use transgenic potato resistant against Colorado potato beetle at the time of registration in Russia [19].

## Safety Assessment of Transgenic Potato Resistant Against Colorado Potato Beetle Conducted in the Russian Federation

Studies were conducted in accordance with the requirements of the Ministry of Health of the Russian Federation authorized for risk and safety assessment of food derived from GM sources [8]. PCR analysis of the potato test and control samples was performed to confirm the identity of the transformation event and its absence in the conventional control line.

**Table 5.337** Registration Status of Superior NewLeaf Potato, Resistant against Colorado Beetle, in Various Countries

| Country | Date of Approval | Scope |
|---|---|---|
| Canada | 1996 | Food |
|  | 1997 | Feed, environmental release |
| USA | 1996 | Food, feed, environmental release |
| Japan | 1997 | Food |

*Note: The current registration status of the transgenic crop is on http://www.biotradestatus.com/*

### 5.4.1 Potato Variety Superior Newleaf Resistant to Damage Caused

**Table 5.338** Protein Content (%) and Amino Acid Composition (g/100 g protein) in Potato

| Ingredient | Conventional Potato | Superior NewLeaf Potato |
|---|---|---|
| Protein | 2.5 | 2.3 |
| *Amino acids* | | |
| Aspartic acid | 14.5 | 13.9 |
| Threonine | 4.56 | 4.32 |
| Serine | 4.09 | 3.88 |
| Glutamic acid | 16.0 | 15.72 |
| Proline | 4.04 | 4.72 |
| Cysteine | 0.35 | 0.32 |
| Glycine | 4.48 | 4.36 |
| Alanine | 4.96 | 4.68 |
| Valine | 6.60 | 6.39 |
| Methionine | 1.40 | 1.17 |
| Isoleucine | 5.26 | 5.12 |
| Leucine | 8.17 | 9.16 |
| Tyrosine | 3.13 | 3.52 |
| Phenylalanine | 5.43 | 5.64 |
| Histidine | 2.17 | 2.04 |
| Lysine | 7.61 | 7.64 |
| Arginine | 5.26 | 4.88 |

#### Biochemical Composition of Superior NewLeaf Potato

The content of proteins in transgenic potato tubers did not significantly differ from that in conventional potato tubers and remained within the range of variations characteristic for potato (0.69–4.63%) [14]. The content and composition of amino acids were similar in both potato varieties (Table 5.338).

The content of carbohydrates remained within the variations characteristic of potato: glucose 0.10–1.30 g/100 g; sucrose 0.10–0.30 g/100 g, and fructose 0.10–1.30 g/100 g (data of long-term studies carried out in the State Research Institute of Nutrition, RAMS). The content of starch, the most important carbohydrate in potato, also varied within the range characteristic for potato: 8.0–29.4 g/100 g [14] (Table 5.339).

Composition of fatty acids in potato tubers was almost identical in both varieties. The differences in the content of lauric, myristic, palmitoleic, stearic, and linolenic acids remained within the range of intraspecific physiological variations characteristic of potato: 0.05–3.00, 0.3–3.0, 0.5–2.0, 5.0–15.0, and 15.2–32.0% of total content of fatty acids, respectively (data of long-term studies carried out in the State Research Institute of Nutrition, RAMS; Table 5.340).

**Table 5.339** Carbohydrate Content (%) in Potato

| Carbohydrate | Conventional Potato | Superior NewLeaf Potato |
|---|---|---|
| Starch | 12.4 | 11.8 |
| Cellulose | 0.73 | 0.69 |
| Fructose | 0.36 | 0.18 |
| Glucose | 0.28 | 0.13 |
| Sucrose | 0.30 | 0.29 |

**Table 5.340** Lipids (%) and Fatty Acids (Rel. %) Content of in Potato

| Ingredient | Conventional Potato | Superior NewLeaf Potato |
|---|---|---|
| Lipids | 0.17 | 0.18 |
| *Fatty acids* | | |
| Lauric 12:0 | 0.7 | 2.6 |
| Myristic 14:0 | 0.9 | 2.7 |
| Palmitic 16:0 | 19.8 | 21.4 |
| Palmitoleic 16:1 | 0.8 | 1.8 |
| Stearic 18:0 | 7.3 | 12.3 |
| Oleic 18:1 | 3.2 | 4.3 |
| Linoleic 18:2 | 42.6 | 38.3 |
| Linolenic 18:3 | 23.9 | 15.2 |
| Arachidic 20:0 | 0.2 | 0.2 |
| Eicosenoic 20:1 | 1.2 | 1.2 |

**Table 5.341** Vitamin Content (mg/100 g product) in Potato

| Ingredient | Conventional Potato | Superior NewLeaf Potato |
|---|---|---|
| Vitamin C | 23.1 | 19.4 |
| Vitamin $B_1$ | 0.088 | 0.089 |
| Vitamin $B_2$ | 0.013 | 0.014 |
| Vitamin $B_6$ | 0.162 | 0.156 |
| β-Carotene | 0.010 | 0.003 |
| Vitamin E (α-tocopherol) | 0.017 | Not detected |

The contents of vitamins in test (transgenic) and control (conventional) potato tubers were almost identical except for the values for vitamin C and β-carotene, which nevertheless remained within the intraspecific range characteristic of potato: 1.0–54.0 and 0.001–0.08 mg/100 g, correspondingly (Table 5.341).

The content of potassium, the most important microelement in potato, varied in the test and control tubers within the range 2294–9400 mg/kg [14].

### 5.4.1 Potato Variety Superior Newleaf Resistant to Damage Caused

**Table 5.342** Mineral Composition (mg/kg) of Potato

| Ingredient | Conventional Potato | Superior NewLeaf Potato |
|---|---|---|
| Iron | 6.40 | 5.96 |
| Sodium | 33.4 | 31.3 |
| Potassium | 4052 | 4203 |
| Calcium | 10.4 | 11.5 |
| Magnesium | 260 | 260 |
| Copper | 0.436 | 0.327 |
| Zinc | 2.63 | 2.23 |

**Table 5.343** Toxic Elements (mg/kg) in Transgenic and Conventional Potato

| Parameter | Conventional Potato | Superior NewLeaf Potato |
|---|---|---|
| Nitrates | 74.5 | 41.6 |
| Nitrites | Not detected | Not detected |
| Patulin | Not detected | Not detected |
| Solanines | 1.0 | 1.1 |
| Chloro-organic pesticides | Not detected | Not detected |
| Lead | 0.035 | 0.023 |
| Cadmium | 0.006 | 0.006 |

There were no significant differences between the two varieties in the content of calcium, iron, or magnesium in the tubers (Table 5.342).

The values of sanitary-chemical parameters for the transgenic and conventional potato tubers did not surpass the limits established in the Sanitary Code of the Russian Federation (Table 5.343).

Thus, the tubers of transgenic and conventional potato did not significantly differ by biochemical composition. The safety parameters of both potato varieties did not surpass the limits established in the regulations valid in Russian Federation [5].

#### Toxicological Assessment of Transgenic Potato

The experiments was carried out on male Wistar rats with an initial body weight of 80–100 g. The test rats were fed diet with boiled transgenic Superior NewLeaf potato resistant to Colorado potato beetle (12 g/day/rat). The control rats were fed diet with an equivalent amount of conventional potato (Table 5.344). The diet was divided into two parts. In the morning, the rats were given the transgenic or conventional potato. The potato intake was determined at 6 h; thereafter the animals received other components of the diet.

**Table 5.344** Daily Rat Diet with Boiled Potato

| Ingredient | Weight, g |
|---|---|
| Potato | 23.2 |
| Grain mix | 38.0 |
| Curd | 4.61 |
| Fish flour | 1.15 |
| Meat of second grade | 9.23 |
| Carrot | 11.5 |
| The greens | 11.5 |
| Cod-liver oil | 0.23 |
| Yeast | 0.23 |
| NaCl | 0.35 |
| Total | 100 |

**Table 5.345** Daily Potato Intake (g/day/rat) in Rats Fed Diet of Conventional Potato (Control) or Transgenic Potato (Test) ($M \pm m$; $n = 10$)

| Duration of Experiment, Weeks | Control | Test |
|---|---|---|
| 1 | 6.1 ± 0.2 | 8.3 ± 0.9* |
| 2 | 6.3 ± 0.2 | 8.8 ± 0.2* |
| 3 | 7.7 ± 0.6 | 8.5 ± 0.2 |
| 4 | 7.1 ± 0.2 | 7.0 ± 0.1 |
| 8 | 8.3 ± 0.8 | 9.0 ± 0.8 |
| 12 | 9.7 ± 0.3 | 9.7 ± 0.4 |
| 16 | 11.5 ± 0.9 | 11.3 ± 1.1 |
| 20 | 11.2 ± 0.5 | 10.7 ± 0.5 |

*$p < 0.05$ in comparison with control.

Throughout the entire duration of the experiment, the general condition of the rats was similar in the control and test groups. No animal mortality was observed during the study in either group. Daily potato intake (per rat) was approximately equal in both groups, although during the first 2 weeks, the intake was significantly higher in the test group (Table 5.345).

Throughout the entire duration of the experiment, there were no differences between the groups in absolute body weight of the rats (Figure 5.16; Table 5.346).

The absolute weights of internal organs did not significantly differ between control and test rats except for a significantly lower absolute weight of the heart on Day 30 and higher absolute weight of kidney on experimental

### 5.4.1 Potato Variety Superior Newleaf Resistant to Damage Caused

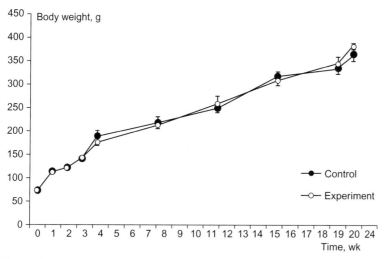

**FIGURE 5.16**
Comparative dynamics of body weight of rats fed diet with transgenic (test) or conventional (control) potato.

**Table 5.346** Comparative Body Weight (g) of Rats Fed Diet of Conventional Potato (Control) or Transgenic Potato (Test) ($M \pm m$; $n = 10$)

| Duration of Experiment, Weeks | Control | Test |
|---|---|---|
| 0 | 72.5 ± 2.8 | 72.7 ± 1.6 |
| 1 | 115.5 ± 4.4 | 113.0 ± 3.2 |
| 2 | 123.5 ± 5.1 | 123.3 ± 1.3 |
| 3 | 144.5 ± 3.7 | 146.0 ± 3.7 |
| 4 | 195.0 ± 11.0 | 181.6 ± 8.3 |
| 8 | 224.9 ± 12.5 | 220.7 ± 9.1 |
| 12 | 258.7 ± 9.5 | 267.9 ± 15.9 |
| 16 | 329.5 ± 11.0 | 320.0 ± 11.8 |
| 20 | 348.7 ± 14.1 | 359.0 ± 14.2 |
| 24 | 379.2 ± 15.7 | 396.7 ± 6.9 |

Note: The differences are not significant ($p > 0.05$).

Day 180 in the test rats (Table 5.347). However, these differences remained within the physiological limits for rats.

### Assessment of Biochemical Parameters

Biochemical examination of blood serum and urine performed on experimental Days 30 and 180 revealed no differences between the test and control group rats (Tables 5.348 and 5.349).

**Table 5.347** Absolute and Relative Weight of Internal Organs of Rats Fed Diet with Conventional (Control) or Transgenic (Test) Potato ($M \pm m$, $n = 6$–$8$)

| Organ | | 30 Days | | 180 Days | |
|---|---|---|---|---|---|
| | | Control | Test | Control | Test |
| Kidneys | Abs.[a], g | 1.43 ± 0.06 | 1.37 ± 0.09 | 2.27 ± 0.13 | 2.70 ± 0.12* |
| | Rel.[b], g/100g | 0.73 ± 0.05 | 0.75 ± 0.02 | 0.60 ± 0.02 | 0.68 ± 0.03 |
| Liver | Abs., g | 7.69 ± 0.23 | 7.34 ± 0.55 | 10.83 ± 0.69 | 11.57 ± 0.20 |
| | Rel., g/100g | 3.94 ± 0.27 | 4.03 ± 0.19 | 2.86 ± 0.16 | 2.92 ± 0.05 |
| Spleen | Abs., g | 1.27 ± 0.08 | 1.15 ± 0.11 | 1.40 ± 0.08 | 1.48 ± 0.07 |
| | Rel., g/100g | 0.65 ± 0.06 | 0.63 ± 0.16 | 0.37 ± 0.02 | 0.37 ± 0.02 |
| Heart | Abs., g | 0.82 ± 0.03 | 0.73 ± 0.03* | 1.15 ± 0.07 | 1.25 ± 0.04 |
| | Rel., g/100g | 0.42 ± 0.025 | 0.40 ± 0.019 | 0.30 ± 0.01 | 0.32 ± 0.02 |
| Testicles | Abs., g | 2.68 ± 0.08 | 2.58 ± 0.08 | 4.89 ± 1.69 | 3.59 ± 0.11 |
| | Rel., g/100g | 1.36 ± 0.065 | 1.43 ± 0.071 | 1.34 ± 0.50 | 0.91 ± 0.03 |
| Hypophysis | Abs., mg | 5.75 ± 0.44 | 4.33 ± 0.74 | 8.4 ± 2.36 | 11.6 ± 2.64 |
| | Rel., mg/100g | 2.91 ± 0.22 | 2.38 ± 0.33 | 2.18 ± 0.61 | 2.90 ± 0.67 |
| Adrenal glands | Abs., mg | 24.50 ± 2.43 | 23.00 ± 1.73 | 26.83 ± 2.79 | 33.5 ± 4.03 |
| | Rel., mg/100g | 12.73 ± 1.70 | 12.91 ± 1.35 | 7.16 ± 0.77 | 8.39 ± 0.91 |
| Seminal vesicles | Abs., mg | 299.7 ± 27.9 | 257.8 ± 35.9 | 635.0 ± 51.56 | 580.0 ± 62.72 |
| | Rel., mg/100g | 150.04 ± 10.33 | 141.52 ± 19.29 | 167.6 ± 12.51 | 146.8 ± 17.19 |
| Prostate | Abs., mg | 135.2 ± 17.6 | 131.2 ± 18.4 | 455.0 ± 71.65 | 483.3 ± 44.2 |
| | Rel., mg/100g | 65.68 ± 11.05 | 70.95 ± 7.37 | 121.26 ± 21.42 | 122.14 ± 15.45 |

[a]Absolute weight of internal organs.
[b]Relative weight of internal organs (per 100g body weight).
*$p < 0.05$ in comparison with control.

**Table 5.348** Biochemical Parameters of Blood Serum of Rats Fed Diet with Conventional (Control) or Transgenic (Test) Potato ($M \pm m$, $n = 6$–$8$)

| Parameter | 30 Days | | 180 Days | |
|---|---|---|---|---|
| | Control | Test | Control | Test |
| Total protein, g/L | 67.7 ± 3.8 | 63.6 ± 4.1 | 92.5 ± 3.2 | 80.9 ± 1.9 |
| Glucose, mM/L | 6.8 ± 0.19 | 6.7 ± 0.48 | 7.6 ± 0.7 | 8.1 ± 0.9 |
| Alanine aminotransferase, μM/min/L | 66.7 ± 4.5 | 73.2 ± 7.3 | 71.8 ± 6.6 | 69.0 ± 4.5 |
| Aspartate aminotransferase, μM/min/L | 69.6 ± 2.9 | 64.3 ± 3.1 | 57.3 ± 4.5 | 66.7 ± 4.0 |
| Alkaline phosphatase, U/L | 438.0 ± 32.1 | 469.9 ± 40.0 | 219.9 ± 33.7 | 222.2 ± 31.3 |

### Assessment of Sensitive Biomarkers

Assessment of activity of the enzymes involved in phases I and II xenobiotic degradation in hepatic microsomes and analysis of the total and nonsedimentable activities of hepatic lysosomal enzymes revealed no differences between the control and test rats (Tables 5.350 and 5.351).

### 5.4.1 Potato Variety Superior Newleaf Resistant to Damage Caused

**Table 5.349** Urinary Biochemical Parameters of Rats Fed Diet with Conventional (Control) or Transgenic (Test) Potato ($M \pm m$, $n = 6{-}8$)

| Parameter | 30 Days | | 180 Days | |
|---|---|---|---|---|
| | Control | Test | Control | Test |
| pH | 6.1 | 6.1 | 7.0 | 7.0 |
| Daily diuresis, mL | 8.8 ± 1.0 | 10.0 ± 2.0 | 9.3 ± 0.5 | 9.0 ± 1.0 |
| Relative density, g/mL | 0.98 ± 0.01 | 0.99 ± 0.01 | 1.00 ± 0.00 | 1.00 ± 0.01 |
| Creatinine, mg/mL | 0.62 ± 0.05 | 0.61 ± 0.11 | 1.01 ± 0.04 | 1.19 ± 0.11 |
| Creatinine, mg/day | 5.24 ± 0.55 | 5.53 ± 0.59 | 9.65 ± 0.40 | 10.20 ± 0.74 |

**Table 5.350** Activity of Enzymes Involved in Metabolism of Xenobiotics in Liver Microsomes in Rats Fed Diet with Conventional (Control) or Transgenic (Test) Potato ($M \pm m$, $n = 6{-}8$)

| Parameter | 30 Days | | 180 Days | |
|---|---|---|---|---|
| | Control | Test | Control | Test |
| Cytochrome P450, nM/mg | 0.79 ± 0.09 | 0.90 ± 0.06 | 0.93 ± 0.02 | 0.89 ± 0.09 |
| Cytochrome $b_5$, nM/mg | 0.78 ± 0.07 | 0.81 ± 0.04 | 0.82 ± 0.02 | 0.77 ± 0.06 |
| 7-ethoxycoumarin-O-deethylation, nM/min/mg | 2.60 ± 0.36 | 2.44 ± 0.23 | 2.0 ± 0.14 | 2.38 ± 0.21 |
| Aminopyrine N-demethylation, nM/min/mg | 9.48 ± 0.55 | 9.56 ± 0.45 | 10.37 ± 0.42 | 10.4 ± 0.27 |
| Acetylesterase, µM/min/mg | 6.63 ± 0.21 | 7.01 ± 0.18 | 11.08 ± 0.36 | 10.15 ± 0.21 |
| Benzpyrene hydroxylase, Fl-units/min/mg | 9.24 ± 0.44 | 9.95 ± 0.75 | 8.12 ± 0.52 | 8.46 ± 1.52 |
| Epoxide hydrolase, nM/min/mg | 5.48 ± 0.38 | 5.36 ± 0.25 | 5.03 ± 0.22 | 4.50 ± 0.39 |
| UDP-Glucuronosil transferase, nM/min/mg | 32.0 ± 2.0 | 34.8 ± 2.2 | 21.9 ± 1.1 | 20.6 ± 0.9 |

**Table 5.351** Total and Non-sedimentable Activity of Hepatic Lysosomal Enzymes of Rats Fed Diet with Conventional (Control) or Transgenic (Test) Potato ($M \pm m$, $n = 6{-}8$)

| Parameters | 30 Days | | 180 Days | |
|---|---|---|---|---|
| | Control | Test | Control | Test |
| *Total activity, µM/min/g tissue* | | | | |
| Arylsulfatase A, B | 2.29 ± 0.04 | 2.28 ± 0.03 | 2.38 ± 0.07 | 2.39 ± 0.07 |
| β-Galactosidase | 2.78 ± 0.07 | 2.76 ± 0.04 | 2.43 ± 0.08 | 2.37 ± 0.06 |
| β-Glucuronidase | 2.13 ± 0.06 | 2.18 ± 0.06 | 2.10 ± 0.02 | 2.06 ± 0.04 |
| *Non-sedimentable activity, % total activity* | | | | |
| Arylsulfatase A, B | 2.72 ± 0.05 | 2.57 ± 0.09 | 3.38 ± 0.15 | 3.43 ± 0.10 |
| β-Galactosidase | 4.92 ± 0.20 | 4.92 ± 0.27 | 6.44 ± 0.22 | 6.43 ± 0.42 |
| β-Glucuronidase | 4.90 ± 0.16 | 5.00 ± 0.14 | 5.04 ± 0.25 | 5.01 ± 0.23 |

**Table 5.352** Hematological Parameters of Peripheral Blood Drawn from Rats Fed Diet with Conventional (Control) or Transgenic (Test) Potato ($M \pm m$, $n = 6\text{--}8$)

| Parameter | 30 Days | | 180 Days | |
|---|---|---|---|---|
| | Control | Test | Control | Test |
| Hemoglobin concentration, g/L | 121.8 ± 6.3 | 109.6 ± 5.8 | 151.12 ± 1.99 | 158.72 ± 2.06 |
| Total erythrocyte count, ×10$^{12}$/L | 5.54 ± 0.18 | 5.67 ± 0.16 | 6.11 ± 0.10 | 6.22 ± 0.16 |
| Hematocrit, vol.% | 48.30 ± 0.52 | 48.80 ± 0.52 | 50.0 ± 0.00 | 50.60 ± 0.38 |
| MCH, pg | 21.96 ± 1.05 | 19.20 ± 0.90 | 24.75 ± 0.76 | 25.56 ± 0.58 |
| MCHC, g/dl | 25.15 ± 1.14 | 22.4 ± 1.18 | 30.22 ± 0.47 | 31.38 ± 0.56 |
| MCV, μm$^3$ | 87.4 ± 1.9 | 86.2 ± 1.69 | 81.81 ± 1.39 | 81.46 ± 1.41 |
| Total leukocyte count, ×10$^9$/L | 11.4 ± 1.34 | 9.6 ± 1.36 | 15.60 ± 1.74 | 13.94 ± 0.81 |

**Table 5.353** Leukogram Parameters of Rats Fed Diet with Conventional (Control) or Transgenic (Test) Potato ($M \pm m$, $n = 6\text{--}8$)

| Parameter | 30 Days | | 180 Days | |
|---|---|---|---|---|
| | Control | Test | Control | Test |
| *Segmentonuclear neutrophils* | | | | |
| rel., % | 19.10 ± 0.35 | 19.0 ± 0.35 | 10.8 ± 1.15 | 16.4 ± 1.5 |
| abs., ×10$^9$/L | 2.20 ± 0.26 | 1.85 ± 0.09 | 1.74 ± 0.77 | 2.31 ± 0.35 |
| *Eosinophils* | | | | |
| rel., % | 0.67 ± 0.35 | 0.50 ± 0.35 | 0.80 ± 0.19 | 0.80 ± 0.38 |
| abs., ×10$^9$/L | 0.09 ± 0.05 | 0.05 ± 0.04 | 0.13 ± 0.036 | 0.102 ± 0.044 |
| *Lymphocytes* | | | | |
| rel., % | 80.1 ± 0.52 | 80.1 ± 0.17 | 88.4 ± 0.77 | 82.8 ± 1.7 |
| abs., ×10$^9$/L | 9.10 ± 1.01 | 7.72 ± 0.17 | 13.79 ± 1.44 | 11.57 ± 0.78 |

## Hematological Assessments

Hematological studies revealed no significant differences between the control and test rats (Tables 5.352 and 5.353).

## Morphological Assessments

The post-mortem macroscopic examination performed on experimental Days 30 and 180 revealed no pathological alterations in the internal organs that could be related to the effect of transgenic potato. A similar conclusion was obtained from the histological examinations of the internal organs of the control and test rats (Table 5.354).

Therefore, the chronic toxicological experiment carried out during 180 days with biochemical, hematological, and morphological examinations revealed no adverse affects of the transgenic potato fed in increased amounts to Wistar rats.

**Table 5.354** Microscopic Examination of Internal Organs in Rats Fed Diet with Conventional (Control) or Transgenic (Test) Potato

| Organ | Control | Test |
|---|---|---|
| Liver | Clear trabecular structure; no alterations in hepatocytes or portal ducts | No differences from control |
| Kidneys | Usual appearance of cortical and medullar substance; no alterations in glomeruli or pelvis epithelium | No differences from control |
| Lung | Alveolar space is air-filled; no alterations in bronchi or blood vessels | No differences from control |
| Spleen | Middle-size or large folliculi with clear wide marginal zones and reactive centers | No differences from control |
| Small intestine | Preserved villous epithelium; usual infiltration in villous stroma | No differences from control |
| Testicle | Usual size and appearance of seminiferous tubules; clearly definable spermiogenesis | No differences from control |

*Note: Generalized data obtained on experimental Days 30 and 180.*

## Assessment of Potential Impact of Transgenic Potato on Immune System in Studies on Mice

### Potential Effect on Humoral Component of Immune System

The immunomodulating effect of transgenic potato on the humoral component of the immune system was examined on two oppositely reacting mice lines, C57Bl/6 and CBA, by determining the level of hemagglutination to sheep erythrocytes (SE). The experimental conditions are described in section 5.1.1.

In the test group of C57Bl/6 mice (low-sensitivity animals), antibodies raised against SE appeared on post-immunization Day 7 at the titer of 1:4. On post-immunization Days 14 and 21, the titers were 1:6 and 1:10, respectively. In the control group of C57Bl/6 mice fed diet with conventional potato, the antibodies appeared on post-immunization Day 7 (1:4) and remained in the blood on Days 14 (1:6) and 21 (1:3).

In the test group of CBA mice (high-sensitivity animals) the antibodies raised against SE also appeared on post-immunization Day 7 at the titer of 1:10; thereafter they were observed on Days 14 (1:8) and 21 (1:16). In the intact control groups of both mice lines (mice were fed potato-free diet), no antibodies against this antigen were detected. Thus, both control and test mice of both lines demonstrated insignificant elevation in the titer of antibodies raised against SE.

### Potential Effect on Cellular Component of Immune System

Examination of the effect of transgenic potato on the cellular component of the immune system, assessed by delayed hypersensitivity reaction, showed that, in C57Bl/6 mice, RI insignificantly increased in both test ($59 \pm 11$) and control (53 ± 10) groups in comparison with the intact control mice not injected with SE (36 ± 13).

In CBA mice, RI was slightly lower in the test (92 ± 12) and control (88 ± 18) groups in comparison with the intact control mice not injected with SE (95 ± 13). Thus, transgenic and conventional potato insignificantly modified RI in mice of both lines.

### Assessment of Potential Sensibilization Effect of Transgenic Potato

The analysis of possible sensitizing action of transgenic potato on the immune response to endogenous metabolic products was carried out by testing mouse sensitivity to histamine. No mortality was observed in either the test or control group of animals, which attests to the absence of sensitizing agent in this potato variety.

### Assessment of Potential Effect of Transgenic Potato on Susceptibility of Mice to *Salmonella typhimurium*

The effect of transgenic potato on susceptibility of mice to infection by salmonella of murine typhus was examined in experiments on mice injected intraperitoneally with various doses of Strain 415 *Salmonella typhimurium*. The following data was obtained.

- In control and test groups, the mortality of mice infected with the greatest dose ($10^4$ microbial cells per mouse) started on post-injection Day 4, and all mice died as early as post-injection Day 6.
- In control and test groups, the smaller doses (10 to $10^3$ microbial cells per mouse) did not induce 100% mortality, and the loss of mice was observed for the duration of the study.
- $LD_{50}$ values in test and control groups were 263 and 512 microbial cells per mouse, correspondingly.

These data showed that *Salmonella typhimurium* produced a typical infection both in control and test mice. Severity of murine typhus was identical in both groups. The differences in $LD_{50}$ values were within the experimental error.

Thus, transgenic potato resistant to Colorado potato beetle produced no effect on humoral or cellular immune system, did not sensitize mice, and did not modify resistance of mice against murine typhus, a typical mouse infection. Therefore, transgenic potato resistant against Colorado potato beetle does not possess immunomodulating properties.

### *Assessment of Potential Impact of Transgenic Potato on Immune System in Studies on Rats*

The study was carried out on male Wistar rats ($n = 50$) weighing initially 180 ± 10 g. After a 7-day adaptation period to the standard vivarium diet,

### 5.4.1 Potato Variety Superior Newleaf Resistant to Damage Caused

**Table 5.355** Severity of Active Anaphylactic Shock in Rats Fed Diet with Conventional (Control) or Transgenic (Test) Potato

| Group of Rats | AI | | Mortality, % |
|---|---|---|---|
| | 3h After Injection | 24h After Injection | |
| Control ($n = 25$) | 2.68 | 3.60 | 84 |
| Test ($n = 25$) | 3.25 | 3.67 | 87 |

Note: The differences are not significant ($p > 0.05$).

**Table 5.356** Parameters of Humoral Immune Response in Rats Fed Diet with Conventional (Control) or Transgenic (Test) Potato ($M \pm m$)

| Group of Rats | $D_{492}$ | Concentration of Antibodies, mg/mL | Logarithm of Antibody Concentration |
|---|---|---|---|
| Control ($n = 25$) | $0.655 \pm 0.027$ | $3.50 \pm 0.58$ | $0.395 \pm 0.097$ |
| Test ($n = 25$) | $0.663 \pm 0.033$ | $4.05 \pm 0.86$ | $0.413 \pm 0.092$ |
| $p$ | $>0.05$ | – | $>0.05$ |

the rats for the following 28 days were fed diet supplemented with conventional (control group) or transgenic (test group) potato in the amount of 12 g/rat/day (Table 5.344).

The model of generalized anaphylaxis was developed according to [8] as described in section 5.1.1.

During the study, the rats of both groups developed normally, which attests to the adequate nutritional value of both diets. On experimental Day 29, the body weights in test and control groups were $252 \pm 6$ and $263 \pm 5$ g, respectively ($p > 0.05$). Assessment of severity of the anaphylactic reaction in control and test rats showed that the differences between all examined parameters were insignificant (Table 5.355).

Table 5.356 shows the mean values of $D_{492}$, the concentration of antibodies, and common logarithm of this concentration in the control and test groups. The differences between these groups assessed by $D_{492}$ and logarithm of antibody concentration were insignificant. Thus, the parameters of humoral immune response did not differ between the control and test group rats.

Overall, these data showed that there were no changes in allergic reactivity and intensity of humoral immune response assessed in the model of generalized anaphylaxis in rats fed diet with transgenic potato in comparison

Table 5.357 Cytogenetic Parameters of Bone Marrow in Mice Fed Diet with Conventional (Control) or Transgenic (Test) Potato ($M \pm m$)

| Number of Cells, % | Control | Test |
| --- | --- | --- |
| with chromosomal aberrations | 0.83 ± 0.23 | 0.90 ± 0.45 |
| with gaps | 1.66 ± 0.45 | 1.59 ± 0.59 |
| with polyploid chromosome set | 2.52 ± 0.67 | 2.05 ± 0.66 |

Note: The numbers of analyzed metaphases were: 440 (test) and 360 (control).

with the control rats fed conventional potato. Thus, the transgenic potato produced no effect on sensitization and allergic reactivity of the laboratory animals.

### Assessment of Potential Genotoxicity of Transgenic Potato

The genotoxic study was carried out on male C57Bl/6 mice weighing 16–18 g. During 30 days, the animals were fed standard vivarium diet isocalorically supplemented with transgenic or conventional potato (5 g/day/mouse). Transgenic or conventional potato tubers were boiled in their skin and peeled before adding to the diet, respectively, of test or control mice. The animals consumed water and feed *ad libitum* (Table 5.344)

The cytogenetic study examined chromosomal aberrations in the bone marrow cells. The experimental conditions are described in section 5.1.1. In each animal, approximately 50–55 cells were analyzed at the metaphasic stage of nuclear division taken from 2-month male C57Bl/6 mice weighing 20–25 g. The genetic alterations in the sex cells were examined by revealing the dominant lethal mutations (see section 5.1.1).

Consumption of transgenic potato did not induce chromosomal aberrations in comparison with conventional potato (Table 5.357). In the basic generation of the control and test mice, the study revealed only chromosomal abnormalities (single segments, rings, and gaps) which are unstable and usually eliminated in subsequent nuclear divisions. Similar abnormalities were also observed in the control mice. There was no difference between the numbers of cells with these abnormalities.

Analysis of genetic alterations in the sex cells of C57Bl/6 mice by counting the dominant lethal mutations revealed no significant differences between control and test groups in pre- and post-implantation mortality and embryonic survival rate. The induced lethality at the post-meiotic stages of spermiogenesis was absent or increased insignificantly (by 3% on Week 2). This increase was not considered to be treatment related, because it was not

**Table 5.358** Dominant Lethal Mutations in Sexual Cells of Mice Fed Diet with Conventional (Control) or Transgenic (Test) Potato

| Parameter, % | Week 1, Mature Spermatozoa | | Week 2, Late Spermatids | | Week 3, Early Spermatids | |
|---|---|---|---|---|---|---|
| | Control | Test | Control | Test | Control | Test |
| Pre-implantation mortality | 7.74 | 13.14 | 7.38 | 6.28 | 5.42 | 5.64 |
| Post-implantation mortality | 6.87 | 7.8 | 8.69 | 6.70 | 7.64 | 7.18 |
| Survival rate | 85.91 | 89.71 | 84.56 | 87.42 | 87.34 | 87.0 |
| Induced mortality | 0 | 0 | 0 | 3 | 0 | 0 |

Note: 857 embryos and 984 corpus lutea from 136 pregnant females were analyzed.

manifested on Week 1, while there was no induced mortality, indicating the absence of any adverse effect on mature spermatozoa (Table 5.358).

Thus, transgenic potato resistant against Colorado potato beetle demonstrated no mutagenic activity in mice.

### Assessment of Technological Parameters

The study of technological parameters was carried out in Moscow State University of Applied Biotechnology (Ministry of Science and Education of the Russian Federation).

To characterize the tubers of Superior NewLeaf potato resistant against Colorado potato beetle in comparison with non-transgenic conventional potato, the technological examination assessed cooking time, starchiness, and adequacy for chip production. This analysis established that the technological parameters of transgenic Superior NewLeaf and conventional potato were identical.

### Conclusions

The results of a complex safety assessment of transgenic potato variety Superior NewLeaf resistant against Colorado potato beetle attest to the absence of any toxic, genotoxic, immunomodulating, or allergenic effects of this potato variety. By biochemical composition, the transgenic and conventional potato varieties were identical.

Based on the results of the studies, the State Sanitation Service of the Russian Federation (Department of State Sanitation and Epidemiological Inspectorate) granted the Registration Certificate which allows the transgenic potato Superior NewLeaf variety to be used in the food industry and placed on the market without restrictions.

# REFERENCES

[1] Chemical composition of the Russian nutrition products: Reference book. Under the editorship of Skurihin IM, Tutelyan VA. [in Russian]. Moscow: DeLi print; 2002. p 236.

[2] Durnev AD, Zhanataev AK, Anisina EA, and others. Application of method of alkaline gel electrophoresis of isolated cells to evaluate genotoxic properties of natural and nonnatural substances: Methodological instructive regulations. Approved by RAMS and RAAS [in Russian]. Moscow; 2006. p 27.

[3] Gubler EV. Numerical Methods to Analyze and Recognize Pathological Processes [in Russian]. Leningrad; 1978. P. 84–6.

[4] Hoot PD. Nutritional value of soybean oil and soybean proteins. Practical Handbook of Soybean Processing and Utilization. Erickson DR, editor, [in Russian]. Moscow; 2002. P. 512–40.

[5] Hygienic safety requirements and nutrition value of nutrition products: Health and hygiene rules and standards (SanPin 2.3.2.1078-01) [in Russian]. Moscow: Ministry of Health of the Russian Federation; 2002. p 164.

[6] Instructions on experimental (non-clinical) study of new pharmaceutical substances. Under the editorship of Habriev R.U. 2nd ed., revised and updated [in Russian]. Moscow: JSC Publishing house "Meditsina" (Medicine); 2005. p 832.

[7] Lushnikov EF, Abrosimov AU. Cell death (apoptosis) [in Russian]. Moscow: Meditsina (Medicine); 2001. [192 p.].

[8] Medical and Biological Assessment of Food Products Derived from Genetically Modified Sources: Methodical Guidelines (MUK 2.3.2.970-00) [in Russian]. Moscow; 2000.

[9] Medico-biologic safety assessment of genetically-engineered and modified organisms of plant origin: Methodological instructive regulations MY 2.3.2.2306-07 [in Russian]. Moscow: Federal Hygienic and Epidemiological Center of Rospotrebnadzor; 2008. p. 21.

[10] Sapronov AR, Zhushman AI, Loseva VA. General technology of sugar and sweetening agents [in Russian]. Moscow: Food Processing Industry; 1979. [p. 353].

[11] Tutelyan VA, Gapparov MMG, Avrenjeva LI, et al. Medico-biologic safety assessment of genetically-engineered and modified corn line MON 88017 [in Russian]. Nutr Issues 2008;77(5):4–12.

[12] Bayer E, Gugel KH, Hagele K, et al. Metabolic products of microorganisms. 98. Phosphinothricin and phosphinothricyl-alanyl-analine. Helv Chim Acta 1972;55(1):224–39.

[13] Brunke KJ, Meeusen RL. Insect control with genetically engineered crops. Trends Biotechnol. 1991;9:197–200.

[14] Consensus document on compositional considerations for new varieties of potatoes: key food and feed nutrients, antinutrients and toxicants. Series on the safety of novel foods and feeds. OECD, Paris; 2002. N 4.

[15] Consensus document on compositional considerations for new varieties of maize (Zea Mays): Key Food and Feed nutrients, antinutrients and secondary plant metabolities. Series on the safety of novel foods and feeds. OECD, Paris; 2002. N 6.

[16] Consensus document on compositional considerations for new varieties of rice (Orysa sativa): key food and feed nutrients and antinutrients. Series on the safety of novel foods and feeds. OECD, Paris; 2004. N 10.

[17] Consensus Document on compositional considerations for new varieties of soybean: key food and feed nutrients and antinutrients. Series on the safety of novel foods and feeds. OECD, Paris; 2001. N 2.

[18] Consensus Document on the Biology of glycine max (L.) merr. (soybean). Series on harmonization of regulatory oversight in biotechnology. OECD, Paris; 2000. N 15.

[19] Database of GM Crop http://www.cera-gmc.org/.

[20] Derelanko MJ, Hollinger MA. Handbook of toxicology, 2nd ed. USA: CRC Press; 2001. [p. 1414].

[21] Gordon-Kamm WJ, Spencer TM, Mangano ML, et al. Transformation of maize cells and regeneration of fertile transgenic plants. Plant Cell 1990;2(7):603–18.

[22] Han Y, Parsons CM, Hymowitz T. Nutritional evaluation of soybeans varying in trypsin inhibitor content. Poultry Sci 1991;70(4):896–906.

[23] Hood RD. Developmental and reproductive toxicology: a practical approach, 2nd ed. USA: CRC Press; 2006. [p. 1168].

[24] Klein TM, Wolf ED, Wu R, Sanford JC. High velocity microprojectiles for delivering nucleic acids into living cells. Nature 1987;327:70–3.

[25] Knowles BH. Mechanism of action of Bacillus thuringiensis insecticidal deltaendotoxins. Adv Insect Physiol 1994;24:275–308.

[26] Kumar PA, Sharma RP, Malik VS. The insecticidal proteins of Bacillus thuringiensis. Adv Appl Microbiol 1996;42:1–43.

[27] Lebrun M, Leroux B, Sailland A. Chimeric gene for the transformation of plants. U.S. patent N 5,510,471. 1996.

[28] Lewi PJ, Marsboom RP. Toxicology reference data – Wistar rat. Amsterdam: Elsevier/North-Holland Biochemical P.; 1981. [p. 358].

[29] McCabe DE, Swain WF, Martinell BJ, Christou P. Stable transformation of soybean (Glicine max) by particle acceleration. Biotech 1988;6:923–6.

[30] McElroy D, Zhang W, Cao J, Wu R. Isolation of an efficient actin promoter for use in rice transformation. Plant Cell 1990;2(2):163–71.

[31] Mensink H, Janssen P. Environmental health Criteria 159. Geneva: World Health Organization; 1994.

[32] Murakami T, Anzai H, Imai S, et al. The bialaphos biosynthetic genes of Streptomyces hygroscopicus: molecular cloning and characterization of the gene cluster. Mol Gen Genet 1986;205:42–50.

[33] Negrotto D, Jolley M, Beer S, et al. The use of phosphomannose-isomerase as a selectable marker to recover transgenic maize plants (Zea maize L.) via Agrobacterium transformation. Plant Cell Reports 2000;19:798–803.

[34] Padgette RS, Kolacz KH, Delannay X, et al. Development, identification and characterization of a glyphosate-tolerant soybean line. Crop Sci 1995;35:1451–61.

[35] Padgette SR, Taylor NB, Nida DL, et al. The composition of glyphosate-tolerant soybean seeds is equivalent to that of conventional soybeans. J Nutr 1996;126(3):702–16.

[36] Pietrzak M, Shillito RD, Hohn T, Potrykus I. Expression in plants of two bacterial antibiotic resistance genes after protoplast transformation with a new plant expression vector. Nucleic Acids Res 1986;14(14):5857–68.

[37] Pryde EH. Composition of soybean oil Erickson DR, Pryde EH, Brekke OL, editors. Handbook of soy oil processing and utilization (5th ed.). St. Louis: American Soybean Association and American Oil Chemists Society; 1980. p. 13–31.

[38] Pullen AH. A parametric analysis of the growing CFHB (Wistar) rat. J Anat 1976;121:371–83.

[39] Reed J, Privalle L, Powell ML, et al. Phosphomannose isomerase: an efficient selectable marker for plant transformation. In Vitro Cell Dev Biol Plant 2001;37:127–32.

[40] Richardson TH, Tan X, Frey G, et al. A novel, high performance enzyme for starch liquefaction. J Biol Chem 2002;277(29):26501–26507.

[41] SCF (Scientific Committee on Food). Opinion on Fusarium toxins – part 3: fumonisin B1 (FB1), expressed on 17 October 2000.

[42] Smith AK, Circle SI. Chemical composition of the seed Smith AK, Circle SI, editors. Soybean: chemistry and technology. Vol. 1. Proteins. : Wesport; 1972. p. 61–92.

[43] Souci SW, Fachmann W, Kraut H. Food composition and nutrition tables, 6th ed. Stuttgart: CRC Press; 2000. <http://www.sfk-online-net>.

[44] Strauch E, Arnold W, Alijah R, et al. Phosphinothricin-resistance gene active in plants and its use. European patent 275957 B1. 1993.

[45] Suckow MA, Weisbroth SH, Franklin CL. The laboratory rat. Burlington, USA: Elsevier Academic Press; 2006. [p. 912].

[46] Taylor NB, Fuchs RL, MacDonald J, et al. Compositional analysis of glyphosate-tolerant soybeans treated with glyphosate. J Agric Food Chem 1999;47(10):4469–73.

[47] Thompson CJ, Movva NR, Tizard R, et al. Characterization of the herbicide-resistance gene bar from Streptomyces hygroscopicus. EMBO J 1987;6(9):2519–23.

[48] Tucker MJ. Diseases of the Wistar Rat. London: Taylor & Francis; 1997. [p 254].

[49] Watson SA. Maize, amazing maize, general properties Wolff IA, editor. CRC handbook of processing and utilization. II. Part. I, Plant products. Florida: CRC; 1982. p. 3–29.

[50] Watson SA. Structure and composition Watson SA, Ramstad RE, editors. Maize: chemistry and technology. St. Paul: American Association of Cereal Chemists; 1987. p. 53–80.

[51] Weber EJ. Lipids of the kernel Watson SA, Ramstad RE, editors. Maize: chemistry and technology. St. Paul: American Association of Cereal Chemists; 1987.

[52] White PJ, Pollak LM. Maize as a food source in the United States. Part. II. Processes, products, composition and nutritional values. Cereal Foods World 1995;40(10):756–62.

[53] WHO. (1994). Glyphosate. World Health Organization (WHO), International Programme of Chemical Safety (IPCS). Environmental Health Criteria N 159. Geneva; 1994.

[54] WHO/FAO. Report of a Joint FAO. WHO Expert Committee on Food Additives: Evaluation of certain mycotoxins in food. Geneva: World Health Organization; 2002.

[55] WHO/IPCS. WHO Food Additives Series 47: Safety evaluation of certain mycotoxins in food. Geneva: World Health Organization; 2001.

[56] Wilcox JR. Breeding soybeans for improved oil quantity and quality. World soybean research conference III: proceeding. Shible R, editor. Boulder; 1985. P. 380–6.

[57] Windels P, Taverniers I, Depicker A, et al. Characterization of the roundup ready soybean insert. Eur Food Res Technol 2002(213):107–12.

[58] Wright M, Dawson J, Dunder E, et al. Efficient biolistic transformation of maize (Zea maize L.) and wheat (Triticum aestivum L.) using the phosphomannose isomerase gene, pmi, as a selectable marker. Plant Cell Rep 2001;20:429–36.

# CHAPTER 6

# Control System of Food Products Derived from Genetically Modified Organisms of Plant Origin

## 6.1 CONTROL OF FOOD PRODUCTS CONTAINING GENETICALLY MODIFIED INGREDIENTS

When treating the use of biotechnology in food production as an extremely promising avenue to solve food supply problems, one should take into consideration the potential possibility of unintentional or intentional production of GMOs by uncontrolled genetic engineering processes that can adversely affect human health in the world. Understandably, legislation in most countries imposes requirements for registration, post-registration surveillance, and labeling of GMO-derived food. It should be stressed that labeling of a GMO food product has nothing to do with its safety: it only informs the consumer of the use of genetic engineering technologies during production of the food product [17].

The USA employs a different control procedure when assessing food/feed safety: once the safety of the GM line and all products derived from it is confirmed, the product is registered and all food products derived from GMOs of this variety (line, event, variety) are placed on the food market under deregulated status (Figure 6.1) [11]. The FDA considers that extra labeling on the mode of production can misguide consumers about the real safety of the food product.

In the European Union, the control of production and life-cycle of GMO-derived food requires post-registration surveillance and obligatory labeling. A control system has been developed which monitors the circulation of GMO products at each marketing stage in order to provide consumers with information on the mode of food production and to ensure adequate labeling (Figure 6.1) [19]. Every batch of food product is accompanied by a certificate on the use of GMO in its production. The threshold for labeling of foodstuff as a GMO-derived product is 0.9%. If a food product contains less than 0.9% GMO per individual ingredient (or per whole product if it consists of a single ingredient), labeling is not required, and the presence of GMO in the

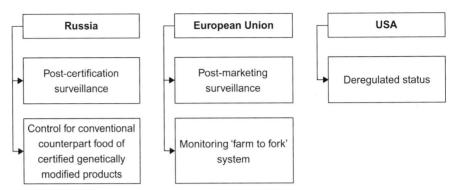

**FIGURE 6.1** Control of food products derived from GMOs in various countries.

foodstuff is considered to be a technologically unavoidable admixture [20]. To provide all types of controls for GMO-derived food products, the applicant must develop and publish qualitative and quantitative methods to detect the transformation events and submit GMO samples with the test reports of the corresponding analyses [20]. The test laboratory of the European Union must examine, test, and certify the method of GMO detection and identification submitted by an applicant, including sampling and identification of transformation events.

An efficient system of state control has been developed in the Russian Federation for control of food products derived from GMOs of plant origin. State control for raw materials and food products derived from GMOs is implemented by the executive bodies authorized to execute sanitary and epidemiologic supervision under the laws of consumer protection, as well as in veterinary and phytosanitary control, in accordance with their respective remits [1]. Based on experience in the control of GMO-derived food accumulated by the responsible parties (Rospotrebnadzor), and taking into account the presence of genetically modified food counterparts in the world food market, the world production of GMO food, and its import into the Russian Federation, a list of food products subject to examination for the presence of GMO of plant origin has been developed [5].

A decision tree has been developed to guide examination of samples of food products to determine the presence and amount of food ingredients derived from GMOs (Figure 6.2).

An interdepartmental commission, on genetic engineering activity, of the Ministry of Science and Education of the Russian Federation in cooperation with RAMS and the scientific council on medical problems of nutrition of the Ministry of Health and Social Services, has developed a database on the world

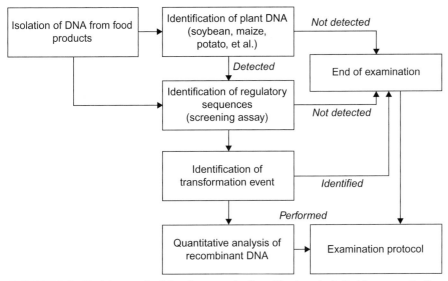

**FIGURE 6.2** Decision tree for laboratory examination of food products that have genetically modified ingredients [5].

production of plant-derived GMO used to produce food and feed, which includes the following data:

- A list of plant-derived GMOs presented and registered on the world food market;
- Regulatory sequences to detect recombinant DNA;
- GMO lines, varieties, and events, which passed through state registration and were included in the public register of food products authorized to be used in the food industry and placed on the market.

To conform to international practice in the control of trade of food containing transgenic ingredients, the Russian Federation established GOST R 53214-2008 "Food products. Analytical methods of detection of genetically modified organisms and the products derived therefrom. General requirements and definitions", a national standard that includes ISO 24276:2006 requirements. This document was registered on December 12, 2008 with the effective date of January 1, 2010. It compiles the general terms and definitions, the requirements and the guidelines to organize GMO-testing laboratories, the technical requirements to confirm the credibility of evidence, a description of the methods and protocols to examine the food products, feed, seed, and plant samples sampled from the environment.

## 6.2 METHODS TO CONTROL FOOD PRODUCTS CONTAINING GENETICALLY MODIFIED INGREDIENTS

Development of methods to control food products derived from plant-derived GMOs accompanied placement of the first GM product into the world food market [12,15].

At present, plant-derived GMOs placed on the market differ from conventional crops by the presence of recombinant DNA in the genome and expressed protein responsible for a novel trait. Both recombinant DNA and the expressed protein can be used to detect the GMO components in a food product. Methods to assay proteins (enzyme immunoassay) demonstrate some advantages related to simple protocol and relatively low cost of analysis in comparison with other methods. At the same time, there are some features limiting their wide use in monitoring of GMO-derived food products. For example, analysis of the foodstuff produced with significant technological processing of basic raw materials (high temperature, acidic medium, enzymatic treatment, etc.) performed with a protein-based technique such as immunodetection, can yield unstable and poorly reproducible results because of denaturation of the proteins [14]. It should be noted that the possibility to detect modified protein in food products is limited by the level of its expression in the plant. In most transgenic varieties placed on the world market, the content of modified protein does not surpass 0.06% in those parts of the plant that are used as food [13,18]. Also, tests based on identification of the modified protein are not specific for a particular transformation event.

The preferred way to monitor GMO-derived food products is provided by methods to detect recombinant DNA [8,10]. DNA structure is identical in all cells of the organism, so any part of the plant can be used to identify GMO, in contrast to the method of detecting the presence of modified protein. The methods based on the assay of recombinant DNA are more sensitive than those based on protein assay. Moreover, they can be used both for screening analyses and for detection of a particular transformation event.

### Identification of Recombinant DNA by Polymerase Chain Reaction (PCR) Method

When considering identification of a GMO by a PCR-based method, it should be taken into account that isolation of DNA is the critical stage that determines feasibility and quality of the analysis. This is explained by the necessity to isolate an adequate amount of DNA suitable for amplification. Transition from raw unprocessed material to a food product is accompanied by technological or heat treatment, which advances the problem to isolate a necessary amount of DNA of the adequate quality. Although DNA is more stable in comparison with the proteins, it also can be degraded by very high temperatures, ultraviolet irradiation, and acidic or enzymatic processing, which specifically affect DNA. For example, amplification may not occur even in a solution with a high DNA content, because of the presence of DNA chains that are too short. The critical length of DNA fragments for the PCR-based method is about 400 nucleotide pairs [15]. DNA cannot be detected in food products that have been subjected to pronounced technological processing (hydrolyzed plant proteins, highly refined oils and starches, soy sauce, and sugar or ethanol derived from transgenic potato or maize) [9,16].

It is important to note that a solution with isolated DNA can also contain proteins, fats, polysaccharides, polyphenols, and other substances capable of inhibiting PCR and impeding the analytical procedures.

At present, three DNA isolation methods are used:

- The CTAB method, based on a standard protocol for plant tissues. A sample of food product is incubated with CTAB detergent (hexadecyltrimethylammonium bromide), thereafter extraction is performed with the use of chloroform followed by sedimentation of DNA with isopropanol;
- The Sorbtion method, based on the use of guanidine thiocyanate salt and a sorbent followed by DNA extraction into solution;
- A method using the Wizard technique, based on specific binding of DNA by a resin (Wizard, Promega) from the solution performed after exposure of the food sample to proteinase K and sodium dodecil sulfate (SDS).

These methods of DNA isolation are authorized in the Russian Federation to examine food products for the presence and amount of recombinant DNA [2–7,22].

A typical genetic construct used in plant genetic engineering contains the target gene responsible for the new trait, the regulatory elements of this gene (the most important are promoter and terminator), and marker genes absent in the genome of most transgenic plants. The use of the same regulatory sequences and marker genes during engineering of transgenic plants makes it possible to carry out screening analyses. For example, the assay for the 35 S promoter of cauliflower mosaic virus and/or NOS terminator from *Agrobacterium tumefaciens* makes it possible to control 98% of food products on the world market, which is one of the most pronounced advantages of the methods based on assay of recombinant DNA.

The PCR-based methods identifying the 35 S promoter and NOS terminator are unified and authorized as the standard assays in 23 countries, the Russian

Federation included [6]. In the Russian Federation, the following primers are used to identify the above-mentioned regulatory sequences:

- 35S-1: 5′ GCT CCT ACA AAT GCC ATC A 3′;
- 35S-2: 5′ GAT AGT GGG ATT GTG CGT CA 3′;
- *nos-1* 5′ GAA TCC TGT TGC CGG TCT TG 3′;
- *nos-2* 5′ TTA TCC TAG TTT GCG CGC TA 3′.

The primers complementary to a site in the target gene are used to identify this gene in a food product. The same gene can be used to engineer a series of transgenic events. An example of such a popular gene is *pat*, which encodes synthesis of phosphinothricin acetyltransferase, which acetylates the free $NH_2$-group of glufosinate ammonium thereby preventing accumulation of ammonia. This gene is present in the genome of all plants tolerant to glufosinate ammonium. The *Cry1Ab* gene from *B. thuringiensis* encoding synthesis of insecticide was used to produce several maize varieties, including MON 810 and Bt11 lines protected against damage by the European corn borer and authorized for use in the food industry and on the market in the Russian Federation.

Identification of a specific genetic construct inserted into the plant genome is performed with construct-specific primers. Because of the high specificity of these methods, detection of recombinant DNA unequivocally indicates the use of GM technologies in the production of the examined foodstuff. However, it should be taken into consideration that the same construct can be used to transform different plants. For example, the genetic constructs pV-ZMBKO7 and pV-ZMGT 10 are present in the genome of maize lines MON 809 (single copy of both constructs), MON 810 (single copy of the first construct), and MON 832 (single copy of the second construct) [14].

Detection of a particular transformation event is based on amplification of a DNA sequence which incorporates a part of the DNA of the inserted genetic construct and a part of the plant genome flanking the inserted sequence. Such amplification unequivocally determines the line of the transgenic plant (Figure 6.3). The methods based on PCR are capable of identifying the most popular transgenic varieties available in the world market. In the Russian Federation, they are unified, authorized, and widely used to control food products not registered in the country [3].

To carry out post-registration monitoring of food containing GM ingredients, the Russian Consumer Protection Agency employs methods and complete protocols to analyze all transgenic lines (transformation events), which are registered in the Russian Federation and authorized for use in the food industry and on the market. To register a novel transgenic foodstuff, the applicant must submit the control method and protocol of the analysis tracing the transgenic event to the Russian Consumer Protection Agency [2–7,22].

## 6.2 Methods to Control Food Products Containing Genetically Modified Ingredients

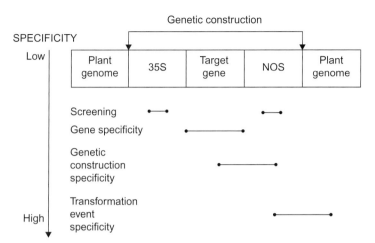

**FIGURE 6.3** Specificity of the methods for detecting recombinant DNA in food and feed.

## Identification of Recombinant DNA with Biological Microchips

Due to rapid development of biotechnology, numerous target genes have become working tools for the development of transgenic plants. According to professional prognosis, production of new transgenic plants will continue, which indicates the need to develop novel approaches for the control of food products derived from the transgenic sources. Considerably promising is the use of DNA technologies with biological microchips. These technologies have a potential for screening of a large amount of plant-derived GMOs by a single assay [4,7,8].

Biochip DNA analysis is based on the complementary principle and specificity of DNA strands. Short sequences of single-strand DNA (probes) complementary to the target DNA are fixed on a very small glass plate area. If the analyzed sample contains the target DNA, it will be bound to the probes when put in contact with the sequence fixed to the glass plate. This process develops at the molecular level with thousands of target DNA sequences in every glass plate. The probe DNA can be designed as complementary to those fragments of the target DNA which incorporate the regulatory sequences (promoter 35 S or terminator NOS), marker genes, target genes, and the regions between the inserted genetic construct and the plant genome. These features can be potentially used to carry out both screening analyses and identification of particular transformation events [1,4,7].

To enhance sensitivity of the method, the target DNA can be amplified by multiplex PCR prior to binding with the probe. The data are read by special devices. The advantageous features of the described method are as follows: (1) the possibility to qualitatively analyze a large number of GMOs simultaneously

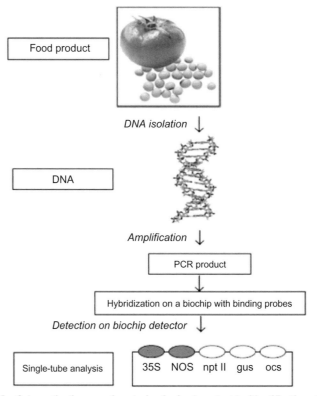

**FIGURE 6.4** Schematic diagram of analysis of a food product by identification of recombinant DNA (screening) using a biological microchip.

in a single assay, (2) high sensitivity in combination with PCR, (3) potential decrease in cost, because of the multiplex character of a single assay, and (4) the possibility of automation.

In the Russian Federation, identification of plant-derived GMOs is performed with the original technique, which employs enzymatic analysis on a biological microchip, including asymmetric multi-primer single-target PCR (MP-ST-PCR), hybridization, and enzymatic analysis of the amplified products on a biological microchip followed by data analysis on a firmware complex Degmigen-001 (Figure 6.4) [4,7]. This method can qualitatively detect the regulatory sequences and marker genes thereby detecting in a single assay transgenic plant varieties, which may be present or absent in the Federal Register of the Russian Federation. The high sensitivity of this method makes it possible to detect 0.1% plant-derived GMO in a foodstuff.

A biological microchip is a regular set of probes fixed on a glass substrate, which ensures specificity of MP-ST-PCR products. Each MP-ST-PCR involves five biotinylated pair primers corresponding to certain transgenic sequences.

On the surface of the biological microchip, fragments of target DNA hybridized with the labeled DNA are detected by enzymatic analysis as colored sites.

Among the sequences available in a biological microchip are DNA fragments that are most frequently used to design transgenic plants: the cauliflower mosaic virus 35 S promoter, transcription terminal sites *nos* and *ocs* from *Agrobacterium tumefaciens*, and the marker genes *nptII* from bacterial transposon Tn5 or *gus* from *Escherichia coli*. Examples of such sequences are shown below.

- The cauliflower mosaic virus 35 S promoter is detected with the following sequences:
  - 35 S_d 5′ CGG CTA CTC CAA GAA TAT CAA AGA TAC AGT TTC AGA A
  - 35 S_r 5′ CCA TTT TCC TTT TTT ATT GTC CTT TCG ATG AAG TGA CA
- Marker gene *gus* from *Escherichia coli* is detected with:
  - gus_d 5′ ACC GTA CCT CGC ATT ACC CTT ACG CTG AAG AGA
  - gus_r 5′ TGC CCG CTT CGA AAC CAA TGC CTA AAG AGA
- Terminator nos from *Agrobacterium tumefaciens* is detected with:
  - nos_d 5′ GGA CAA GCC GTT TTA CGT TTG GAA CTG ACA GA
  - nos_r 5′ GCC TGA CGT ATG TGC TTA GCT CAT TAA ACT CCA GA
- Marker gene *nptII* from bacterial transposon Tn5 is detected with:
  - npt_d 5′ GTG ACC CAT GGC GAT GCC TGC TTG C
  - npt_r 5′ ACC CAG CCG GCC ACA GTC GAT GAA TCC AGA
- Promoter *ocs* from *Agrobacterium tumefaciens* is detected with:
  - ocs_d 5′ AAA AAG TGG CAG AAC CGG TCA AAC CTA AAA GA
  - ocs_r 5′ CGT TAT TAG TTC GCC GCT CGG TGT GTC GTA GA

Here 35 S_d, gus_d, nos_d, npt_d, and ocs_d are the direct primers and 35 S_r, gus_r, nos_r, npt_r, and ocs_r are the reverse primers labeled with biotin.

These DNA fragments are incorporated in most transgenic plants placed on the world market, corresponding to 98% of transgenic food and feed products.

## Quantitative Detection of Recombinant DNA by Real-Time PCR Analysis

To determine quantitative content of GMO in food products, most countries use real-time PCR analysis of the data. This method is an established standard for such analyses. To reveal the amplification products during real-time analysis, DNA probes containing a fluorescent label and a quencher are used. During the course of PCR, the label and quencher are disconnected, resulting in fluorescence of intensity correlated with the amount of amplicons. Intensity of fluorescence is proportional to the amount of recombinant DNA in the analyzed sample. This is determined from the level of fluorescence intensity and the number of circles shown on the computer's monitor. Several approaches have been used to detect changes in the number of amplicons in the course of PCR with individual probe design (the linear probes

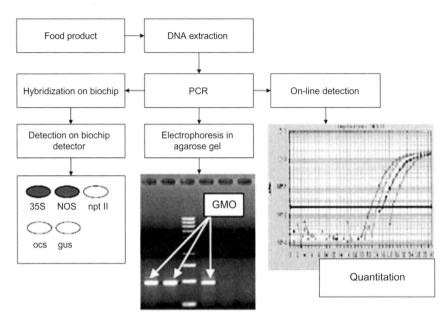

**FIGURE 6.5** Schematic diagram of examination of a food product for the presence and content of plant-derived GMOs.

of TaqMan, the real-time PCR hairpin-forming probes of Molecular Beacons, etc.) [2,9,21].

A special feature of this method is determination of PCR products directly during the reaction. The method is characterized by high sensitivity and specificity, absence of contamination by PCR products (analysis is performed in a closed tube without an electrophoretic stage), saving of laboratory space, and short duration of the analysis. In the Russian Federation, real-time PCR-based methods are unified and authorized for quantitative assay of plant-derived GMOs in food products [2].

Thus, examination of foodstuffs for the presence of plant-derived GMOs is carried out in the Russian Federation through three methodical approaches (Figure 6.5).

Examination of multi-component food products for the presence of plant-derived GMOs may include several tests based on different methods. As an example, consider analysis of boiled sausage for the presence of GMOs (Figure 6.6). Analysis of this product with screening methods (PCR and microchip technique) detected the presence of recombinant DNA. Further analysis of ingredients of the product showed that the source of this DNA was a flavor aromatic additive, which has incorporated soybean proteins. Then PCR with corresponding primers revealed glyphosate-tolerant transgenic

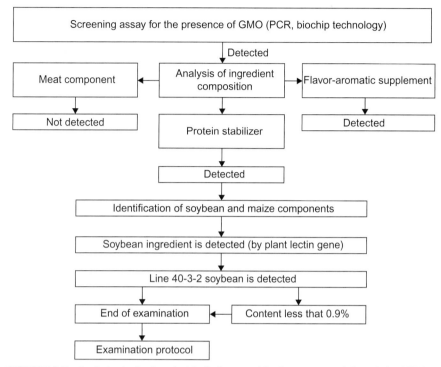

**FIGURE 6.6** Analysis of a food product (boiled sausage) for the presence of plant-derived GMOs.

soybean line 40-3-2 registered for use in the food industry in the Russian Federation since 1999. Quantitative analysis with PCR technique and on-line data analysis determined the content of recombinant DNA as less than 0.9%.

In 2007, the Russian Federation established the minimum content of GMO ingredients in a food product that must be labeled as GMO-derived. This triggered the development, approbation, and introduction of the methods of quantitative tracing of recombinant DNA and production control for all GMO events admitted in the Russian Federation [22]. Extensive introduction of modern biotechnology into food production is a potent stimulus to develop and improve the technical approaches to control transgenic food.

# REFERENCES

[1] Genetically modified food products. Organization and results of supervision. The data of State Sanitary and Epidemiological Agency of Russian Federation in 2003–2004. Belyaev EN, editor. Moscow, 2005.

[2] Technical guides: Methods for quantitation of plant-derived genetically modified sources (GMS) in food products (MUK 4.2. 1913-04). Moscow, 2004.

[3] Technical guides: Tracing of plant-derived genetically modified sources (GMS) with polymerase chain reaction (MUK 4.2. 1902-94). Moscow, 2004.

[4] Technical guides: Identification of plant-derived genetically modified organisms (GMO) with enzymatic analysis on biological microchips (MUK 4.2. 2008-05). Moscow, 2005.

[5] Technical guides: Order and organization of the control for food products derived from (or produced with the use of) plant-derived raw materials that have genetically modified counterparts (MUK 2.3.2. 1917-04). Moscow, 2004.

[6] National Standards of Russian Federation: Method to identify plant-derived genetically modified sources (GMS). GOST R 52173-2003. Moscow, 2004.

[7] National Standards of Russian Federation: Method to identify plant-derived genetically modified sources (GMS) with biological microchips. GOST R 52174-2003. Moscow, 2004.

[8] Sorokina EYu. Transgenic food under supervision. Science in Russia 2005(3) C. 37–40.

[9] Anklam E. Analytical methods for detection and determination of genetically modified organisms in agricultural crops and plant-derived food products. Eur Food Res Technol 2002;214(1):3–26.

[10] Bonfini L, Heinze P, Kay S. Van den Eede Review of GMO Detection and Quantification Techniques EUR 20384 EN European Communities. 2002. <http://www.osservaogm.it/pdf/IRCReview.pdf>.

[11] List of Completed Consultations on Bioengineered Foods. US Food and Drug Administration. <http://www.fda.gov/default.htm>.

[12] James C. Cultivation areas of transgenic (GM) varieties the world over: ISAAA report N 34-2005. 2005. <http://www.isaaa.org>.

[13] Harrison LA, Bailey MR, Naylor MW, et al. The expressed protein in glyphosate-tolerant soybean, 5-enolpyruvylshikimate-3-phosphate synthase from agrobacterium sp. strain CP4, is rapidly digested *in vitro* and is not toxic to acutely gavaged mice. J Nutr 1996;126:728–40.

[14] Holst-Jensen A, Ronning SB, Lovseth G, Berdal KG. PCR technology for screening and quantification of genetically modified organisms (GMOs). Anal Bioanal Chem 2003;375(8):985–93.

[15] Kuiper HA. Summary report of the ILSI Europe workshop on detection methods for novel foods derived from genetically modified organisms. Food Control 1999;10(6):339–49.

[16] Meyer R, Jaccaud E. Detection of genetically modified soya in processed food products: development and validation of PCR assay for specific detection of glyphosate-tolerant soybeans, Proceedings of the EURO FOOD CHEM IX Conference, Interlaken, Switzerland, event N 220. 1997. N 1. P. H23–8.

[17] Modern food biotechnology, human health and development: an evidence-based study. Food Safety Department World Health Organization, Geneva, 2005. <http://www.who.int/foodsafety/biotech/who_study/en/index.html>.

[18] Padgette SR, Taylor NB, Nida DL, et al. The composition of glyphosate-tolerant soybean seeds is equivalent to that of conventional soybeans. J Nutr 1996;126(3):702–16.

[19] Regulation (EC) N 1829/2003 of the European Parliament and of the Council of 22 September 2003 on genetically modified food and feed. Off J Eur Union 2003; L. 268. P. 1–23.

[20] Regulation (EC) N 1830/2003 of the European Parliament and of the Council of 22 September 2003 concerning the traceability and labeling of genetically modified organisms and the traceability of food and feed products produced from genetically modified organisms and amending Directive 2001/18/EC. Off J Eur Union 2003; L. 268. P. 24–8.

[21] Hubner P, Waiblinger HU, Pietsch K, Brodmann P. Validation of PCR methods for quantitation of genetically modified plants in food. J AOAC Int 2001;84(6):1855–64.

[22] Collected articles on the methodical guidelines 'Execution of Supervision on Production and Turnover of GMO-derived Food. Part 1'. Moscow, 2008.

# CHAPTER 7

# Monitoring of Food Products Derived from Genetically Modified Organisms of Plant Origin in the Russian Federation

Post-registration surveillance of food products derived from GMOs has been carried out in the Russian Federation since 2001. During 2001 the State Research Institute of Nutrition of RAMS and the Center of State Sanitary and Epidemiological Inspectorate (Federal State Agency) analyzed 1,505 food products in Moscow that were selected from the food distribution network and food manufacturing facilities. The analysis showed that transgenic food products amounted to about 20% of total food products that contained substances derived from genetically modified crops. All GM products contained processed glyphosate-tolerant soybean line 40-3-2, which is approved for use as food in Russia (Table 7.1).

In 2002, the State Research Institute of Nutrition and the Federal Service for Consumer Rights Protection and Public Wellbeing analyzed 2,004 food products made of soybean or containing the products of soybean processing. It was established that food products derived from transgenic soybean amounted to 16.2% of all such products sold in the domestic market of the Russian Federation (Table 7.2). Analysis of soybean varieties revealed the glyphosate-tolerant line (transformation event 40-3-2) that is admitted to the market in Russia. Quantitative analysis of GMO content (i.e., the content of soybean line 40-3-2) showed that 14% of food products derived from this soybean line contained less than 1.0% recombinant DNA, while only 2.2% of the food products contained a greater percentage.

In 2003, the laboratories of the Federal Inspectorate for Protection of Consumers and Human Wellbeing carried out 4,300 analyses of food products present on the domestic market in the Russian Federation in order to detect GMO (Table 7.3). Selection of products from the food distribution network and food manufacturing facilities was based on the list of food products subjected to analysis for the presence of GMO by the methods authorized in the Russian Federation [2–7].

**Table 7.1** Analysis of Food Products Containing Substances Derived from Genetically Modified Plants in 2001 (Data of the State Research Institute of Nutrition and the Russian Consumer Supervision Agency)

| Food Products | Number of Examined Samples | GM Not Detected | GM Detected |
|---|---|---|---|
| Products of soybean processing | 606 | 455 | 151 (24.9%) |
| Sausages | 300 | 200 | 100 (33.3%) |
| Dairy products | 55 | 44 | 11 (20%) |
| Baby food products | 300 | 298 | 2 (0.66%) |
| Food supplements | 244 | 207 | 37 (15.2%) |
| Total | 1505 | 1204 | 301 (20.0%) |

**Table 7.2** Analysis of Food Products Derived from Soybean or Containing Processed Soybean in 2002 (Data of the State Research Institute of Nutrition and the Russian Consumer Supervision Agency)

| Food Products | Number of Examined Samples | GMO Not Detected | GMO Detected |
|---|---|---|---|
| Products of soybean processing (soybean proteins and flour) | 720 | 502 | 218 (30.3%) |
| Sausages | 405 | 350 | 55 (13.5%) |
| Dairy products | 75 | 65 | 10 (13.3%) |
| Baby food products | 250 | 250 | 0 |
| Food supplements | 360 | 335 | 25 (6.9%) |
| Biologically active supplements | 115 | 100 | 15 (13.0%) |
| Confectionery | 79 | 77 | 2 (2.5%) |
| Total | 2004 | 1679 | 325 (16.2%) |

The highest number of food samples containing plant-derived GMOs was detected in meat products (272 in absolute score), predominantly in sausages containing ingredients derived from soybean, and in the "others" group (129) mostly composed of processed soybean products. In the Russian Federation, 11.88% of food products contained GMOs [1].

In 2004, the laboratories of the Federal Service for Consumer Rights Protection and Public Wellbeing carried out 12,953 analyses of food products for the presence of plant-derived GMO (Table 7.4).

**Table 7.3** Analysis of Food Products Containing GM-Derived Substances in 2003 (Data of the Federal Service for Consumer Rights Protection and Public Wellbeing [1])

| Food Products | Number of Examined Samples | Detected GMO, % |
|---|---|---|
| Meat products (containing ingredients of plant origin) | 1535 | 17.72 |
| Poultry products (containing ingredients of plant origin) | 44 | 29.5 |
| Others[a] | 715 | 16.41 |
| Bakery and flour products | 162 | 16.67 |
| Fish and other seafood products | 23 | 26.0 |
| Baby food products | 123 | 6.50 |
| Grain and grain products | 252 | 13.49 |
| Fruitage and berries | 6 | 16.7 |
| Dairy products | 240 | 1.67 |
| Vegetable fat products | 203 | 0.99 |
| Tinned food | 662 | 1.51 |
| Vegetables and watermelons | 182 | 1.65 |
| Sugar and confectionery | 69 | 2.9 |
| Potato | 84 | 3.6 |
| Total | 4300 | 11.88 |

[a]This group contained vegetable proteins, including products of soybean processing.

In 2004, the greatest number of samples containing plant-derived GMO was detected in meat and "other" (Table 7.4) products: respectively, 946 and 466. In addition, the following groups of food products were examined in 2004: honey and apiculture products, alcoholic beverages, beer, nonalcoholic beverages, and wild plant products, but analysis revealed no plant-derived GMO. On average, the number of food products containing GMOs in the Russian Federation was 11.98%, approximately equal to that in 2003.

According to the data of the Federal Service for Consumer Rights Protection and Public Wellbeing, 19,795 analyses of food products were carried out in the Russian Federation in 2005 for the presence of GMOs [8]. In this year, the number of food products containing substances derived from plant GMOs in the Russian Federation was 6.8%, which was smaller by 43.4% than that in 2003–2004.

According to the State Research Institute of Nutrition of RAMS, in 2003–2005 the amount of food containing substances derived from plant GMOs was 11.9% of the total amount of products containing genetically modified

**Table 7.4** Analysis of Food Products Containing GMO-Derived Substances in 2004 (Data of the Federal Service for Consumer Rights Protection and Public Wellbeing [1])

| Food Products | Number of Examined Samples | Detected GMO, % |
|---|---|---|
| Meat products (containing ingredients of plant origin) | 4609 | 20.53 |
| Poultry products (containing ingredients of plant origin) | 188 | 15.43 |
| Others[a] | 2689 | 16.75 |
| Bakery and flour products | 653 | 6.74 |
| Fish and other seafood products | 93 | 6.45 |
| Baby food products | 230 | 5.65 |
| Grain and grain products | 348 | 2.30 |
| Fruitage and berries | 145 | 1.38 |
| Dairy products | 686 | 1.31 |
| Vegetable fat products | 642 | 1.09 |
| Preserved food | 1325 | 0.98 |
| Vegetables and watermelons | 513 | 0.98 |
| Sugar and confectionery | 640 | 0.63 |
| Potato | 195 | 0 |
| Total | 12 956 | 11.98 |

[a]This group contained vegetable proteins, including products of soybean processing.

counterparts (Table 7.5), which corresponds with the data of the Federal Service for Consumer Rights Protection and Public Wellbeing.

The studies carried out in 2003–2005 showed that plant-derived GMOs were represented in food products by transgenic soybean variety 40-3-2 allowed for use in the food industry and on the market in the Russian Federation. About 4.4% of food products derived from maize contained ingredients from maize line MON 810 resistant to the European corn borer and glyphosate-tolerant NK 603 and GA21 lines, which are allowed for processing in the food industry and on the market in the Russian Federation. However, in some cases, ingredients with transgenic maize event Bt 176 were detected, which was not registered in the Russian Federation.

Analysis of food products derived from wheat, tomato, vegetable marrow, papaya, and melon sampled from the food distribution network in 2004–2006, which were not registered in the Russian Federation but present in the world market, revealed neither recombinant DNA nor modified proteins in all examined samples (Table 7.5). This corresponds with information from

**Table 7.5** Analysis of Food Products Containing GMO-Derived Substances Sampled in 2004–2006 (Data of the State Research Institute of Nutrition, RAMS)

| Food Products | Number of Examined Samples | Not Detected | Detected |
|---|---|---|---|
| Products of soybean processing | 620 | 380 | 240 |
| Meat products | 385 | 228 | 157 |
| Baby food products | 660 | 654 | 6 |
| Food supplements | 590 | 546 | 44 |
| Dietary food | 26 | 25 | 1 |
| Biologically active supplements | 352 | 342 | 10 |
| Maize and products of maize processing | 489 | 467 | 22 |
| Rice and products of rice processing | 250 | 250 | 0 |
| Wheat and products of wheat processing | 80 | 80 | 0 |
| Tomato and products of its processing | 380 | 380 | 0 |
| Products of vegetable marrow processing | 75 | 75 | 0 |
| Melon and products of its processing | 35 | 35 | 0 |
| Papaya | 25 | 25 | 0 |
| Starches | 50 | 50 | 0 |
| Total | 4017 | 3537 | 480 (11.9%) |

WHO and other international bodies that these crops are produced on a small scale in a few countries, predominantly for internal use [9].

Quantitative analysis of the content of plant-derived GMO in food products available on the domestic market showed that virtually all imported products derived from soybean and labeled "GMO-free" do contain recombinant DNA at the level of technically unavoidable presence (0.1–0.9%). Sampling analyses were carried out to test the isolates and food concentrates of soybean protein, soybean flour, and sausages that contained soybean proteins (Table 7.6).

Analysis of food products labeled as "GMO-derived" revealed the presence of recombinant DNA characteristic of glyphosate-tolerant soybean line 40-3-2 in the content of 0.9–37% (Table 7.7).

Thus, according to the data of the Federal Service for Consumer Rights Protection and Human Wellbeing [1] and the State Research Institute of Nutrition, the average share of transgenic food placed on the domestic market in the Russian Federation during 2004–2006 was 6.8–12% of the total amount of tested products. Depending on the region, they included the products of soybean processing (15–35% of the total amount of products derived

**Table 7.6** Content of Recombinant DNA in Products Containing Soybean Labeled as GMO-Free in the Food Distribution Network and in Food Manufacturing Facilities in 2006

| Food Products | Number of Examined Samples | Content of Recombinant DNA, % |
|---|---|---|
| Isolate of soybean proteins | 50 | 0.1–0.8 |
| Concentrate of soybean protein | 25 | 0.05–0.7 |
| Soybean flour | 15 | 0.1–0.7 |
| Sausages | 40 | 0.05–0.6 |

**Table 7.7** Content of Recombinant DNA in Products Containing Soybean Labeled as "Derived from GMO" in the Moscow Distribution Network and Food-Producing Factories in 2006

| Food Products | Number of Examined Samples | Content of Recombinant DNA, % |
|---|---|---|
| Isolate of soybean proteins | 15 | 0.9–20 |
| Concentrate of soybean protein | 10 | 0.9–25 |
| Soybean flour | 15 | 1.0–30 |
| Sausages | 15 | 1.0–37 |

from soybean) and those of maize processing (similarly, 0.5–3%). As a rule, detection of recombinant DNA in complex multi-component products such as sausages and confectionery products revealed the presence of ingredients derived from transgenic soybean and maize.

State control over food products containing GMO ingredients led by the Federal Service for Consumer Rights Protection and Human Wellbeing resulted in analysis of 30,966, 44,441, 47,935, 38,655, and 33,423 samples of food products in 2006, 2007, 2008, 2009, and 2010, respectively. In these years, the proportion in which GMO-derived food was detected was 2.74%, 1.13%, 0.62%, 0.23%, and 0.16%, respectively (Table 7.8).

A reduction in the use of GMOs in food was observed. This trend indicates a refusal of producers to use biotechnology-derived raw materials and the replacement of vegetable protein (soy) used in production of meat and sausage products with poorly digestible connective proteins or starchy components which reduce the nutritional value of products by 20%. Avoiding the use of biotechnological materials reduces consumption of valuable protein, resulting in a greater imbalance of protein / fat / carbohydrates in the diet of the population of the Russian Federation.

**Table 7.8** Analysis of Food Products Containing GMO-Derived Ingredients in 2006–2010

| Food Products | 2006 | | 2007 | | 2008 | | 2009 | | 2010 | |
|---|---|---|---|---|---|---|---|---|---|---|
| | Total Number of Examined Samples | GMO Detected, % | Total Number of Examined Samples | GMO Detected, % | Total Number of Examined Samples | GMO Detected, % | Total Number of Examined Samples | GMO Detected, % | Total Number of Examined Samples | GMO Detected, % |
| Food products, total | 30 966 | 2.74 | 44 441 | 1.13 | 47 935 | 0.62 | 38 655 | 0.23 | 33 423 | 0.16 |
| Meat products (containing plant-derived ingredients) | 7 641 | 6.30 | 11 545 | 2.47 | 10 726 | 1.08 | 7 771 | 0.36 | 6 238 | 0.20 |
| Bakery, flour-grinding, and confectionery | 3808 | 0.67 | 3965 | 0.13 | 4518 | 0.07 | 4 107 | 0.12 | 3 706 | 0.08 |
| Grain and grain-derived food | 1087 | 0.55 | 1018 | 0.88 | 1287 | 0.78 | 771 | 0.13 | 928 | 0.21 |
| Baby food products | 564 | 0.71 | 884 | 0.23 | 807 | 0 | 863 | 0.12 | 937 | 0.10 |
| Dairy products (containing plant-derived ingredients) | 1763 | 0.40 | 2379 | 0.76 | 3205 | 0 | 2 599 | 0.00 | 1 949 | 0.05 |
| Vegetables, melons and gourds, greens | 1996 | 0.40 | 3491 | 0.2 | 2643 | 0 | 2 549 | 0.00 | 2 346 | 0.00 |
| Preserved food | 3632 | 0.80 | 5962 | 0.6 | 6714 | 0.01 | 6 408 | 0.00 | 6 253 | 0.09 |

*(Continued)*

**Table 7.8** (Continued)

| Food Products | 2006 | | 2007 | | 2008 | | 2009 | | 2010 | |
|---|---|---|---|---|---|---|---|---|---|---|
| | Total Number of Examined Samples | GMO Detected, % | Total Number of Examined Samples | GMO Detected, % | Total Number of Examined Samples | GMO Detected, % | Total Number of Examined Samples | GMO Detected, % | Total Number of Examined Samples | GMO Detected, % |
| Vegetable fat products | 3614 | 1.11 | 2495 | 0.33 | 2800 | 4.61 | 1 014 | 0.10 | 911 | 0.00 |
| Cereals | | | 1066 | 0.47 | 2010 | 0.05 | 2 093 | 0.24 | 1816 | 0.33 |
| Potato | 637 | 1.26 | 742 | 0.54 | 965 | 0 | | | | |
| Poultry products (containing plant-derived ingredients) | 496 | 3.43 | 923 | 2.82 | 1122 | 0.36 | 681 | 0.29 | 597 | 0.50 |
| Seafood (containing plant-derived ingredients) | 279 | 2.15 | 457 | 2.19 | 575 | 0.35 | 486 | 0.00 | 303 | 0.00 |
| Other | 5449 | 4.4 | 9514 | 1.62 | 10563 | 0.47 | 9 313 | 0.46 | 7 439 | 0.54 |

*Note:* Data of the Federal Service for Consumer Rights Protection and Human Wellbeing. State Report, 2007–2010.

# REFERENCES

[1] Genetically modified food products. Organization and results of supervision. The data of State Sanitary and Epidemiological Agency of Russian Federation in 2003–2004. In: Belyaev EN, editor. Moscow: 2005.

[2] Technical guides: Methods for quantitation of plant-derived genetically modified sources (GMS) in food products (MUK 4.2. 1913-04). Moscow: 2004.

[3] Technical guides: Tracing of plant-derived genetically modified sources (GMS) with polymerase chain reaction (MUK 4.2. 1902-94). Moscow: 2004.

[4] Technical guides: Identification of plant-derived genetically modified organisms (GMO) with enzymatic analysis on biological microchips (MUK 4.2. 2008-05). Moscow: 2005.

[5] Technical guides: Order and organization of the control for food products derived from (or produced with the use of) plant-derived raw materials that have genetically modified counterparts (MUK 2.3.2. 1917-04). Moscow: 2004.

[6] National Standards of Russian Federation: Method to identify plant-derived genetically modified sources (GMS). GOST R 52173-2003. Moscow: 2004.

[7] National Standards of Russian Federation: Method to identify plant-derived genetically modified sources (GMS) with biological microchips. GOST R 52174-2003. Moscow: 2004.

[8] Regulation No. 2 of Chief State Sanitary Physician of Russian Federation of 29.08. 2006 'On strengthening supervision on production and trade turnover of food products'.

[9] Modern food biotechnology, human health and development // World Health Organization, <http://www.who.int/foodsafety/biotech/who_study/en/index.html>; 2005.

# CHAPTER 8

# Information Service for the Use of Novel Biotechnologies in the Food Industry

The beginning of the 21st century has been characterized by profound qualitative changes in the clinical sciences, based upon accumulation and integration of sophisticated theoretical and technological methods over the entire range of medical and biological subjects. Deciphering of the human genome and the development of such sciences as genomics, proteomics, transcriptomics, and metabolomics have provided tools to reveal the mechanisms underlying health and illnesses, longevity or death, and made it possible to operate in clinics at the molecular level. In fact, the development of all modern branches of biological science is based on genetic technologies and their wide use in human life.

After spectacular victory over such devastating diseases as plague and smallpox, the modern world met another challenge, of bioterrorism, fraught with the appearance of new diseases. Widening the instrumental basis to manipulate the genome not only strengthens confidence in the future fate of the Earth, but also necessitates precautionary defense against new threats. Evidently, almost any new scientific breakthrough can be considered as a direct or indirect threat to human life, but also as a factor promoting the development of civilization. The most salient examples were provided by nuclear physics with its "military" and "peaceful" atoms. Intellectual expansion without restrictive moral principles could inflict devastating consequences, which explains the extreme responsibility of scientific society in the modern world.

The ability of human beings to accept new technologies depends on culture, education, and free access to related impartial information. In this respect, the role of the mass media that form social views on problems cannot be overestimated. Balanced description of the situation, weighed judgments, and comprehensive study of issues are the cornerstones of progress. Precious time should not be traded for momentary profit, which could be followed by negative impact on the enterprise and long-term stagnation.

One of the challenges of modern life is the parallel but distant coexistence of science and common society, which limits scientific knowledge to a narrow circle of technocrats. The conceptions elaborated in limited scientific circles numbering thousands of researchers cannot attract the attention of the public. This discrepancy provokes the development of pseudoscience and distorts scientific results. Instead of scientists, charlatans attract the public's attention and squander the common wealth on useless or downright dangerous projects.

A striking example of such irresponsibility is demonstrated by the discussion about plant-derived GMOs (so-called "Frankenstein's food"). It focused on propaganda of fears, ignoring the evident fact that food supply to mankind is rapidly becoming a global problem characterized by increased demand for various raw materials used in the food industry. This real challenge can be met only with the tools provided by new technologies and especially by genetic engineering.

Despite the fact that the probative evidence on safety of GM food sources is reliably based on decades of scientific research preceding the placing of the first GMO on the market (in the period 1981–1995 alone, more than 500,000 scientific papers were published on biotechnology in the USA, Great Britain, Germany, and France, which are the leaders in applied biotechnology [1]), numerous newspapers and magazines appeared with flashy headlines about the harm of "Frankenstein's food".

By contrast, the Russian scientific community considers the problems of biotechnology synthesis of GMOs and other substances with due account for the necessity to develop modern technologies. In 2003, the Russian Academy of Sciences, Russian Academy of Medical Sciences, and Russian Academy of Agriculture resolved: "Development of fundamental and applied principles of GMO food production must be considered as the priority and strategically important scientific guidelines…. The works on creating GM plants resistant against pathogens, insects, and stress abiotic factors must be supported as well as the search for new forms of plants needed for production of high quality food, fodder, and biomaterials. The works on creating GMOs capable of producing medical preparations, enzymes, vaccines, and other biologically active compositions are of primary importance" [2–4].

Manufacturing and production of GM plants for food and feed purposes is very promising for socially important areas of modern biotechnology. This point is emphasized by the statement of the President of the Russian Federation Federal Assembly (May 2006) that the economic policy of Russia in 2007–2012 should focus on "the development of the scientific and technological potential of the Russian Federation for the implementation of priority directions of science development, technology and engineering". Fundamental and applied research in the life sciences, safety assessment of new sources of food and food ingredients, the introduction of innovative technologies, including bio- and nanotechnology, should be consistent with the objective of the state policy in the area of food security made by the Food Security Doctrine of the Russian Federation.

In the 21st century, overpopulation and exhaustion of resources will increase. Scientific progress is the only safeguard of mankind's survival. Thus, fundamental and applied research should not weaken its pace, because stagnation and even delay in this work could be disastrous.

## REFERENCES

[1] Patel P, Bousios A, Senker J. UK Performance in Science related to Biotechnology: An Analysis of Publications data. Report prepared for the UK Department of Trade and Industry, December 2003.

[2] Resolution of Bureau of Life Sciences Department of Russian Academy of Medical Sciences. No. 137, 25.11.2003.

[3] Resolution of Presidium of Russian Academy of Medical Sciences. No. 162, 17.09.2003.

[4] Decree of the Presidium of Russian Agricultural Academy of 20.11.2003, "On the use of genetically modified plants in order to protect plants".

# Index

*Note*: Page numbers followed by "*f*" and "*t*" refers to figures and tables, respectively.

## A

A2704-12, *see* Soybean line A2704-12
A5547-127, *see* Soybean line A5547-127
Aflatoxin B1
  maize line 3272 259*t*
  maize line 88017 225*t*
  maize line Bt11 194*t*
  maize line GA 21 129*t*
  maize line MIR604 242*t*
  maize line MON 810 146*t*
  maize line MON 863 179*t*
  maize line NK 603 162*t*
  maize line T25 209*t*
  rice line LLRICE62 276*t*
  soybean line 40-3-2 50*t*
  soybean line A2704-12 73*t*
  soybean line A5547-127 95*t*
*Agrobacterium*, plant cell transformation mediation principles 1–3, 5, 6*f*, 7
  soybean line 40-3-2 44–45
Agrolistic transformation 8
Amino acid score, soybean line 40-3-2 59*t*
a-Amylase, *see* Maize line 3272
Anaphylaxis, *see specific plant lines*

## B

*Bacillus thuringiensis*
  mechanism of toxin action 11–12
  safety 43–44
  transgenic plants, *see* Maize line 88017; Maize line Bt11; Maize line MIR604; Maize line MON 810; Superior NewLeaf potato
Biolistic transformation 7–8, 8*f*
Bt11, *see* Maize line Bt11

## C

Cadmium
  maize line 3272 259*t*
  maize line 88017 225*t*
  maize line Bt11 194*t*
  maize line GA 21 129*t*
  maize line MIR604 242*t*
  maize line MON 810 146*t*
  maize line MON 863 179*t*
  maize line NK 603 162*t*
  maize line T25 209*t*
  rice line LLRICE62 276*t*
  soybean line 40-3-2 50*t*
  soybean line A2704-12 73*t*
  soybean line A5547-127 95*t*
  Superior NewLeaf potato 293*t*
Carbohydrate content
  maize line 3272 258*t*
  maize line 88017 224*t*
  maize line Bt11 193*t*
  maize line GA 21 127*t*
  maize line MIR604 241*t*
  maize line MON 810 144*t*
  maize line MON 863 177*t*
  maize line NK 603 160*t*
  maize line T25 207*t*
  rice line LLRICE62 274*t*
  soybean line 40-3-2 47*t*
  soybean line A2704-12 71*t*
  soybean line A5547-127 93*t*
  Superior NewLeaf potato 292*t*
Colorado potato beetle resistance, *see* Superior NewLeaf potato
Comet assay, safety assessment of genetically modified crops 40
Corn borer, *see* European corn borer
Corn rootworm, resistant maize, *see* Maize line 88017; Maize line MON 863
Cotton, genetically modified crop percentage 21*f*

## D

Decrees, Russian Federation 27–28
*Diabrotica*, *see* Corn rootworm
DNA methylation, transformation effects in plants 9
DNA microarray, control system for genetically modified plant food products 313–315, 314*f*, 316*f*

## E

EU, *see* European Union
European corn borer, resistant maize, *see* Maize line Bt11; Maize line MIR604; Maize line MON 810
European Union (EU)
  approved genetically modified crops 22*t*
  control system for genetically modified plant food products 307–308, 308*f*
  planted area of genetically modified crops 20*t*

## F

Fatty acid composition
  maize line 3272 258*t*
  maize line 88017 224*t*
  maize line Bt11 193*t*
  maize line MIR604 241*t*

Fatty acid composition (*Continued*)
  maize line MON 810 145t
  maize line MON 863 178t
  maize line NK 603 161t
  maize line T25 208t
  rice line LLRICE62 275t
  soybean line 40-3-2 48t
  soybean line A2704-12 72t
  soybean line A5547-127 93t
  Superior NewLeaf potato 292t
Federal Law No. 29-FZ 26
Federal Law No. 86-FZ 25
40-3-2, *see* Soybean line 40-3-2

# G

GA 21, *see* Maize line GA 21
Gene gun, *see* Biolistic transformation
Genetic engineering, principles for plants 1
Glufosinate
  maize tolerant lines, *see* Maize line Bt11; Maize line T25
  mechanism of action 11
  rice tolerant line, *see* Rice line LLRICE62
  soybean tolerant lines, *see* Soybean line A2704-12; Soybean line A5547-127
Glyphosate
  maize tolerant lines, *see* Maize line 88017; Maize line GA 21; Maize line NK 603
  mechanism of action 10–11
  soybean tolerant lines, *see* Soybean line 40-3-2; Soybean line MON 89788

# H

Heavy metals, *see* Cadmium; Lead
8-Hydroxy-2-deoxyguanosine (8-OxodG), safety assessment of genetically modified crops 39

# I

Immune response, *see specific plant lines*

# L

Lead content
  maize line 3272 259t
  maize line 88017 225t
  maize line Bt11 194t
  maize line GA 21 129t
  maize line MIR604 242t
  maize line MON 810 146t
  maize line MON 863 179t
  maize line NK 603 162t
  maize line T25 209t
  rice line LLRICE62 276t
  soybean line 40-3-2 50t
  soybean line A2704-12 73t
  soybean line A5547-127 95t
  Superior NewLeaf potato 293t
Lipid peroxidation (LPO)
  maize line GA 21 studies 133t
  maize line MON 810 studies 150t
  maize line MON 863 studies 183t
  maize line NK 603 studies 166t
  maize line T25 studies 214t
  rice line LLRICE62 studies 283t
  soybean line A2704-12 studies 80t
  soybean line A5547-127 studies 100t
Lipids, *see* Fatty acid composition
LLRICE62, *see* Rice line LLRICE62
LPO, *see* Lipid peroxidation

# M

Maize, genetically modified crop percentage 21f
Maize line 3272
  a-amylase expression 255
  registration status by country 256, 256t
  safety assessment in Russian Federation
    biochemical parameters 265t
    biomarkers 266t, 267t
    composition 257, 258t, 259t
    genotoxicity 266–268, 268t
    hematology 264t
    immune response studies
      mice 268–269, 269t
      rats 269–271, 270t
    morphology 262t, 263t, 264t
    overview 257–272
    proximate parameters 260f, 261t, 262t
  technological parameter assessment 271
  transformation 256
Maize line 88017
  registration status by country 222, 223t
  safety assessment in Russian Federation
    composition 222–223, 224t, 225t
    genotoxicity 234t, 273
    immune response studies
      mice 235t, 273–289
      rats 236–237, 236t, 237t
    overview 222–238
    toxicological assessment in rat
      biochemical parameters 231t, 232t
      biomarkers 232t, 233t
      hematology 230t
      morphology 228t, 229t, 230t
      overview 223–231
      proximate parameters 226f, 227t, 228t
  technological parameter assessment 237–238
  transformation 221–222
Maize line Bt11
  registration status by country 191, 192t
  safety assessment in Russian Federation
    composition 192–194, 193t, 194t
    genotoxicity 203, 204t
    immune response studies
      anaphylaxis 203t
      cell-mediated immunity 201
      humoral immunity 200, 203t
      rat studies 202
      *Salmonella* infection susceptibility 201–202
      sensitization 186
    overview 192–205
    toxicological assessment in rat
      biochemical parameters 197, 198t
      biomarkers 197, 198t, 199t
      hematology 197, 199t
      morphology 200, 200t
      overview 195–200
      proximate parameters 195f, 196t
    *Streptomyces viridochromogenes* pat gene 191
    technological parameter assessment 204–205
  transformation 191
Maize line GA 21

registration status by country 126, 126t
safety assessment in Russian Federation
  composition 127–129, 127t, 128t, 129t
  genotoxicity 139–141, 140t, 141t
  immune response studies
    anaphylaxis 138t
    cell-mediated immunity 137
    humoral immunity 136–137, 139t
    rat studies 138–139
    *Salmonella* infection susceptibility 137–138
    sensitization 137
  overview 126–142
  toxicological assessment in rat
    biochemical parameters 132–134, 132t
    biomarkers 133t, 134t, 135t
    hematology 134–135, 135t, 136t
    morphology 135, 136t
    overview 129–136
    proximate parameters 130f, 130t, 131t
    rat diet 129t
  technical parameter assessment 141–142
  transformation 126
Maize line MIR604
  registration status by country 239, 240t
  safety assessment in Russian Federation
    composition 240, 241t, 242t
    genotoxicity 250–251, 250t, 251t
    immune response studies
      mice 251–252, 252t
      rats 253, 253t, 254t
    overview 240–255
    toxicological assessment in rat
      biochemical parameters 105t
      biomarkers 249t, 250t
      hematology 247t
      morphology 245t, 246t
      overview 240–250
      proximate parameters 243t, 244f, 244t
    technological parameter assessment 253–255
  transformation 239

Maize line MON 810
  registration status by country 143, 143t
  safety assessment in Russian Federation
    composition 143–146, 144t, 145t, 146t
    genotoxicity 156–157, 157t
    immune response studies
      anaphylaxis 155t
      cell-mediated immunity 154
      humoral immunity 153–154, 156t
      rat studies 155–156
      *Salmonella* infection susceptibility 154–155
      sensitization 154
    overview 143–158
    toxicological assessment in rat
      biochemical parameters 149, 149t
      biomarkers 149–151, 150t, 151t
      hematology 151, 152t, 153t
      morphology 151t, 152
      overview 146–153
      proximate parameters 146–149, 147f, 147t, 148t
    technical parameter assessment 157–158
  transformation 142–143
Maize line MON 863
  registration status by country 176, 176t
  safety assessment in Russian Federation
    composition 176–179, 177t, 178t, 179t
    genotoxicity 188–189, 189t
    immune response studies
      anaphylaxis 187t
      cell-mediated immunity 186
      humoral immunity 185–186, 188t
      rat studies 187–188
      *Salmonella* infection susceptibility 186–187
      sensitization 186
    overview 176–190
    toxicological assessment in rat
      biochemical parameters 181, 182t
      biomarkers 181–184, 182t, 183t, 184t

hematology 184, 184t, 185t
      morphology 184, 185t
      overview 179–185
      proximate parameters 180f, 180t, 181t
    technological parameter assessment 189–190
  transformation 175–176
Maize line NK 603
  registration status by country 159, 159t
  safety assessment in Russian Federation
    composition 156t, 159–162, 161t, 162t
    genotoxicity 173–174, 173t, 174t
    immune response studies
      anaphylaxis 172t
      cell-mediated immunity 170–171
      humoral immunity 170, 172t
      rat studies 172–173
      *Salmonella* infection susceptibility 171
      sensitization 171
    overview 159–175
    toxicological assessment in rat
      biochemical parameters 165, 166t
      biomarkers 165–167, 166t, 167t, 168t
      hematology 168, 169t
      morphology 169, 170t
      overview 162–170
      proximate parameters 164f, 164t, 165t
      rat diet 163t
    technological parameter assessment 174–175
  transformation 158–159
Maize line T25
  registration status by country 206, 206t
  safety assessment in Russian Federation
    composition 207–209, 207t, 208t, 209t
    genotoxicity 219–220, 219t, 220t
    immune response studies
      anaphylaxis 218t
      cell-mediated immunity 217
      humoral immunity 216–217, 219t

Maize line T25 (*Continued*)
  rat studies 218–219
  *Salmonella* infection susceptibility 217–218
  sensitization 217
  overview 206–221
  toxicological assessment in rat
    biochemical parameters 212, 212*t*
    biomarkers 213–215, 213*t*, 214*t*
    hematology 215, 215*t*, 216*t*
    morphology 215–216, 216*t*
    overview 209–216
    proximate parameters 210*f*, 211*t*
  *Streptomyces viridochromogenes pat* gene 205–206
  technological parameter assessment 220–221
  transformation 206
Mineral composition
  maize line 3272 259*t*
  maize line 88017 225*t*
  maize line Bt11 194*t*
  maize line GA 21 128*t*
  maize line MIR604 242*t*
  maize line MON 810 145*t*
  maize line MON 863 178*t*
  maize line NK 603 162*t*
  maize line T25 209*t*
  rice line LLRICE62 275*t*
  soybean line 40-3-2 49*t*
  soybean line A2704-12 73*t*
  soybean line A5547-127 94*t*
  Superior NewLeaf potato 293*t*
MIR604, *see* Maize line MIR604
MON 810, *see* Maize line MON 810
MON 89788, *see* Soybean line MON 89788

# N

Net protein efficiency ratio (NPER)
  soybean line 40-3-2 57–58
  soybean line A2704-12 83
NK 603, *see* Maize line NK 603
NPER, *see* Net protein efficiency ratio

# O

*Ostrinia nubilalis*, *see* European corn borer
8-OxodG, *see* 8-Hydroxy-2-deoxyguanosine

# P

PCR, *see* Polymerase chain reaction
PER, *see* Protein efficiency ratio
Phospholipid composition, soybean line 40-3-2 48*t*
Planted area, genetically modified crops
  country distribution 18*t*, 19*t*
  European Union 20*t*
  growth trends 17*f*, 18*t*
Polymerase chain reaction (PCR), control system for genetically modified plant food products
  DNA microarray 313–315, 314*f*, 316*f*
  principles 310–312
  real-time polymerase chain reaction 315–317, 317*f*
  specificity of methods 313*f*
Potato, *see* Superior NewLeaf potato
Potato beetle, *see* Colorado potato beetle resistance
Protein composition
  maize line 3272 258*t*
  maize line 88017 224*t*
  maize line Bt11 193*t*
  maize line GA 21 127*t*
  maize line MIR604 241*t*
  maize line MON 810 144*t*
  maize line MON 863 177*t*
  maize line NK 603 160*t*
  maize line T25 207*t*
  rice line LLRICE62 274*t*
  soybean line 40-3-2 47*t*
  soybean line A2704-12 71*t*
  soybean line A5547-127 92*t*
  Superior NewLeaf potato 291*t*
Protein efficiency ratio (PER)
  soybean line 40-3-2 57–58
  soybean line A2704-12 83
Proteomics, safety assessment of genetically modified crops 38–39

# R

Raffinose, soybean line 40-3-2 49*t*
Rapeseed, genetically modified crop percentage 21*f*
Reporter genes 9
Rice line LLRICE62
  registration status by country 273, 273*t*
  safety assessment in Russian Federation
    composition 269*t*, 273–274, 275*t*, 276*t*
    genotoxicity 286–287, 287*t*, 288*t*
    immune response studies
      anaphylaxis 286*t*
      cell-mediated immunity 284
      humoral immunity 283–284, 286*t*
      rat studies 285–286
      *Salmonella* infection susceptibility 284–285
      sensitization 284
    overview 273–289
    toxicological assessment in rat
      biochemical parameters 277–281, 279*t*
      biomarkers 280*t*, 281*t*
      hematology 282–283, 282*t*
      morphology 283, 283*t*
      overview 276–283
      proximate parameters 277*f*, 278*t*
    rat diet 276*t*
  technological parameter assessment 288
  transformation 272–273
Russian Federation
  control system for genetically modified plant food products
    decision tree 309*f*
    overview 307–309, 308*f*
    techniques
      DNA microarray 313–315, 314*f*, 316*f*
      polymerase chain reaction 310–312
      real-time polymerase chain reaction 315–317, 317*f*
      specificity of methods 313*f*
  food product monitoring
    content of recombinant DNA 324*t*
    genetic modification detection rates 320*t*, 321*t*, 322*t*, 323*t*, 325*t*–326*t*
    regulatory agencies 319–320
  human health safety assessment of genetically modified crops, *see also specific crops*
    allergenicity potential 36–37, 37*t*

biomarkers 39
classification of assessments 32
data evaluation 37
genotoxicity potential 36, 40
nutritional assessment
  compositional equivalence analysis 38
  proteomics 38–39
  overview 31, 32, 33f
  toxicity potential 33–36, 35f
information service for use of novel biotechnologies in food industry 329
legislation and regulation of genetically modified crops 25
registered genetically modified crops 23t

## S

*Salmonella typhimurium*, infection susceptibility studies in mice
  maize line Bt11 201–202
  maize line GA 21 137–138
  maize line MON 810 154–155
  maize line MON 863 186–187
  maize line NK 603 171
  maize line T25 217–218
  rice line LLRICE62 284–285
  soybean line 40-3-2 61
  soybean line A2704-12 86
  soybean line A5547-127 103–104
  Superior NewLeaf potato 300
Selection markers 8–9
Sensitization, *see specific plant lines*
Soybean
  genetically modified crop percentage 21f
  genetic modification detection rates in food products 320t
Soybean line 40-3-2
  *Agrobacterium*-mediated transformation 44–45
  registration status by country 45, 46t
  safety assessment in Russian Federation
    composition 46–50, 47t, 47t, 48t, 49t, 50t
    dietary and nutritional value of protein 56–59, 58t, 59t
    genotoxicity studies in mice 64–66, 65t, 66t
    immune response studies in mice and rats
      anaphylaxis in rats 62, 62–63, 63–64, 63f, 63t
      cell-mediated immunity 60
      humoral immunity 59–60, 64t
      *Salmonella typhimurium* susceptibility 61
      sensitization 60–61
      overview 45–69
    reproduction studies in rats 66–68, 67t, 68t
    toxicological assessment in rat
      biomarkers 54, 54t, 55t
      hematology 55, 55t, 56t
      morphology 56, 57t
      proximate parameters 51, 52f, 52t, 53t, 54t
      rat diet 50–56, 51t
Soybean line A2704-12
  registration status by country 70, 70t
  safety assessment in Russian Federation
    biological value and digestibility of protein 82–85, 83t, 84t
    composition 70–74, 71t, 72t, 73t
    genotoxicity 88–89, 89t
    immune response studies
      humoral immunity 85–87, 88t
      cell-mediated immunity 85–86
      sensitization 86
      *Salmonella typhimurium* susceptibility 86
      anaphylaxis 87t
    overview 70–90
    toxicological assessment in rats
      biochemical parameters 74–78, 78t
      biomarkers 79t, 80t
      hematology 81, 81t, 82t
      morphology 76f, 77t, 81–82
      overview 74–82
      rat diet 75t, 76t
    *Streptomyces viridochromogenes pat* gene 69
    technological parameter assessment 89–90
    transformation 69–70
Soybean line A5547-127
  registration status by country 91, 91t
  safety assessment in Russian Federation
    composition 91–95, 92t, 93t, 94t, 95t
    genotoxicity 106–107, 107t
    immune response studies
      anaphylaxis 105t
      cell-mediated immunity 103
      humoral immunity 103
      rat studies 104–105, 105t
      *Salmonella typhimurium* susceptibility 103–104
      sensitization 103
    overview 91–108
    toxicological assessment in rats
      biochemical parameters 97, 98t, 99t
      biomarkers 97–101, 99t, 100t, 100t, 101t
      hematology 101, 101t, 102t
      morphology 101–102, 102t
      overview 95–103
      proximate parameters 95–97, 96f, 97t, 98t
      rat diet 96t
    *Streptomyces viridochromogenes pat* gene 91
    technological parameter assessment 56
    transformation 91
Soybean line MON 89788
  quality assessment 110t
  registration status by country 109, 109t
  safety assessment in Russian Federation
    composition 110, 111t
    genotoxicity 120–121, 122t
    immune response studies
      mice 121–123, 123t
      rats 123–124, 124t
    overview 110–125
    toxicological assessment in rat
      biochemical parameters 118t
      biomarkers 119t, 120t
      hematology 117t
      overview 110–120
      proximate parameters 113t, 114f, 115t, 116t
      rat diet 112t
    technological parameter assessment 124–125
    transformation 109

Stachyose, soybean line
    40-3-2 49t
*Streptomyces hygroscopicus, bar* gene
    272
*Streptomyces viridochromogenes, pat*
        gene
    maize line Bt11 191
    maize line T25 205–206
    soybean line A2704-12 69
    soybean line A5547-127 91
Superior NewLeaf potato
    registration status by
        country 290, 290t
    safety assessment in Russian
            Federation
        composition 291–293, 291t,
            292t, 293t
        genotoxicity 302–303, 302t, 303t
        immune response studies
            anaphylaxis 301t
            cell-mediated immunity
                299–300
            humoral immunity
                299, 301t
            rat studies 300–302

*Salmonella* infection
    susceptibility 300
    sensitization 300
    overview 273–289
    toxicological assessment in rat
        biochemical parameters 295,
            296t
        biomarkers 296, 297t
        hematology 298, 298t
        morphology 298, 299t
        overview 293–299
        proximate parameters 295f,
            295t, 296t
        rat diet 294t
    technological parameter
        assessment 303
    transformation 289–290

# T

T25, *see* Maize line T25
T-DNA, *see* Transfer DNA
Ti-plasmid 6–7
Transfer DNA (T-DNA)
    delivery to plant cells 5–6
    gene insertion 7

# U

United States
    approved genetically modified
        crops 22t
    control system for genetically
        modified plant food
        products 307, 308f

# V

Vitamin composition
    maize line 3272 259t
    maize line 88017 225t
    maize line Bt11 194t
    maize line GA 21 128t
    maize line MIR604 242t
    maize line MON 810 145t
    maize line MON 863 178t
    maize line NK 603 161t
    maize line T25 208t
    rice line LLRICE62 275t
    soybean line 40-3-2 49t
    soybean line A2704-12 72t
    soybean line A5547-127 94t
    Superior NewLeaf potato 292t